LIQUID CRYSTALS

SOLID STATE PHYSICS

Advances in
Research and Applications

Editors
Henry Ehrenreich
Division of Applied Sciences, Harvard University
Cambridge, Massachusetts

Frederick Seitz
Rockefeller University, New York, New York

David Turnbull
Division of Applied Sciences, Harvard University
Cambridge, Massachusetts

The following supplements are published within the framework of the series:

LIQUID CRYSTALS

Guest Editor:

L. LIEBERT

Laboratoire de Physique des Solides
Université Paris-Sud
Orsay, France

ACADEMIC PRESS · New York San Francisco London ·
1978

A Subsidiary of Harcourt Brace Jovanovich, Publishers

ACADEMIC PRESS, INC.
111 Fifth Avenue, New York, New York 10003

United Kingdom Edition published by
ACADEMIC PRESS, INC. (LONDON) LTD.
24/28 Oval Road, London NW1 7DX

LIBRARY OF CONGRESS CATALOG CARD NUMBER: 55-12299

ISBN 0-12-607774-6

PRINTED IN THE UNITED STATES OF AMERICA

Contents

Macromolecules and Liquid Crystals: Reflections on Certain Lines of Research

P. G. DE GENNES

Liquid-Crystal Synthesis for Physicists

P. KELLER AND L. LIEBERT

Elasticity of Nematic Liquid Crystals

H. J. DEULING

The Dielectric Permittivity of Liquid Crystals

W. H. DE JEU

Instabilities in Nematic Liquid Crystals

E. DUBOIS-VIOLETTE, G. DURAND, E. GUYON, P. MANNEVILLE, AND P. PIERANSKI

Lyotropic Liquid Crystals: Structures and Molecular Motions

JEAN CHARVOLIN AND ANNETTE TARDIEU

Liquid Crystals and Their Analogs in Biological Systems

Y. BOULIGAND

Contributors to Supplement 14

Numbers in parentheses indicate the pages on which the authors' contributions begin.

Y. BOULIGAND, *E.P.H.E. et Centre de Cytologie Experimentale, C.N.R.S., 94200 Ivry-sur-Seine, France* (259)

JEAN CHARVOLIN, *Laboratoire de Physique des Solides, Laboratoire associe au C.N.R.S., Université Paris-Sud, 91405 Orsay, France* (209)

P. G. DE GENNES, *Collège de France, Paris Cedex 05, France* (1)

W. H. DE JEU, *Philips Research Laboratories, Eindhoven, The Netherlands* (109)

H. J. DEULING, *Gesamthochschule Kassel, D 3500 Kassel, West Germany* (77)

E. DUBOIS-VIOLETTE, *Laboratoire de Physique des Solides, Université Paris-Sud, 91405 Orsay, France* (147)

G. DURAND, *Laboratoire de Physique des Solides, Université Paris-Sud, 91405 Orsay, France* (147)

E. GUYON, *Laboratoire de Physique des Solides, Université Paris-Sud, 91405 Orsay, France* (147)

P. KELLER, *Laboratoire de Physique des Solides, Université Paris-Sud, 91405 Orsay, France* (19)

L. LIEBERT, *Laboratoire de Physique des Solides, Université Paris-Sud, 91405 Orsay, France* (19)

P. MANNEVILLE, *Centre d'Etudes Nucléaires de Saclay, Gif sur Yvette, France* (147)

P. PIERANSKI, *Laboratoire de Physique des Solides, Université Paris-Sud, 91405 Orsay, France* (147)

ANNETTE TARDIEU, *Centre de Génétique Moléculaire, C.N.R.S., 91190 Gif-sur-Yvette, France* (209)

Preface

The characterization and understanding of the liquid crystalline, or ordered fluid, states of matter have been long-standing challenges. In the past decade there has been a sharp resurgence in the efforts to meet these challenges. These efforts have been marked by fruitful collaboration of physicists and chemists and have led to important advances in the understanding of liquid crystals. This supplement presents a series of reviews of some of these advances by a group of European scientists who have made distinguished contributions to the field.

We are indebted to L. Liebert of the University of Paris-Orsay for coordinating the preparation of the volume and also to Robert B. Meyer for useful advice.

HENRY EHRENREICH
FREDERICK SEITZ
DAVID TURNBULL

LIQUID CRYSTALS

Macromolecules and Liquid Crystals: Reflections on Certain Lines of Research

P. G. DE GENNES

Collège de France, Paris, France

I. Introduction

There are many relations between liquid crystals and polymers: (a) It has been known for a long time that rod-like molecules can form nematic or cholesteric phases.[1] (b) Polymerization of mesomorphic phases built with suitable vinyl esters has led to the preparation of various frozen mesomorphic textures.[2] (c) More recently various groups have studied phase equilibria between mesomorphic monomers and the corresponding polymers.[3] (d) Effects related to the polymerization and cross-linking of

[1] C. Robinson, *Discuss. Faraday Soc.* **25**, 29 (1958); for a general review, see "Liquid Crystalline Order in Polymers" (A. Blumstein, ed.). Academic Press, New York, 1978.

[2] L. Strzelecki and L. Liebert, *Bull. Soc. Chim. Paris* **2**, 597,603, and 605 (1973); Y. Bouligand, P. E. Cladis, L. Liebert, and L. Strzelecki, *Mol. Cryst. Liq. Cryst.* **25**, 223, (1974).

[3] B. Kronberg *et al.*, *J. Phys. Chem.* (to be published); A. Dubault, M. Casagrande, and M. Veyssié, *Mol. Cryst. Liq. Cryst.* (to be published).

a solute species in a cholesteric matrix have been considered theoretically.[4] (e) The possible existence of nematic rubbers (with strong
coupling between stress and nematic order) has been mentioned.[5] No
chemical program has been launched—to the author's knowledge—in
these last two directions.

The present paper discusses two other situations of interest: The first
is concerned with *lyophilic systems*.[6] The phase diagrams of lipid–water
systems and also of detergent–water systems are of a fascinating
complexity. However, most theoretical discussions on lipid conformations suffer from a basic difficulty: Each chain (with a number N of
aliphatic carbons ranging between ~12 and 20) has a very large number
of degrees of freedom. But the values of N are still not large enough to
permit picturing the chain as a simple flexible polymer; end effects are
essential. Also, as is usual in chemical physics, some of the most
valuable information should come from comparisons between different
N values; but with lipids the requirements of hydrophilic/hydrophobic
balance severely limit the span of accessible values of N. These
difficulties are removed to a large extent when we deal with flexible
block copolymers (AB), together with a solvent which associates selec-
tively with B. The phase diagrams and structures of these systems have
been a subject of active study.[7]

We shall not attempt here any theoretical discussion of the phase
diagrams. We shall discuss only the *size* of some typical structures, and
sheets and micelles for these sizes; scaling laws can be predicted in
rather simple terms (all our examples are restricted to systems where all
chains are *fluid*). These questions have already been actively studied.[8,8a]
Our approach here is simplified: We consider only the limit of *strongly
incompatible systems* with *sharp boundaries*, for which comparatively
simple scaling laws can be established.[8b]

[4] P. G. de Gennes, *Phys. Lett. A* **28**, 725 (1969).

[5] P. G. de Gennes, *C. R. Hebd. Seances Acad. Sci., Ser. B* **281**, 101 (1975).

[6] V. Luzzati, *Biol. Membr.* **2**, 1 (1968); J. M. Vincent and A. E. Skoulios, *Acta
Crystallogr.* **20**, 432 (1966); B. Gallot and A. Skoulios, *Kolloid-Z.* **209**, 164; **210**; 143
(1966); J. Charvolin and A. Tardieu, this volume, Chapter 6.

[7] For a general survey of block copolymers, see *J. Polym. Sci., Part C* **26** (1969); S.
Aggawal, ed., "Block Copolymers." Plenum, New York, 1970; D. C. Allport and W. H.
James, eds., "Block Copolymers." Halstad Press, New York, 1973; S. Tuzar and P.
Kratovchil, *Adv. Colloid Interface Sci.* **6**, 201 (1976).

[8] D. Meier, *J. Polym. Sci., Part C* **26**, 81 (1969); *Polym. Prepr., Am. Chem. Soc., Div.
Polym. Chem.* **11**, 400 (1970); "Block and Graft Copolymers," p. 105. Syracuse Univ.
Press, Syracuse, New York, 1973; S. Tuzar and P. Kratovchil, *Adv. Colloid Interface
Sci.* **6**, 201 (1976).

[8a] E. Helfand, *in* "Recent Advances in Blends, Grafts, and Blocks" (L. Sperling, ed).
Plenum, New York, 1974.

[8b] The opposite limit, with weak interfacial energies, is discussed in a forthcoming paper
by J. Rault.

PURE
SOLVENT ——▶

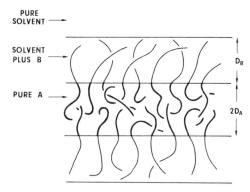

SOLVENT
PLUS B ——▶ D_B

PURE A ——▶

                                                               $2D_A$

FIG. 1.   Model for one fluid sheet of block copolymers AB. The solvent is assumed compatible with B and strictly incompatible with A. The interfacial energy is assumed to be large. Then, the B coils are strongly stretched, while the A coils are densely packed and only weakly stretched.

a. The problem of a *single sheet* in a lamellar phase of copolymer plus solvent is analyzed qualitatively in Section II. Our model for the sheet (Fig. 1) essentially assumes an A "core" (of half-thickness D_A) with constant density—equivalent to that of an A homopolymer melt. Outside this region we have B chains, mixed with some solvent, and extending up to a length D_B. The basic contributions included in the free energy are:

An A–A interaction which is saturated inside the core, and which we need not know explicitly, replacing its effects by a constraint on the core density.

An entropy of confinement of the A chains inside the core.

An effective repulsion between B units in the solvent.

A drop in entropy due to the stretching of B chains.

An interface energy proportional to the area of contact between the core and the solvent.

We give in Section II a qualitative discussion of this free energy and of the sizes D_A, D_B using scaling methods[9] which are particularly appropriate for these many chain problems. We do *not* discuss the total repeat period D of the structure (which depends on the overall solvent concentration) but restrict our considerations to the domain where $D \geq 2(D_A + D_B)$; no direct overlap between sheets.

b. In Section III we consider *one spherical micelle* with an A core and a "corona" of B chains (Fig. 2). Here for simplicity we omit all coefficients and discuss only the power laws: Our results sometimes

[9] For a review, see P. G. de Gennes, *Nuovo Cimento Rivista* 7, 363 (1977).

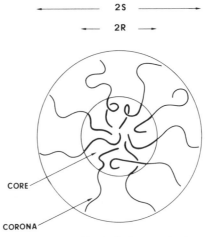

FIG. 2.   Model for an AB micelle in a good solvent of B.

differ significantly from the powers quoted in the existant theoretical literature.[8] Note that in the present (exploratory) study we do not try to predict the relative stabilities of various phases, involving micelles, rods, or sheets. With the list of contributing energies which has been given above, a comparison of this sort is feasible in principle, but requires heavy numerical calculations. Furthermore, phase stability may be affected by various weak forces which are not included in our discussion—in particular by the long-range components of the van der Waals forces.[10]

In a second part of this paper we turn to a completely different class of problems, where macromolecules are inserted as *solutes* in a conventional mesomorphic matrix. The lamellar phases (Lα in the crystallographic classification[6]) of lipid–water systems can be prepared as single domains of reasonable size. It may be possible to dissolve in these phases certain flexible polymer chains which enter preferentially into one of the components.[10a] As mentioned in Daoud *et al.*,[11] these systems may become of interest for neutron studies. In the present text we discuss some of their theoretical properties, assuming essentially that the thicknesses (D_L, D_W) of the lipid and water layers are large when compared with the monomer size, so that a comparatively simple "continuum" analysis may be applied. In Section IV we cover the case

---

[10] A. Parsegian, *Chem. Phys.* **56**, 4393 (1972).

[10a] Throughout this work, for definiteness, we speak of a hydrophobic chain, confined mainly to the lipidic sheets—although the inverse situation is equally interesting.

[11] M. Daoud *et al.*, *Macromolecules* **8**, 804 (1975).

of one ideal chain, and we discuss in particular the average number of consecutive units (g) which are trapped in one particular layer. When this number becomes as large as the polymerization index N, the chain becomes strictly two-dimensional. In Section V we propose a tentative extension of these ideas to the more realistic case in which excluded volume effects become important. In Section VI we discuss the feasibility of certain related experiments.

II. Sheet Structures of AB Copolymers

The situation is pictured in Fig. 1. The chains have an A portion with N_A monomers, and a mean square end-to-end distance $N_A a^2$. Similarly, the B portion has N_B monomers and a square size $N_B a^2$ (for simplicity we take the same size per monomer for both A and B). To make things definite, as in the Flory–Huggins theory, we describe our chains as walks on a simple cubic lattice of unit length a. In the A regions the chains fill the lattice completely. In the B regions they occupy only a small fraction ρ of the sites, the rest being occupied by solvent molecules. We treat the B chains as an athermal system with nonintersecting chains, but no other interaction between them. (In the notation of Flory[12] $\chi = 0$.) All these simplifications are not essential provided that the solvent is good for B and poor for A. Our free energy per chain (apart from irrelevant contributions) then splits into three parts:

$$F = \sigma\Sigma + F_A + F_B,\qquad \text{(II.1)}$$

where Σ is the interface area per chain and σ is an interfacial energy coefficient. F_A and F_B represent contributions from both sides, and will be constructed below. Note that with our assumption of complete saturation on the A side the half-thickness D_A of the A region is related to Σ by the relation

$$\Sigma D_A = N_A a^3$$

or

$$a^2/\Sigma = D_A/N_A a = x.\qquad \text{(II.2)}$$

We shall always be concerned with situations in which $x \ll 1$ or $\Sigma \gg a^2$. Then (and only then) is it indeed correct to describe the interface energy by a single extensive term of the form $\sigma\Sigma$: Using for instance the language of lipids, the special interactions between polar heads become

[12] P. Flory, "Principles of Polymer Chemistry." Cornell Univ. Press, Ithaca, New York, 1953.

weak when $\Sigma > a^2$, and we are left with an oil/water interfacial energy proportional to Σ.[12a]

The thickness D_B of the other region is related to Σ and ϕ by the relation

$$N_B a^3 = \rho D_B \Sigma. \tag{II.3}$$

1. THE DENSE SIDE (A)

In the energy F_A we need *not* write explicitly the A–A attractions. Their contribution is large but independent of D_A, and thus not interesting here. The only relevant term is the stretching free energy: Each A chain is stretched over a thickness D_A. We assume the following inequalities:

$$N_A^{1/2} < D_A < N_A a.$$

The first will turn out to be satisfied for large values of σ. The second inequality is equivalent to $x < 1$ or $\Sigma \gg a^2$, as discussed after Eq. (II.2). When both inequalities hold we may write the stretching free energy simply as

$$F_A = \frac{3T}{2N_A a^2} D_A{}^2 = \frac{3T}{2} N_A \left(\frac{a^2}{\Sigma}\right)^2 = \frac{3T}{2} N_A x^2 \tag{II.4}$$

(we choose temperature units which make Boltzmann's constant equal to unity).

2. THE SEMIDILUTE SIDE (B)

The situation on the B side is more complicated for the following reason: in a solution, with concentrations $\rho \ll 1$, the chains cannot be treated as ideal at all scales. We know now that there are qualitative deviations from the mean field (Flory–Huggins) theory for such a system.[11] Here we shall proceed in two steps: (1) rederive the Flory–Huggins results, which are very simple; (2) go to a more rigorous scaling approach which was recently applied to this particular problem by S. Alexander.[11a]

a. Mean Field Approximation

If we treat the chains as ideal and uncorrelated, we expect an interaction energy per chain:

$$F_{int} = \tfrac{1}{2} T N_B \rho = \tfrac{1}{2} T N_B{}^2 a^2 / D_B \Sigma. \tag{II.5}$$

[12a] Also the entropy decrease due to the localization of the block joints[8] is of order $\ln \Sigma$ and the resulting contribution to the free energy is negligible when Σ/a^2 is large.

[11a] S. Alexander, *J. Phys. (Paris)* **38**, 977, 983 (1977).

The stretching energy is for ideal chains

$$F_{\text{stretch}} = 3TD_B{}^2/2N_B a^2. \tag{II.6}$$

Minimizing the sum $F_B = F_{\text{int}} + F_{\text{stretch}}$ at fixed Σ with respect to D_B, we arrive at

$$D_B = \text{const } N_B a x^{1/3} \tag{II.7}$$

and an energy

$$F_B = \text{const } TN_B x^{2/3} \tag{II.8}$$

b. Scaling Approach

In reality, the chains are not ideal at small scales; also they are correlated (they tend to avoid each other). For this reason both (II.5) and (II.6) are overestimates. The correct scaling description of semidilute chains is given in Daoud et al.[11] The central concept is that of a correlation length $\xi = a\rho^{-3/4}$. At distances $r < \xi$ the chains show excluded volume effects. But for $r > \xi$ they behave as ideal (stretched) chains. The basic unit is not the monomer, but a "blob" of size ξ containing a number of monomer $g_\xi = (\xi/a)^{5/3} = \rho(\xi/a)^3$. The repulsive energy, including correlation effects, is of order T/ξ^3 per cm^3, or per chain

$$F_{\text{int}} \sim (N_B/g_\xi)T. \tag{II.5'}$$

The stretching energy (for a sequence of N/g_ξ blobs of size ξ) is

$$F_{\text{stretch}} \sim TD_B{}^2/(N_B/g_\xi)\xi^2. \tag{II.6'}$$

In (II.5') and (II.6') the numerical prefactors are unknown at present. Minimizing the sum $F_{\text{int}} + F_{\text{stretch}}$ at fixed Σ, Alexander finds that, as usual, both terms become of the same order of magnitude. Thus he obtains

$$\begin{aligned} D_B{}^2 g_\xi/N_B\xi^2 &\sim N_B/g_\xi \\ D_B &\sim (N_B/g_\xi)\xi. \end{aligned} \tag{II.10}$$

The concentration is

$$\rho = N_B/D_B\Sigma \cong g_\xi/\xi\Sigma.$$

From the scaling laws[11] we also know that $\rho \cong (g_\xi/\xi^3)$. Thus we are led to the essential property

$$\Sigma = \xi^2. \tag{II.11}$$

The resulting picture is very simple: each chain can be visualized as a succession of *blobs* of size $\xi = \Sigma^{1/2}$. The projected area of any blob on the interface is Σ. Different blobs do not overlap very much, and each chain is very nearly a linear stacking of blobs, as expressed by (II.10): D_B is the product of the number of blobs per chain (N_B/g_ξ) by the size of each blob (ξ). However, the reader should keep in mind that some transverse fluctuations are allowed: the end-to-end distance, projected on the interface plane $(R_\perp)$, may be shown to be

$$R_\perp \cong (N_B/g_\xi)^{1/2}\xi, \tag{II.12}$$

i.e., the transverse path is a random walk.

Comparing (II.10) and (II.11), we find that $D_B(\Sigma)$ is *still given by the mean field equation* (II.7). On the other hand, what is not correctly given by mean field is the energy. The Alexander result, derived from (II.5′) and (II.6′),

$$F_B \cong T(N_B/g_\xi) = CTN_B(a^2/\Sigma)^{5/6} = CTN_B x^{5/6} \tag{II.8′}$$

where C is an unknown constant. This differs from (II.8). As pointed out by Alexander, the mean field method gives us an incorrect energy, but a correct size D_B for this copolymer problem, just as it does for the single chain excluded volume problem.

3. The Balance between Bulk and Surface Energies

The total energy per chain of our block copolymer system is obtained from (II.1), (II.4), and (II.8′)

$$\frac{F}{N_A T} = \frac{\eta}{x} + \frac{3}{2}x^2 + \nu x^{5/6} \tag{II.13}$$

where

$$\eta = \sigma a^2/N_A T \qquad \nu = CN_B/N_A. \tag{II.14}$$

All equilibrium properties can now be obtained by minimizing (II.13) with respect to x, i.e., by finding the optimum area $\Sigma = a^2 x^{-1}$. We shall not discuss this in great detail, but simply consider the major limiting cases.

a. Very Small B Chains ($\nu \to 0$)

Here F_B is negligible and, after minimization of $F_A = \sigma\Sigma$, we find

$$x = (\eta/3)^{1/3}$$

or

$$D_A = N_A^{2/3}(\sigma/3T)^{1/3}a^{5/3}. \tag{II.15}$$

This remarkable power law should be comparable to experiment, but we have not found relevant data in the literature.[7] For lipids, all the X-ray results[6] point toward a different law (D_A linear in N_A or Σ independent of N_A). But, as explained in the introduction, lipids are not amenable to our discussion (N_A is too small and also Σ/a^2 is of order unity).

b. Large B Chains

Here the dominant terms in the energy balance are the surface energy and F_B. Minimization with respect to x gives

$$x = \left(\frac{6}{5} \cdot \frac{\eta}{\nu}\right)^{6/11} \tag{II.16}$$

from which all characteristic lengths (D_A, D_B, etc.) can be derived. For example, using (II.7), we obtain

$$D_B = \text{const } N_B a (\eta/\nu)^{2/11}. \tag{II.17}$$

It may be useful at this atage to define the limits of validity of the "large B chain" approximation: this holds only when $F_A \ll F_B$ or when $x^2 \ll \nu x^{5/6}$. From (II.16), this is equivalent to the condition

$$\eta < \nu^{18/7}. \tag{II.18}$$

For practical use it is helpful to rewrite (II.17) in an explicit form:

$$D_B \cong a N_B^{9/11} (\sigma a^2/T)^{2/11}. \tag{II.19}$$

c. Intermediate Case

In between the two extreme regimes described above, there is of course a cross over region where the full equation (II.13) should be used. But this region will not be too helpful for the verification of the model, and we shall not discuss it in detail.

III. Dilute Micelles of AB Copolymers

We now consider the conformations in a single micelle, assuming that it is strictly spherical (Fig. 2). The physics is basically the same as in the previous case. The complete self-consistency equations would, in principle, require a rather complicated calculation of variations along the radial direction, which we have not carried out. Our approach is a simplified one, in which we consider that the A chains are stretched with an end-to-end distance equal to the core radius R. Similarly, we treat the B chains as stretched over an interval $S - R$, equal to the thickness of the "corona" (the B region). We take the density as saturated in the core, and this is rigorous within our model. But we also take ρ = constant (<1) in the corona; this would be at its worst when we have a small core and a large diffuse corona. The situation in this limit, however, resembles that of the excluded-volume problem for a single chain, where the assumption of a constant ρ inside the coil does not spoil the characteristic exponent derived by Flory,[12] (although it changes numerical coefficients). Thus we can expect that our results remain qualitatively correct all along. We shall introduce the number p of chains per micelle related to R by our assumption of saturated core density:

$$pN_A a^3 = 4\pi/3R^3 \qquad (\text{III.1})$$

and the interface area per chain

$$\Sigma = 4\pi R^2/p \qquad (\text{III.2})$$

To avoid lengthy discussions, we shall report only the results for the limit of *small B chains*. Here we predict a core of radius

$$R = \text{const} \cdot aN_A^{2/3}(\sigma a^2/T)^{1/3} \qquad (\text{III.3})$$

scaling like the thickness of one layer in the same regime for sheet structures [Eq. (II.8)]. The number of chains per micelle is *linear in N_A*:

$$p = \text{const} \cdot N_A \sigma a^2/T. \qquad (\text{III.4})$$

Equations (III.3) and (III.4) differ from the predictions of Meier for the same reason.[8] Working with ideal chains trapped in a sphere of radius R, Meier concluded that to reach a uniform density profile one should have $R \cong R_0 = N_A^{1/2}a$. The difference stems from the fact that we include the interface energy in our calculation of R. We find that the micelle prefers to increase p and stretch all its chains to a radius $R > R_0$ (still keeping a saturated density), the increase in R diminishing the surface ratio.[12b]

Of course the Meier result is expected to hold when the surface energy is very small. In fact when

$$\sigma a^2/T < N_A^{-1/2} \qquad (\text{III.5})$$

our expression for the A-chain stretching energy (which held only for strong stretching $R > R_0$) breaks down: After inclusion of a more complete form for this term, one can check that when the inequality (III.6) prevails, the calculation of Meier[8] applies. Using his procedure (i.e., fixing $R = \text{const} \cdot R_0$) Meier has obtained a rather good numerical agreement for the description of various block copolymer structures.

[12b] On the other hand, we agree with Meier[8] as regards the use of ideal chain statistics in the core: The core is a medium of high polymer concentration, where the chains tend to become ideal (as first noticed by Flory[12] for melts).

where (III.5) does not hold. This agreement may reveal some flaw in our argument; alternatively it may be that the difference between both predictions (involving a factor $N_A^{1/6}$) was too weak to show up in the available experiments. Clearly what is needed is a systematic study of sizes as a function of N_A covering large N_A values.

IV. One Ideal Chain in a Layered System

We now turn to the second problem mentioned in the introduction, and discuss first the conformations of one ideal chain dissolved in a smectic matrix. The word "ideal" amounts to assuming that all self-interactions of the chain are weak. Then, to a certain extent, the chain can be pictured as a simple random walk. However, a further distinction must be introduced already at this stage. Let us think, for definiteness, of our smectic as being a lamellar lipid–water system, the chain being hydrophobic. (1) If this hydrophobic character is not too strong, the chain will tend to have shorter portions in the water layers, but will still behave like a flexible coil in these regions. (2) On the other hand, if the monomer–water repulsion is very strong, the chain portions in the water will be mainly straight (and normal to the layers) to minimize the contacts. In this second case, a somewhat different approach, allowing for strong stretching, is required.

4. WEAK STRETCHING IN ADVERSE REGIONS

When spatial variations are not too fast, the statistical weight $G_N(r_0, r)$ for one chain extending from point r_0 to point r is ruled by an equation of the Schrödinger form:[12c]

$$-(a^2/6)\nabla^2 G_N + V(r)G_N = -\partial G_N/\partial N, \qquad (IV.1)$$

where N is the number of units (each with a mean square length a^2) and $TV(r)$ is the local potential seen by one monomer. The qualitative aspect of V across the layers may be represented by square-well potentials. With a convenient choice of origin we would write

$$\left.\begin{array}{l} V = 0 \quad \text{in the lipid regions;} \\ V = +W \equiv \Delta F/T \quad \text{in the water regions.} \end{array}\right\} \qquad (IV.2)$$

In fact many of the equations to be written below allow for more general shapes of V: The essential features are the periodicity along the normal (z) to the layers (period $D = D_L + D_W$), and the existence of a repulsive barrier in the water regions.

[12c] See, for instance, P. G. de Gennes, *Rep. Prog. Phys.* **32**, 187 (1969).

One further simplification which has been put into Eq. (IV.1) concerns the length of the statistical element a^2. We have ignored anisotropy effects (i.e., the fact that the mean square averages $a_x{}^2$ and $a_z{}^2$ are different) and solvent effects (a^2 might be different in the L and W regions). All these corrections are easy to include but complicate the notation severely.

The solution G_N to Eq. (IV.1) splits into two separate factors: a "transverse" factor (for x and y coordinates) with the usual Gaussian form, and a more interesting z-dependent part:

$$G_N(\mathbf{r_0}, \mathbf{r}) = G_N{}^0(x - x_0, y - y_0)G_N{}^z(z_0, z). \qquad (IV.3)$$

It is useful to expand $G_N{}^z$ in terms of eigenfunctions[12c] $u_{k,n}$ defined by

$$-(a^2/6)(d^2u/dz^2) + V(z)u = \epsilon u, \qquad (IV.4)$$

where ϵ is an eigenvalue: Eq. (II.4) coincides with the Schrödinger equation for one electron propagating in a one-dimensional periodic potential. The eigenfunctions have the Bioch form[13]

$$u_{kn} = e^{ikz}v_{kn}(z), \qquad (IV.5)$$

where $v(z + D) \equiv v(z)$. The wavevector k is restricted to the first Brillouin zone $(-\pi/D < k < \pi/D)$ and $n = 0, 1, \ldots$ is a band index. The weight factor $G_N{}^z$ is explicitly[3]

$$G_N{}^z(z_0 z) = \sum_{nk} u_{nk}^*(z_0)u_{nk}(z)e^{-\epsilon_{nk}N}. \qquad (IV.6)$$

For $N \to \infty$ the sum is dominated by the bottom of the lowest band (minimum of ϵ_{nk}) for which we have

$$\epsilon_{0k} = \epsilon_0 + (a^2/6)c^2k^2 + 0(k^4) \qquad (IV.7)$$

where c^2 is a coefficient, dependent on the periodic potential $V(z)$, and to be discussed below. Inserting (IV.7) into (IV.6) we have

$$G_N{}^z(z_0 z) \to u_{00}(z_0)u_{00}(z)(2\pi)^{-1/2}R_z^{-1}\exp{-\frac{1}{2}\left(\frac{z - z_0}{R_z}\right)^2}, \qquad (IV.8)$$

where the size R_z along z is given by

$$R_z = c(Na^2/3)^{1/2} = cR_0. \qquad (IV.9)$$

Thus the rms end-to-end size of the coil in this regime is x_0 along x and y, and is cx_0 along z. We may call c the *contraction factor*. A qualitative feeling for c may be obtained from the following picture: On the average

---

[13] N. Ashcroft and D. Mermin, "Solid State Physics," Chapters 8–11. Holt, New York, 1976.

the chain has g consecutive units in one lipid layer, then a few in a water layer, then again g in the next lipid layer, etc. The average size along z is thus $R_z{}^2 \sim D^2N/g$, from which we extract

$$g \sim (D/a)^2 c^{-2}. \qquad \text{(IV.10)}$$

For large water thickness c is small and g is large. In particular, when g becomes of order N, the chain is confined in a single lipid layer (and then the approximation IV.8 breaks down).

Methods of calculation of the "effective mass factor" c^2 for arbitrary periodic potentials $V(z)$ can be found in textbooks on band theory.[13] Here we shall consider only the limiting case of a repulsive barrier of the square form (IV.2). Then, in most cases of interest, the effective mass is controlled by *tunneling* through the barrier, and we have

$$c^2 = \text{const} \cdot e^{-KD_W} \qquad (KD_W \gg 1)$$

$$a^2K^2/6 = W - \epsilon_{00} = W - (\pi^2/6)(a/D_L)^2 \qquad \text{(IV.11)}$$

$$\cong W.$$

Using Eq. (IV.10), we then find that the criterion for localization in one layer ($g > N$) can be written in the form

$$\alpha(W) = (6W)^{1/2}Dw/a \gtrsim \ln N \qquad \text{(IV.12)}$$

(where we have omitted all nonessential factors adding a constant to $\ln N$).

5. STRONG STRETCHING IN ADVERSE REGIONS

It is important to realize that Eqs. (IV.1) through (IV.12) have a strong intrinsic limitation (which has no counterpart in the analog quantum mechanical problem); namely Eq. (IV.1) applies only when the space variations of $G_N(\mathbf{r}_0\mathbf{r})$, on a length equal to the monomer size a, are weak.[14] This is always correct for large N in the lipid phase, where the chain is smoothly coiled. But it may become incorrect in the water phase, where $G^{-1}\partial G/\partial r \sim K$, K being defined in Eq. (IV.11). Thus for Eq. (IV.12) to hold we must have $Ka < 1$ or $6W < 1$ [still maintaining $W > (a/D_L)^2$ and $KD_W > 1$].

What happens when $6W \gtrsim 1$? In this case we expect the chain to become completely stretched in the water portions, and normal to the layers. The number of units involved in such a "connection" is D_W/a

---

[14] This has not always been recognized sufficiently in theoretical discussions on block copolymers.

and the corresponding Boltzmann factor is

$$e^{-\alpha(W)} = e^{-D_W/aW} \qquad (6W \gtrsim 1).$$

In this limit we expect Eq. (II.10) for g to be replaced by

$$g \cong (D_L/a)^2 e^{+\alpha(W)}, \qquad \text{(IV.10')}$$

where $(D_L/a)^2$ is the number of consecutive units in one lipid layer in the absence of any barriers. The criterion for confinement in one layer is now

$$\alpha(W) = (D_W/a)W \gtrsim \ln N. \qquad \text{(IV.12')}$$

Taking, for instance, $\Delta F = 1$ kHz/monomer ($W \sim 2$), $a = 5$ Å, and $D_W = 20$ Å, this implies that chains with $N \gtrsim 10^3$ will be localized in one layer.

V. A Real Chain in a Layered System

Let us now assume that the lipid is a good solvent for the chains, so that there is an excluded volume v per monomer which is large. More specifically let us take $v = a^3$ (or $\chi = 0$ in the Flory notation[12]). For definiteness we also restrict our attention to the limit $6W > 1$ where the connecting portions between sheets are straight, as in the second part of Section IV.

We expect the number of consecutive units g inside one layer to become now

$$g = (D_L/a)^{1/\nu_3} e^\alpha, \qquad \text{(V.1)}$$

where $\nu_3 \sim 3/5$ is the excluded volume exponent for three-dimensional coils[12] and α is given by Eq. (II.12'). Equation (III.1) ensures that for $\alpha \to 0$ g reaches the correct scaling form. For $g \gtrsim N$ we expect confinement in a single layer. For $g \ll N$ we must discuss first the shape of the chain portion associated with g monomers in one layer. This is a flat pancake, or "spot." The lateral radius of the spot is, from ref. 15:

$$\rho_\perp = g^{\nu_2}(a/D_L)^{(\nu_2/\nu_3 - 1)}, \qquad \text{(V.2)}$$

where $\nu_2 \cong 3/4$ is the excluded volume exponent for two dimensions. The reader may check that when D_L increases and becomes equal to $\rho_\perp$, the normal three-dimensional scaling law holds: This confirms the

[15] M. Daoud and P. G. de Gennes, *J. Phys. (Paris)* **38**, 85 (1977).

exponent of a/D_L in Eq. V.2. Inserting (V.1) into (V.2) we get

$$\rho_\perp = e^{\alpha\nu_2}D_L. \qquad (V.3)$$

Let us now look at the statistics of the chain on a larger scale. We have N/g "spots," each occupying a size $\rho_\perp$ along x and y, D along z; the corresponding volumes are mutually exclusive. We may qualitatively picture the situation by putting the centers of the spots on a periodic (tetragonal) lattice of mesh sizes $(\rho_\perp, \rho_\perp, D)$. This can be reduced to a cubic lattice of mesh unity by the transformations

$$x' = x/\rho_\perp; \quad y' = y/\rho_\perp; \quad z' = z/D. \qquad (V.4)$$

On this lattice the three-dimensional scaling law[11,12] gives us a size

$$R' = (N/g)^{\nu_3}, \qquad (V.5)$$

and finally we obtain the real size:

$$\left. \begin{array}{l} R_z = R'D = (N/g)^{\nu_3}D; \\ R_\perp = R'\rho_\perp = (N/g)^{\nu_3}\rho_\perp. \end{array} \right\} \qquad (V.6)$$

Of particular interest is the contraction ratio, which is now

$$R_z/R_\perp \cong D/\rho_\perp \cong (D/D_L)e^{-\alpha\nu_2}. \qquad (V.7)$$

VI. Conclusions

6. SCALING LAWS FOR BLOCK COPOLYMER/SOLVENT SYSTEMS

We hope to have shown that block copolymers can give us the best model of amphiphilic behavior. For very large molecular masses (as opposed to those which are found in common lipids and nonionic detergents) the sizes of the building blocks should obey some relatively simple scaling laws. The essential notion is that to test a model completely one should vary systematically the length of both subchains. Of course there are limitations on the ratio $\nu = N_B/N_A$, similar in origin to those which are imposed by hydrophilic/hydrophobic balance in conventional detergents.[16] But even at fixed ν, varying the overall molecular mass over more than one decade is a crucial experiment. Another interesting parameter is the quality of the solvent: Here we

[16] See, for instance, P. Becher, "Emulsions," pp. 232-255. Van Nostrand-Reinhold, Princeton, New Jersey, 1965.

have assumed an excellent solvent for B, and no solubility for A, but many other cases can be of interest[8,8a,8b]. Also, for certain values of N_A/N_B, the micelles may become non spherical.

We might at this point add a word of comment on size determinations. For lamellar structures, X-ray scattering is the best choice. For micelles, inelastic light scattering (giving the diffusion coefficient and hydrodynamic radius) is most convenient; this gives us the outer radius S. To obtain the inner radius R we wish to point out the interest of a physicochemical method: To the dilute solution of micelles let us add a few chains of an A homopolymer, with N_A' units per chain; these will dissolve readily in the core, provided that they are not too large, i.e., that their unperturbed size $N_A'^{1/2}a$ is smaller than R. In the opposite limit they will dissolve less readily, and the fraction coefficient f will have the form[17]

$$f(N_A') \cong f(0)\exp\left(-\frac{\pi^2}{6}N_A'a^2/R^2\right).$$

Thus from a study of $f(N_A')$ as a function of N_A' one should be able to extract the radius R.

Finally we must emphasize once more that all our size considerations refer to separate units (e.g., dilute micelles). When the B chains from different units begin to overlap, the problem is modified; for flat sheets it is still tractable in simple terms, but for micelles (or rods) the self-consistent field equations become quite complex.

7. FUTURE EXPERIMENTS ON CHAINS TRAPPED IN A SMECTIC MATRIX

A number of experiments on chains trapped in an $L\alpha$ phase may become feasible in the near future. Conventional light-scattering experiments are not too tempting, because a smectic phase contains defects (dislocations, focal conics, etc.) which may dominate the signal. Neutron scattering from deuterated polymers may be more efficient. In particular, it is conceivable to use mixtures of heavy and light water having a specific scattering amplitude equal to that of the lipid matrix—in which case most of the scattering due to large defects will be eliminated. The neutron experiments could deal with various regimes:

a. Dilute coils confined in a single layer would give a very interesting example of two-dimensional behavior.

b. Semidilute coils, always confined in a single layer, would provide a

---

[17] For similar discussions of the confinement entropy, see E. Casassa *J. Polym. Sci., Part B* **5**, 773 (1967); E. Casassa and Y. Tagami, *Macromolecules* **2**, 14 (1969).

good realization of the slit problem discussed in ref. 15. From this reference we expect two different semidilute regimes, with distinct scaling laws.

c. Finally, for very large N we may observe the regimes $N > g$ where each coil is a prolate ellipsoid. In the dilute limit, one of the most interesting probes for this case would be the intrinsic viscosity $[\eta]$ for shear flows *parallel to the smectic layers*. A detailed discussion of $[\eta]$ including backflow is delicate in a smectic matrix, but clearly $[\eta]$ depends strongly on R_z.

In another direction we may ask whether solutions of chains in lamellar lipids can reach complete equilibrium in a reasonable amount of time. For instance, one might try to start from some high temperature isotropic solution involving lipid, water, and dilute chains, the chains being possibly dissolved by micellar action. Then one would cool the system into the lamellar phase, and prepare a single domain by mechanical means. At this stage the chains may still be out of equilibrium, e.g., with an excess of connecting portions going through the water sheets. However, there appears to be no special barrier for the elimination of connecting portions: When a chain "reptates" along its average path, the number of units in contact with the water need not increase; the relaxation times connected with this process should remain reasonably short. On the other hand, there is an obvious barrier for the *creation* of new connecting portions: If the chain is initially confined into one layer, although its equilibrium shape involves more than one layer, it may take a long time to expand along the normal (the expansion will take place mainly through both ends of the chain). These kinetic problems are obviously complex, but they do not seem to forbid the equilibrium studies which have been discussed here.

Finally we should mention that lipid–water systems are not the only candidate for these experiments. *Thermotropic smectics* give us layers with alternate aromatic and aliphatic character, and a given polymer chain will often dissolve preferentially in one of these regions. The main difference is that the two sublayers are relatively thin; in this case the segregation effects which were discussed here may be less essential than the anisotropy effects ($a_x{}^2 \neq a_z{}^2$) which were ignored in Eq. (II.1): For instance when $N \to \infty$ the chain could be oblate rather than prolate.

8. CHAINS CONFINED IN ROD STRUCTURES

It should be also possible to study hydrophilic flexible chains dissolved in an inverse hexagonal lipid–water structure.[6] Here the water regimes form separate rods, and if a certain criterion (similar to IV.2') holds, the chains should be confined to one tube, as first suggested by F.

Brochard.[18] The geometrical properties of these one dimensionally confined chains are analyzed in ref. 15. Some of the related dynamical problems are studied in Brochard and de Gennes.[18]

ACKNOWLEDGMENT

The writer is greatly indebted to S. Alexander, who pointed out a serious weakness in the first version of Section II.

·

[18] F. Brochard and P. G. de Gennes, *J. Chem. Phys.* **67,** 52 (1977).

Liquid-Crystal Synthesis for Physicists

P. KELLER AND L. LIEBERT

Laboratoire de Physique des Solides, Université Paris-Sud, Orsay, France

I. Introduction

Most chemistry reviews on liquid crystals are of little interest for physicists. In these books an accumulation of chemical formulae* and transition temperatures corresponding to the various mesophases are found. Our experience as chemists working in permanent contact with physicists has encouraged us to consider a new presentation. Our purpose is to examine several classes of liquid crystals, and to specify their synthesis and their yields, (the time of reaction is not given because it is too difficult to evaluate) only from a practical point of view. Our ambition is to make accessible for an isolated physicist, with no possible help from a nearby chemistry laboratory, a certain number of syntheses for the purpose of preparing good purity samples required for his physical studies. A little equipment, some work, and much "good will" are necessary. To take a realistic and humorous analogy, this recipe collection should be similar to the cookbook found in all good houses, which allows the neophyte to cook fine meals after some attempts have been relegated to the garbage can.

Very often, the difficulties of organic synthesis will appear through the miscarried experiment. Never in our laboratory has the debate been more open and richer than when a physicist has tried to make a synthesis and has thus experienced all the traps involved.

More generally, it is hoped that this kind of article will improve the relationships between chemists and physicists. The latter often seem to consider the organic molecules as convenient tools, forgetting about the difficulties of synthesis. A physicist well-informed about chemistry will avoid certain experimental pitfalls. Also, he will use more understandable language when dealing with a chemist. Moreover, the two scientists will then be in a better position to find new ideas that could be profitable for both.

We describe here the synthesis of very simple families of liquid crystals of the types currently required by the physicists of our labora-

---

* It is worth emphasizing that the extensive use by physicists of chemical designations like (p-methoxybenzylidène-p'-butylaniline) or even more synthetic forms like (505, CBOOA, . . .) is often a cause of ambiguity, as several names can be given to the same structure. In addition it does not provide much enlightenment concerning the relation between the chemical structure and the properties. We strongly recomend the use of developed chemical formulae in the papers (they can be inserted in the first figure of any paper).

tory. These liquid crystals have a formula of the type

$$R_1 -\!\!\left\langle\!\!\bigcirc\!\!\right\rangle\!\!- X -\!\!\left\langle\!\!\bigcirc\!\!\right\rangle\!\!- R_2$$

where

$R_1 = R_2 = -OC_nH_{2n+1}$		alkoxy chain
$R_1 = R_2 = -C_nH_{2n+1}$		alkyl chain
$R_1 = R_2 = -CH{=}CH-COOC_nH_{2n+1}$		cinnamate
$R_1 \neq R_2$		
$X = -CH{=}N-$		Schiff's bases
$X = -N{=}N- \atop \downarrow \atop O$		azoxy compounds
$X = -\underset{\overset{\|}{O}}{C}-O-$		ester compounds

From this list it can be seen that R_1 and R_2 and the core part can be worked on at will according to the physical problem studied or the desired mesophase.

In most cases we have actually synthesized the products mostly by following the literature, often with slight modifications of the procedures. We apologize to the authors for these modifications, which were made with a regard for simplification. To avoid repetitions, in Section VI we have selected some methods of purification used in the different syntheses (or to purify commercially available liquid crystals).

We were interested only in a small number of series because we thought these simple syntheses were accessible to a physicist. After a certain amount of success with these preparations, a physicist reader may want to try syntheses with a more specific phase or molecule in mind. In this case, he must refer to the classical chemical literature of liquid crystals. In fact, for this second level the presence of a chemist seems necessary.

Perhaps, before embarking on a synthesis, a physicist should spend a short time in a chemical laboratory to see the apparatus used and the different manipulations.

II. A General Presentation

All the syntheses described here will be preceded by different symbols, following the way in which a climbing guide indicates the different levels of difficulty. These symbols will be:

E = Easy
LD = A Little Difficult
D = Difficult
VD = Very Difficult

But, as in a mountain climb, where the difficulty of the hike depends on the snow, the weather, and other conditions sometimes unexpected, nothing can replace experience.

Sometimes we give the general name of the reaction (Friedel and Crafts, Schmidt, . . .), but in this article we do not intend to give the basic theory of the reactions. If the reader wants to learn some details, he should refer to any standard organic chemistry book.

Our plan for all the syntheses quoted will be the following:

1. Procedure with symbols of difficulty, names of authors, and formulae.
2. Reagents and solvents required for the synthesis.
3. Apparatus used.
4. Experiment, with the details of the synthesis.
5. Purification.
6. *Remarks* specifying the "*hazards*," and the observations leading to an increase of the yield and a reduction of the time of reaction. Read these remarks carefully before starting the practical work. References to remarks are given in parentheses.

The following abbreviations will be used:

c.p. = commercial product
C = crystal
S_B = smectic B
S_H = smectic H
S_C = smectic C
S_A = smectic A
N = nematic
I = isotropic liquid
Bp = boiling point
Mp = melting point
mmHg = pressure in millimeters of mercury
°C = degree Celsius
$n_D^{20°C}$ = refractive index at 20°C for the D sodium line.

Before all the syntheses, we shall specify the scheme of the reaction as:

$$A + B \xrightarrow[[Y]]{(X)} C + D;$$

(X) means X is the solvent used; [Y] means Y participates in the reaction as catalyst or intermediary complex.

III. Schiff's Bases

Schiff's bases are the most numerous mesomorphic compounds in the literature because they are easily prepared by a reaction between aromatic aldehydes and amines as follows:

where R_1 and R_2 are various substituents as seen in the introduction.

These compounds represent a very important class because they are used most often in physical experiments. Their synthesis is always easy but their purification is sometimes difficult. The most important thing to know about Schiff's bases is that they are unstable substances, very sensitive to moisture and acids.

Let us examine now the different series of Schiff's bases and especially the most often required:

= p-alkyloxybenzylidene-p'-alkylanilines

= p-substituted benzylidene-p'-alkyloxyanilines

= p-substituted benzylidene-p'-alkylaminocinnamates

= terephthalydene-di-(p-substituted aniline)

1. Preparation of *p*-Alkyloxybenzylidene-*p'*-Alkylanilines

The most famous representative of this series is MBBA (*p*-methoxy-benzylidene-*p'*-butylaniline),[1] the first room-temperature nematic:

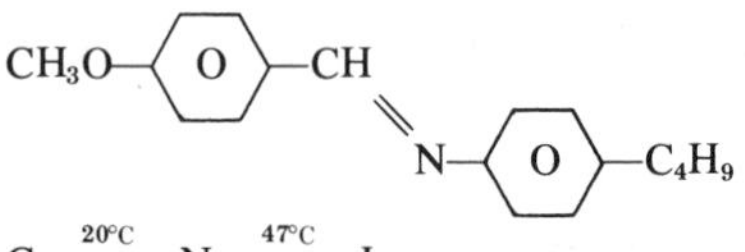

$$C \xrightarrow{20°C} N \xrightarrow{47°C} I.$$

The compounds required for the preparation of *p*-alkyloxybenzylidene-*p'*-alkylanilines are *p*-*n*-alkyloxybenzaldehydes and *p*-alkylanilines.

a. Preparation of p-n-Alkyloxybenzaldehydes

They are obtained by a one-step reaction from commercial products. The reaction is

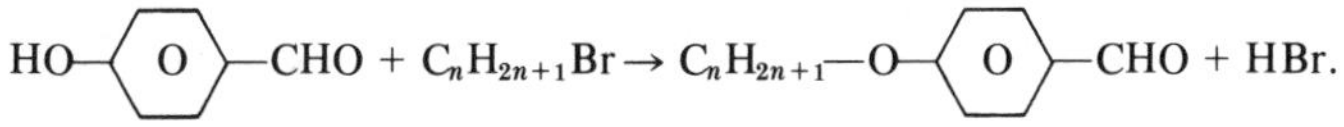

Procedure: LD, after Gray and Jones.[2]

Reaction:

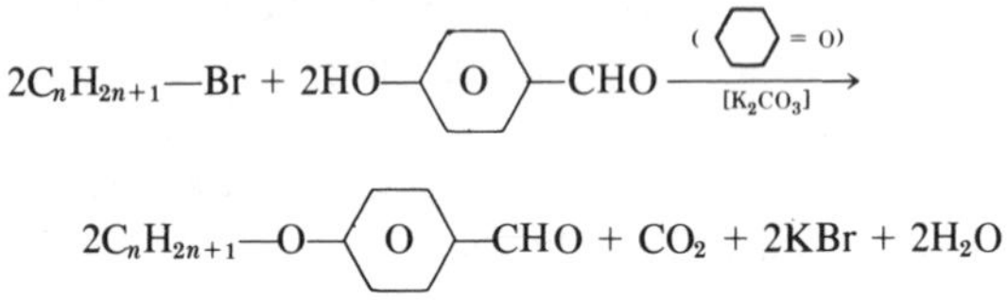

Reagents and solvents:

n-alkylbromide:	$C_nH_{2n+1}Br$	(c.p.)
p-hydroxybenzaldehyde:	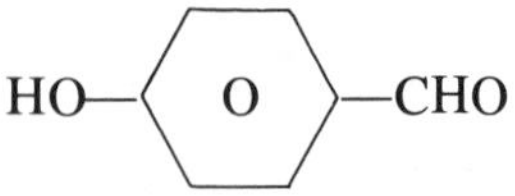 HO—⟨O⟩—CHO	(c.p.)
anhydrous potassium carbonate:	CO_3K_2	(c.p.)
cyclohexanone:	⟨ ⟩ = O	(c.p.)

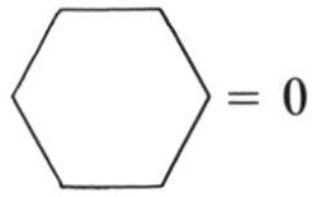

[1] Kelker, H., and Scheurle, B., *Angew. Chem., Int. Ed. Engl.* **8**, 884, (1969).
[2] Gray, G. W., and Jones, B., *J. Chem. Soc.* p. 1467 (1954).

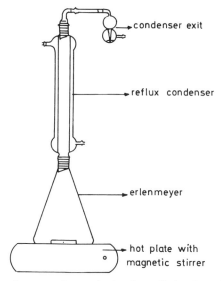

FIG. 1.   Apparatus for synthesis of *p-n*-alkyloxybenzaldehydes.

Apparatus: The apparatus for this synthesis is shown in Fig. 1.

Experiment: 12.2 g (0.1 mole) of *p*-hydroxybenzaldehyde, 60 g (0.4 mole) of anhydrous potassium carbonate, 80 ml of cyclohexanone (as solvent), and 0.16 mole of *n*-alkylbromide are introduced into an erlenmeyer (500 ml) fitted out with a reflux–condenser (1). The mixture is then stirred vigorously and heated until reflux (2). A significant volume of gas (CO_2 formed during the reaction) is evolved. This can be controlled by bubbling at the condenser exit. After refluxing for 3 hr, the reaction is finished.

Purification: The reaction mixture is cooled to room temperature and then filtered off to remove excess of K_2CO_3 and KBr formed during the reaction. The precipitate is washed with 50 ml of ether (3). This filtrate is then evaporated on a rotovapor: Ether is removed first, but for cyclohexanone elimination the temperature of the water bath must be 90–100°C and the vacuum about 10–20 mmHg (a water pump is sufficient). The residue obtained is distilled under vacuum (see Section VI). Yield: 75–85%.

Remarks: 1. Warning: Do not forget to lubricate the ground glass joint of the condenser with an extremely thin film of vacuum grease; it can be attacked by the alkali, which causes the joint to stick.

2. A magnetic stirrer is sufficient.

3. Warning: The temperature of the precipitate must be lower than 36°C (Bp of ether = 36°C).

4. The following *p-n*-alkyloxybenzaldehydes are common commercial products:

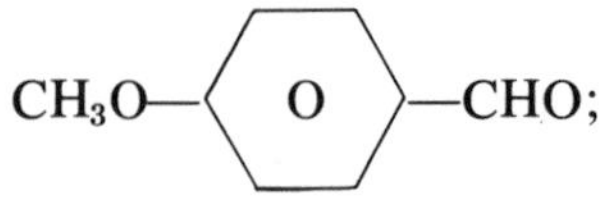

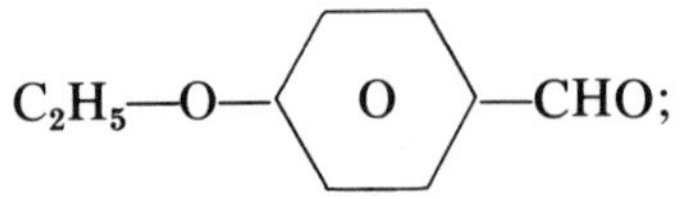

$$CH_3O\!-\!\langle\ O\ \rangle\!-\!CHO;$$

$$C_2H_5\!-\!O\!-\!\langle\ O\ \rangle\!-\!CHO;$$

and for $n = 3$ to $n = 8$, only one company sells them.

These commercial aldehydes must be redistilled before use because they are never pure: Aromatic aldehydes are quickly oxidized in acids which are unsuitable for the following reaction.

5. For the branched *p*-alkyloxybenzaldehydes of general formula

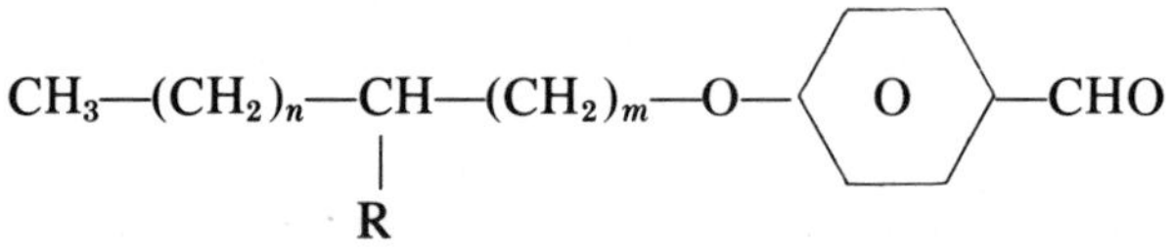

$$CH_3\!-\!(CH_2)_n\!-\!CH\!-\!(CH_2)_m\!-\!O\!-\!\langle\ O\ \rangle\!-\!CHO \quad \text{with} \begin{cases} 0 \leqslant n \\ 0 \leqslant m \end{cases}$$
$$\underset{R}{\vert}$$

in the previous reaction, you must replace

$$C_nH_{2n+1}Br \quad \text{by} \quad CH_3(CH_2)_n\!-\!\underset{\underset{R}{\vert}}{CH}\!-\!(CH_2)_m\!-\!Br.$$

b. Preparation of p-n-Alkylanilines

p-n-alkylanilines are obtained by the following three reactions from commercial products.

a. Friedel and Crafts reaction:

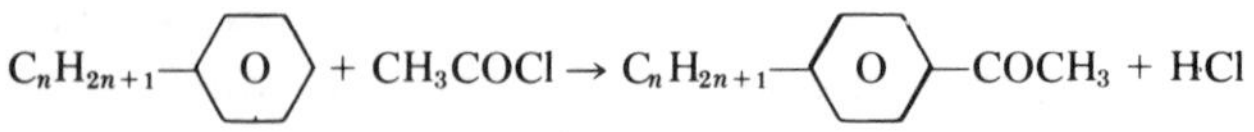

$$C_nH_{2n+1}\!-\!\langle\ O\ \rangle + CH_3COCl \rightarrow C_nH_{2n+1}\!-\!\langle\ O\ \rangle\!-\!COCH_3 + HCl$$

b. Schmidt reaction on acetophenone:

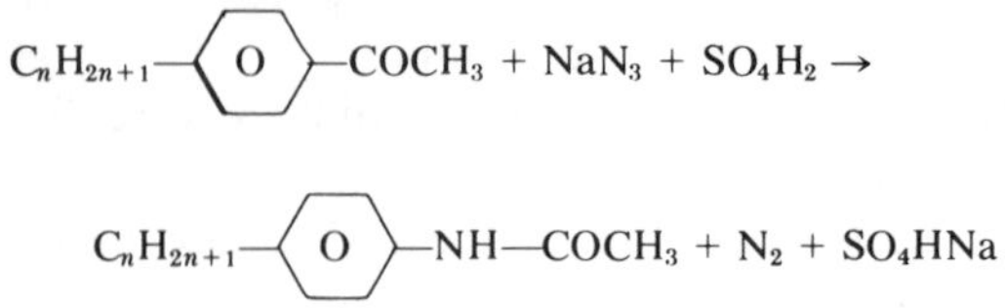

$$C_nH_{2n+1}\!-\!\langle\ O\ \rangle\!-\!COCH_3 + NaN_3 + SO_4H_2 \rightarrow$$

$$C_nH_{2n+1}\!-\!\langle\ O\ \rangle\!-\!NH\!-\!COCH_3 + N_2 + SO_4HNa$$

c. Hydrolysis reaction of acetanilide:

$$C_nH_{2n+1}\!\!-\!\!\left\langle O \right\rangle\!\!-\!\!NHCOCH_3 + H_2O \rightarrow C_nH_{2n+1}\!\!-\!\!\left\langle O \right\rangle\!\!-\!\!NH_2 + CH_3COOH$$

i. Preparation of p-n-Alkylacetophenone. Procedure: D, after Van der Veen *et al.*[3]

Reaction:

$$C_nH_{2n+1}\!\!-\!\!\left\langle O \right\rangle + CH_3COCl \xrightarrow[\text{[AlCl}_3\text{]}]{\text{(CCl}_4\text{)}} C_nH_{2n+1}\!\!-\!\!\left\langle O \right\rangle\!\!-\!\!COCH_3 + HCl$$

Reagents:

n-alkylbenzene:	$C_nH_{2n+1}\!\!-\!\!\left\langle O \right\rangle$	(c.p.)
dried aluminum chloride:	AlCl$_3$	(c.p.)
acetylchloride:	CH$_3$COCl	(c.p.)
carbon tetrachloride:	CCl$_4$	(c.p.)
hydrochloric acid:	HCl	(c.p.)
dried sodium hydrogen carbonate:	NaHCO$_3$	(c.p.)
molecular sieves (4 Å):		(c.p.)

Apparatus: The apparatus for this synthesis is shown in Figs. 2 and 3.

Experiment: p-n-alkylbenzene (0.5 mole) is added dropwise through a dropping funnel to a stirred mixture of 80 g (0.6 mole) of AlCl$_3$ and 42.6 ml (0.5 mole) of acetylchloride in 350 ml of CCl$_4$ (as solvent) cooled by an ice bath in a reactor of 2 l; during the addition, the temperature of the reactional mixture must be kept between 0 and 5°C (1). The ice bath is removed and the reactional mixture is stirred for an additional hour at room temperature. The mixture is then poured cautiously (2) while stirring into a solution of concentrated hydrochloric acid (250 ml) and ice (500 g).

Purification: This purification is an extraction (see Section VI). The resulting heterogeneous solution is then poured into a separatory funnel (2 l). Two layers are formed: The lower one is the organic layer (organic products in CCl$_4$) and the higher one is the aqueous layer. The organic layer is kept and the aqueous layer discarded.

The organic layer is washed by extraction twice with 250 ml of 2 *N*

[3] Van der Veen, J., de Jeu, W. H., Grobben, A. H., and Boven, J., *Mol. Cryst. Liq. Cryst.* **17**, 291 (1972).

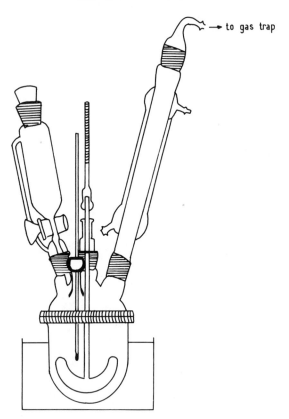

FIG. 2.   Apparatus for synthesis of p-n-alkylacetophenone.

hydrochloric acid (removal of aluminum salts), once with 250 ml of sodium hydrogen carbonate solution (removal of the acid traces), and finally with 250 ml of water (at each stage, the organic layer is always the lower). The resulting organic solution is then dried on molecular sieves (Merk 4 Å) for 3 hr. The solvent is removed under reduced pressure with a rotovapor and the residue is distilled in vacuum (see Section VI). Yield: 70–90%.

Remarks: 1. Warning: This temperature is important because you may bring about the formation of the ortho derivative at high temperature and this derivative is very difficult to separate from the desired para-derivative.

2. Warning: the hydrolysis of the complex formed during this Friedel and Crafts reaction is an exothermic reaction.

3. The boiling points and refractive indices for *n*-alkylacetophenone[3] are given in Table I.

GAS TRAP

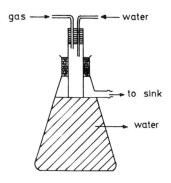

FIG. 3.   Apparatus for synthesis of *p-n*-alkylacetophenone.

ii. Preparation of p-n-Alkylacetanilide. Procedure: VD, after Van der Veen *et al.*[3]

Reaction:

$$C_nH_{2n+1}\!-\!\langle O \rangle\!-\!COCH_3 + NaN_3 + SO_4H_2 \xrightarrow{(CH_2Cl_2)}$$

$$C_nH_{2n+1}\!-\!\langle O \rangle\!-\!NHCOCH_3 + N_2 + SO_4HNa$$

Reagents and solvents:

p-n-alkylacetophenone: $C_nH_{2n+1}\!-\!\langle O \rangle\!-\!COCH_3$

sulfuric acid: H_2SO_4 (c.p.)
sodium azide: NaN_3 (c.p.)

TABLE I.   BOILING POINTS AND REFRACTIVE INDICES FOR *n*-ALKYLACETOPHENONE

n	Bp(°C)	mmHg	$n_D^{20°C}$
3	85	1	1.5246
5	108	1	1.5152
6	120	1.1	1.5126
7	130	1	1.5106
8	140	1	1.5070
9	150	1	1.5044
10	162	1	solid Mp at 37–37.5°C

dichloromethane: CH_2Cl_2 (c.p.)
sodium hydrogen $NaHCO_3$ (c.p.)
  carbonate:
petroleum ether (60–80°C),   (c.p.)
  generally mixture of
  C_6H_{14} and C_7H_{16}

Apparatus: The same as that described in Section III,1,*b,i* with a reactor of 1 l and the dropping funnel replaced by a stopper.

Experiment: 7 g (0.11 mole) of sodium azide (1) is added by small portions to a stirred solution of *p-n*-alkylacetophenone (0.1 mole) in 146 ml of 70% sulfuric acid (2). To prevent foaming, 50 ml of dichloromethane is added before addition of NaN_3 (3). During the addition, the temperature of the reactional mixture must be kept between 15 and 20°C by an ice–water bath. During the reaction nitrogen is released and the mixture becomes more viscous. After stirring an additional half-hour, the reactional mixture is slowly poured into a mixture of 200 ml of chilled water and 100 ml of dichloromethane.

Purification: This purification is an extraction (see Section VI). The resulting heterogeneous solution is poured into a separatory funnel (2 l). Two layers are formed: The lower is the organic layer (organic products in solution in dichloromethane) and the higher is the aqueous layer. The two layers are separated; the organic layer is kept and the aqueous layer is extracted twice with 50 ml of dichloromethane. Each time the organic layer is the lower one.

The combined organic layers are washed by extraction with 50 ml of a sodium hydrogen carbonate solution and water, respectively. A part (150–200 ml) of dichloromethane is removed under reduced pressure with a rotovapor and the residual solution is poured into 200 ml of petroleum ether (60–80°C). The resulting white precipitate is filtered off and dried in vacuum. Yield: 75–85%.

Remarks: 1. Warning: You must pour 8–10% of sodium azide first. After a quarter of an hour you will observe nitrogen evolving and you can add by small portions the remainder of the sodium azide. Sodium azide may explode if temperature is elevated. *Use goggles and rubber gloves for this manipulation, which is hazardous.*

2. The 70% sulfuric acid solution is made as follows: to 150 ml of water are added slowly 350 g of sulfuric acid (about 192 ml if the sulfuric acid used has a density of 1.83). *Warning: the acid is poured into the water, never the contrary.*

3. The addition of dichloromethane is important because it regulates the contact between SO_4H_2 and NaN_3.

4. The melting points of *p-n*-alkylacetanilide are given in Table II.

TABLE II. MELTING POINT OF p-n-
ALKYLACETANILIDE

n	Mp (°C)
3	99
5	102
6	91
7	91
8	94
9	97
10	101

iii. Preparation of p-n-Alkylaniline. Procedure: LD, after Van der Veen *et al.*[3]

Reaction:

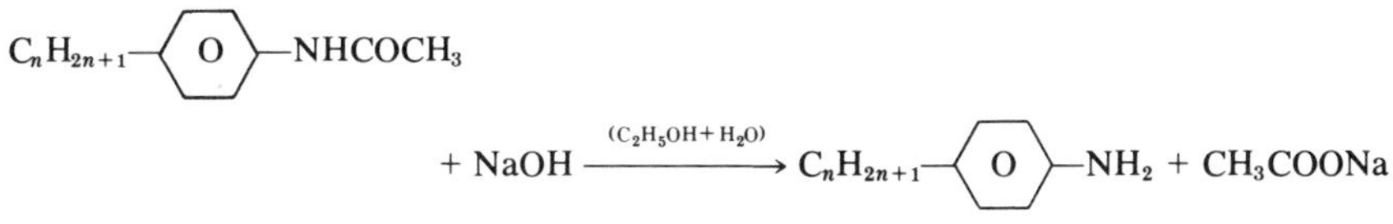

$$C_nH_{2n+1}\!-\!\!\left\langle O \right\rangle\!-\!NHCOCH_3$$

$$+ \, NaOH \xrightarrow{(C_2H_5OH + H_2O)} C_nH_{2n+1}\!-\!\!\left\langle O \right\rangle\!-\!NH_2 + CH_3COONa$$

Reagents and solvents:

p-n-alkylacetanilide: $C_nH_{2n+1}\!-\!\!\left\langle O \right\rangle\!-\!NHCOCH_3$

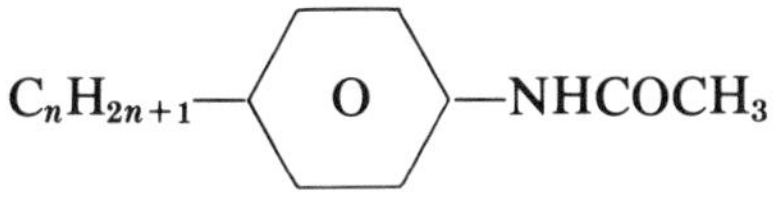

sodium hydroxide:	NaOH	(c.p.)
ethanol:	C_2H_5OH	(c.p.)
benzene:	C_6H_6	(c.p.)
dried potassium carbonate:	CO_3K_2	(c.p.)

Apparatus: The same as that of Section III,1,*a*, but the condenser exit is not required.

Experiment: A mixture of p-n-alkylacetanilide (0.1 mole), 160 ml of ethanol, and 92 g of sodium hydroxide previously dissolved in 60 ml of water is refluxed for $5\frac{1}{2}$ hr (1).

Purification: The mixture is cooled and a part of the solvent is removed by distillation with rotovapor until the volume of residue is about 40 to 50 ml. This solution is poured into a mixture of 150 ml of water and 200 g of ice. The heterogeneous mixture is then poured into a separatory funnel (1 l), and extracted (see Section VI) twice with 100 ml of benzene (2). Each time the organic layer is the higher. The combined organic layers are extracted twice with 200 ml of water and dried with

TABLE III.   BOILING POINTS OF *p-n-*
ALKYLANILINE

n	Bp (°C)	mmHg
3	61	0.6
5	84	0.6
6	96	0.6
7	107	0.6
8	118	0.6
9	126	0.6
10	139	0.6

potassium carbonate. The benzene is distilled off with rotovapor and the residue is distilled in vacuum (see Section VI). Yield: 80–85%.

Remarks: 1. See Remark 1 of Section III,1,*a*.

2. Warning: Benzene is a dangerous solvent because it may disturb the blood content and can bring about leukemia. *It is advisable to work under a hood.*

3. The *p-n*-alkylanilines with $n = 1, 2, 4$ are commercial but must be distilled before use.

4. The anilines are classified as carcinogenic products and special care is recommended, including use of rubber gloves, noninhalation of vapors, and storage of waste before burning by a specialist company.

5. The boiling points of *p-n*-alkylanilines are given in Table III.

c. Preparation of p-n-Alkyloxybenzylidene-p'-n-Alkylaniline

Formula:

$$C_n H_{2n+1}O\text{—}\hexagon\text{—}CH\text{=}N\text{—}\hexagon\text{—}C_{n'} H_{2n'+1}$$

with $1 \leq n$ and $4 \leq n'$

for which the mesophases exist.

In fact, for these syntheses two procedures are commonly available, and we will describe successively the two methods on two famous products most demanded by physicists: MBBA[1] and 40.8.[4] MBBA has served actually for dynamic scattering, thermal conductivity, disclinations studies and 40.8 for the near second-order transition between SmA and N.

[4] Smith, G. W., Gardlung, Z. G., and Curtis, R. J., *Mol. Cryst. Liq. Cryst.* **19**, 327 (1973).

i. Preparation of MBBA[1] *by Benzene Method. Procedure:* LD.
Reaction:

$$CH_3O-\langle O \rangle-CHO + H_2N-\langle O \rangle-C_4H_9 \xrightarrow[\text{[CH}_3-\langle O \rangle-SO_3H]}{\text{(C}_6H_6)}$$

$$CH_3O-\langle O \rangle-CH$$
$$N-\langle O \rangle-C_4H_9 + H_2O$$

Reagents and solvents:

p-methoxybenzaldehyde (redistilled before use):	$CH_3O-\langle O \rangle-CHO$	(c.p.)
p-n-butylaniline (redistilled before use):	$C_4H_9-\langle O \rangle-NH_2$	(c.p.)
benzene:	$\langle O \rangle$	(c.p.)
p-toluene sulfonic acid:	$CH_3-\langle O \rangle-SO_3H$	(c.p.)

Apparatus: The apparatus for this synthesis is shown in Fig. 4.

Experiment: In an erlenmeyer of 2 l, 29.8 g (0.2 mole) of *p*-butylaniline, 27.2 g (0.2 mole) of *p*-methoxybenzaldehyde, 0.5 g of *p*-toluene sulfonic acid (as acid catalyst), and 800 ml of benzene (1) (as solvent) are refluxed for $3\frac{1}{2}$ hours. The water formed during the reaction is removed azeotropically with a Dean–Stark apparatus. In fact, the azeotropic water–benzene mixture distills, and in the water condenser this azeotropic mixture separates into two immiscible layers. Thus one can see the volume of water formed and know when the reaction is over (see the apparatus).

Purification: The mixture is filtered off warm (1), about 60 to 70°C. After cooling the benzene is removed by distillation on rotovapor and the residue is distilled under reduced pressure (see Section VI).

10 g of distilled MBBA is dissolved in 1 l of hexane and crystallized in

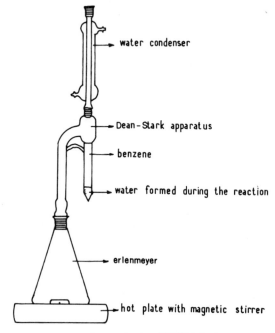

FIG. 4.   Apparatus for synthesis of MBBA by benzene method.

a freezer ($-18°$). The white crystals obtained are filtered off very rapidly and distilled, in general, twice under reduced pressure. Ebullition: 170–180° at 3–4 mmHg; yield: approximately 50%.

The time of reaction is difficult to determine because the purity of distilled MBBA, verified by the value of isotropic transition, sometimes requires three or four distillations in vacuum.

After this purification, the transition temperature must be:

$$ C \xrightarrow{21°C} N \xrightarrow{47°C} I. $$

Remarks: 1. Warning: For benzene see Remark 2 of Section III,1,*a,iii*.

2. If the product has a transition temperature above room temperature, the repeated crystallizations (see Section V) are more practical.

3. Do not forget to purify the starting materials extremely well, which will make the number of distillations smaller.

ii. Preparation of 40.8[4] by Ethanol Method. Procedure: E, after Smith *et al.*[4]

Reaction:

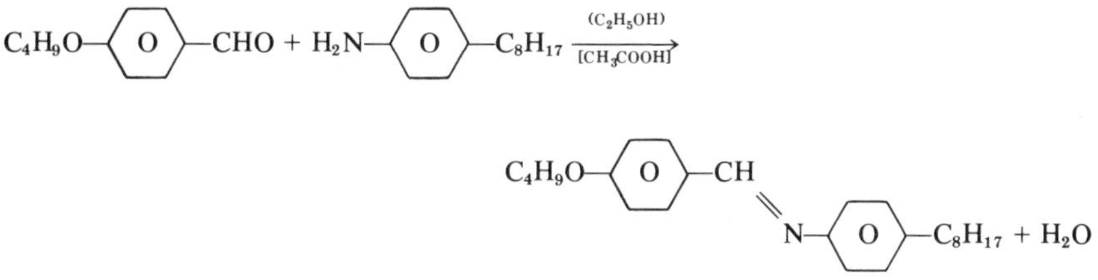

$$C_4H_9O-\langle O \rangle-CHO + H_2N-\langle O \rangle-C_8H_{17} \xrightarrow[\text{[CH}_3\text{COOH]}]{\text{(C}_2\text{H}_5\text{OH)}}$$

$$C_4H_9O-\langle O \rangle-CH$$
$$\| N-\langle O \rangle-C_8H_{17} + H_2O$$

Reagents:

p-n-butoxybenzaldehyde: $C_4H_9O-\langle O \rangle-CHO$

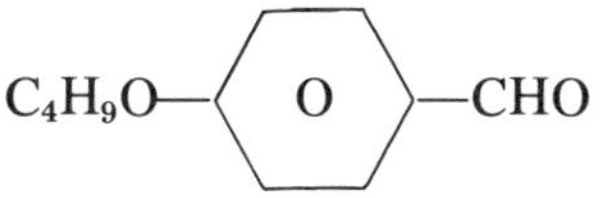

p-n-octylaniline: $C_8H_{17}-\langle O \rangle-NH_2$

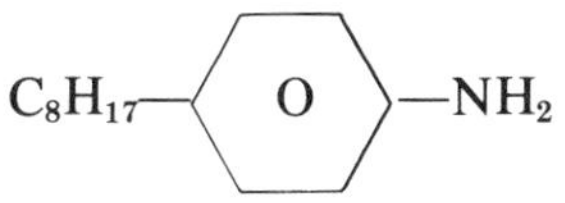

ethanol: C_2H_5OH (c.p.)
acetic acid: CH_3COOH (c.p.)

Apparatus: 250 ml erlenmeyer; magnetic stirrer.

Experiment: In an 250 ml erlenmeyer a mixture of 3.56 g (0.02 mole) of p-n-butoxybenzaldehyde, 4.10 g of p-n-octylaniline, one or two drops of acetic acid as catalyst, and 100 ml of ethanol as solvent is stirred with a magnetic stirrer. After about 20 or 30 min of stirring, the product formed during the reaction reprecipitates. The reaction is practically complete after 3 hr.

Purification: The precipitate is collected by filtration and washed twice with 25 ml of cold ethanol. The 40.8 is then recrystallized several times in ethanol until the transition temperatures remain constant (see general chapter on the purification). Yield: 60–65%.

The transition temperatures of 40.8 must be:

$$C \xrightarrow{33°} S_B \xrightarrow{49.5°} S_A \xrightarrow{63.5°} N \overset{79.0°}{\underset{\ell}{\rightarrow}} I$$

Thus physicists have studied the transition between smectic A and nematic. The purity of the material required for this experiment is very high, but the instability of 40.8 does not permit great accuracy on $T_{c_{A\rightarrow N}}$ if long-term experiments are required.

Remarks: 1. This experiment is very easy, but a very pure product is difficult to obtain.

2. We have seen 40.8 absolutely dark after certain physical experiments of several days. The impure product had to be discarded.

3. 40.8 decomposes rapidly in the presence of air and heat but in an experiment made in a sealed tube under argon its transition temperatures have been stable for a very long time of experiment.

We have seen the general preparation of *p-n*-alkyloxybenzylidene-*p-n*-alkylanilines; let us see now a new series of Schiff's bases.

2. PREPARATION OF *p*-SUBSTITUTED BENZYLIDENE-*p'n*-ALKYLOXYANILINE

Formula:

$$\begin{cases} 1 \leq n \\ R_1 \text{ may be any substituent.} \end{cases}$$

The raw materials are the *p*-substituted benzaldehydes:

$$\begin{cases} R_1 = OC_nH_{2n+1} & \text{(see Section III,1,}a\text{)} \\ R_1 = -CN & \text{(c.p.)} \\ R_1 = -Br, -Cl & \text{(c.p.)} \\ R_1 = \end{cases}$$

and *p-n*-alkyloxyanilines, $C_nH_{2n+1}O-$$-NH_2$, of which we will describe the preparation.

a. Preparation of p-n-Alkyloxyaniline

p-n-alkyloxyanilines are obtained by the following two-step reaction.

a. O-alkylation reaction:

b. Hydrolysis of p-alkyloxyacetanilide:

$$C_nH_{2n+1}O\!-\!\!\left<\!O\!\right>\!\!-\!HNCOCH_3 + KOH \xrightarrow{\text{(C}_2\text{H}_5\text{OH)}} C_nH_{2n+1}O\!-\!\!\left<\!O\!\right>\!\!-\!NH_2 + CH_3COOK$$

These two reactions seem to be similar, but in the first one we form the potassium salt of the phenol derivative at room temperature that permits the O-alkylation of acetanilide, and in the second the potassium hydroxide at an increased temperature hydrolyzes the acetanilide into aniline. Thus the physicist can see that the conditions of the experiment are very important and he must follow the "recipe" carefully.

i. Preparation of p-n-Alkyloxyacetanilides. Procedure: E, after Buu-hoi *et al.*[5]
Reaction:

$$CH_3CONH\!-\!\!\left<\!O\!\right>\!\!-\!OH + KOH \xrightarrow{\text{(C}_2\text{H}_5\text{OH)}} CH_3CONH\!-\!\!\left<\!O\!\right>\!\!-\!OK + H_2O$$

$$CH_3\!-\!CO\!-\!NH\!-\!\!\left<\!O\!\right>\!\!-\!OK + C_nH_{2n+1}\!-\!Br \rightarrow CH_3\!-\!CO\!-\!NH\!-\!\!\left<\!O\!\right>\!\!-\!OC_nH_{2n+1} + KBr$$

Reagents:

p-acetaminophenol: $CH_3CONH\!-\!\!\left<\!O\!\right>\!\!-\!OH$ (c.p.)

ethanol: C_2H_5OH (c.p.)
potassium hydroxide: KOH   (c.p.)
n-alkylbromide: $C_nH_{2n+1}Br$ (c.p.)

Apparatus: The apparatus for this synthesis is shown in Fig. 5.
Experiment: In an erlenmeyer (250 ml) equipped with a dropping funnel and a water condenser, 15.1 g (0.1 mole) of p-acetaminophenol is dissolved into 75 ml of hot ethanol with magnetic stirring. After cooling, 7 g (0.125 mole) of potassium hydroxide dissolved in a minimum of water (about 5 ml) is added slowly through the dropping funnel (1, 2). Then, using the same dropping funnel, n-alkylbromide (0.11 mole), dissolved in 25 ml of ethanol, is added dropwise to the mixture.
    The mixture is refluxed for 1 hr after the end of the addition and then 25 ml of water is poured into it. After cooling of the mixture, the p-n-alkyloxyacetanilide crystallizes.

[5] Buu-Hoï, N. P., Gautier, M., and Dat Xuong, N., *Bull. Soc. Chim. Fr.* p. 2154 (1962).

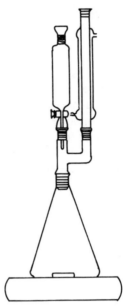

FIG. 5.   Apparatus for synthesis of *p-n*-alkyloxyacetanilides.

Purification: The solid obtained is filtered off, washed twice or three times with water until the filtrate is neutral, and recrystallized (see general section on purification) from ethanol/water (about 75–25%). Yield: 70–90%.

Remarks: 1. The dissolution of potassium hydroxide by a minimum of water is dangerous and exothermic: *Use goggles and rubber gloves.*

2. Do not forget to lubricate the ground glass joint, for the same reasons as in Section III,1,*a*.

    ii. Preparation of p-n-Alkyloxyaniline. Procedure: LD, after Buu-Hoï *et al.*[5]

Reaction:

$$C_nH_{2n+1}O-\left\langle O \right\rangle-NHCOCH_3 + KOH \xrightarrow{(C_2H_5OH)}$$

$$C_nH_{2n+1}-O-\left\langle O \right\rangle-NH_2 + CH_3COOK$$

Reagents and solvents:

p-n-alkyloxyac-
     etanilide: $C_nH_{2n+1}-O-\left\langle O \right\rangle-NHCOCH_3$

ethanol:	C_2H_5OH	(c.p.)
potassium hydroxide:	KOH	(c.p.)
anhydrous sodium sulfate:	SO_4Na_2	(c.p.)

benzene: (c.p.)

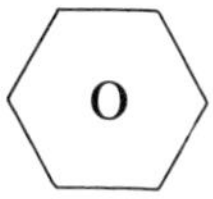

Apparatus: The apparatus is identical to that used in Section III,2,*a,i*.

Experiment: In an erlenmeyer equipped with a dropping funnel and a water condenser, a mixture of *p-n*-alkyloxyacetanilide (0.05 mole) and ethanol (50 ml) is refluxed.

Then, through the dropping funnel, 20 *N* potassium hydroxide (12.5 ml) (1) is added slowly. After the end of the addition, the reaction mixture is refluxed for 5 hr.

Purification: Ethanol is removed by distillation using rotovapor, and the *p-n*-alkyloxyaniline is recovered by filtration if the amine is solid or by extraction with benzene (2) (see Section VI) if it is liquid. This last procedure is the more convenient purification. The residue from the distillation is extracted twice with 50 ml of benzene. The combined organic layers are washed with water until they are neutral and then dried on anhydrous sodium sulfate.

The benzene is removed by distillation using rotovapor and the residue is then distilled in vacuum (see Section VI). Yield: 80–85%.

Remarks: 1. The 20 *N* potassium hydroxide solution is prepared by dissolving 22.4 g of KOH (0.4 mole) in 20 ml of water. *Eyes should be protected with goggles for this operation.*

2. Benzene is a dangerous solvent and also the extraction with benzene can be difficult. Emulsions are very often formed and the separation into two layers can take some time.

b. Preparation of p-Substituted Benzylidene-p'-n-Alkyloxyaniline

Formula:

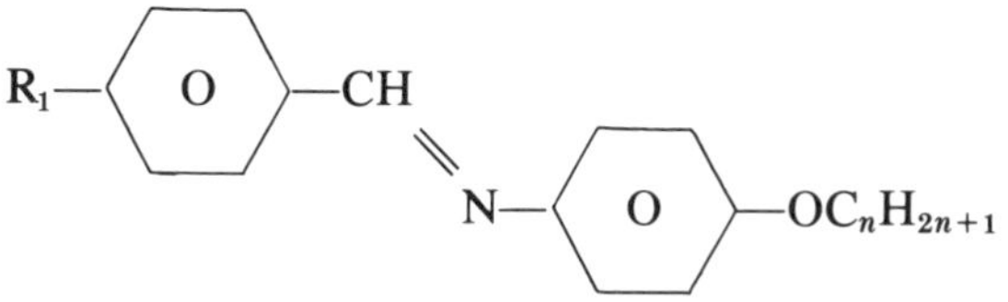

We will describe only the preparation of *p*-cyanobenzylidene-*p-n*-octyloxyaniline or CBOOA which is required for thermal conductivity and structure studies. The synthesis for the other members of the series may be made by a similar procedure.

i. Preparation of CBOOA. Procedure: E, ethanol method (see Section III,1,*c,ii*).
Reaction:

$$NC-\langle O \rangle-CHO + H_2N-\langle O \rangle-OC_8H_{17} \xrightarrow[\text{[CH}_3\text{COOH]}]{\text{(C}_2\text{H}_5\text{OH)}}$$

$$NC-\langle O \rangle-CH$$
$$\begin{array}{c} \\ N-\langle O \rangle-OC_8H_{17} + H_2O \end{array}$$

Reagents:

p-cyanobenzaldehyde:	$NC-\langle O \rangle-CHO$	(c.p.)
p-n-octyloxyaniline:	$H_2N-\langle O \rangle-OC_8H_{17}$	
ethanol:	C_2H_5OH	(c.p.)
acetic acid:	CH_3COOH	(c.p.)
heptane:	C_7H_{15}	(c.p.)

Apparatus: The apparatus for this synthesis is the same as that in Section III,1,*c,ii*.

Experiment: In an erlenmeyer, *p*-cyanobenzaldehyde (2.62 g), *p-n*-octyloxyaniline (4.10 g), and one or two drops of acetic acid as catalyst in 100 ml of absolute ethanol are stirred for 3 hr at room temperature.

Purification: The precipitate formed during the reaction is filtered off, washed two or three times with cold ethanol (25 ml portion), and recrystallized twice in ethanol and several times in heptane until the transition temperatures remain constant. Yield: 70–80%.

For the other members of the series the same synthesis may be used; to optimize the solvent of crystallization one must see Section VI.12.

3. PREPARATION OF p-SUBSTITUTED BENZYLIDENE-p'-AMINOCINNAMATE

Formula:

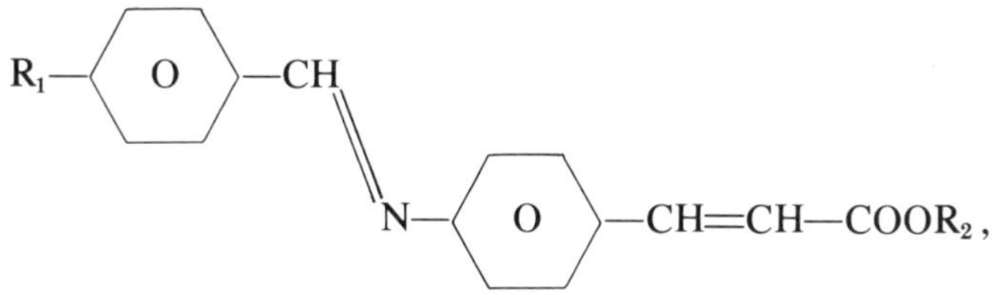

$$R_1 - \langle O \rangle - CH$$
$$N - \langle O \rangle - CH = CH - COOR_2,$$

where R_1 is a variable substituent (see Section III.2) and R_2 generally are linear or branched aliphatic chains. These Schiff's bases show different smectic mesophases, and when R_2 contains an asymmetric carbon we have cholestric mesophases and sometimes chiral smectic C.

a. Preparation of p-Aminocinnamates

For us, the best synthesis is the one proposed by Leclercq et al.,[6] using the following sequences of reactions.

a. Preparation of p-nitrocinnamoylchloride:

$$NO_2 - \langle O \rangle - CH = CHCOOH + SOCl_2 \rightarrow$$
$$O_2N - \langle O \rangle - CH = CH - COCl + SO_2 + HCl$$

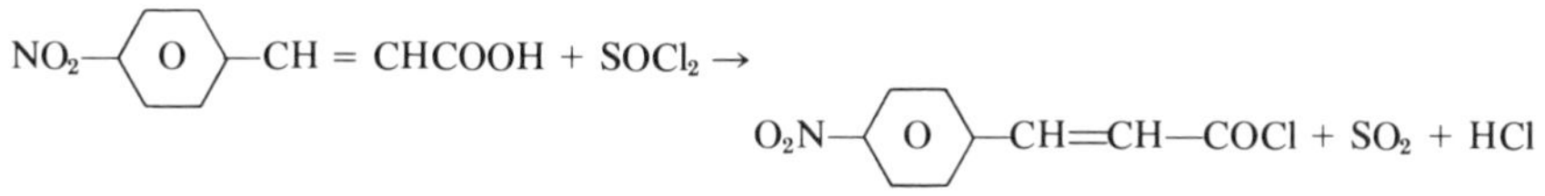

b. Esterification of p-nitrocinnamoylchloride:

$$O_2N - \langle O \rangle - CH = CH - COCl + HOR \rightarrow O_2N - \langle O \rangle - CH = CH - COOR + HCl$$

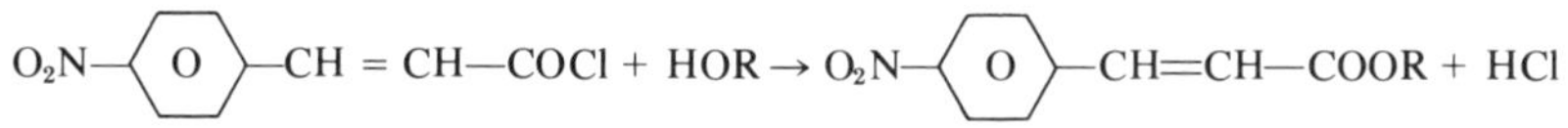

c. Reduction of p-nitrocinnamate:

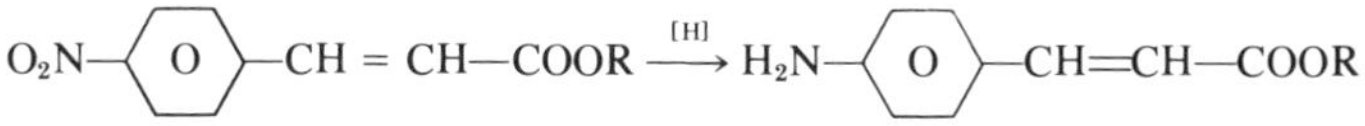

$$O_2N - \langle O \rangle - CH = CH - COOR \xrightarrow{[H]} H_2N - \langle O \rangle - CH = CH - COOR$$

i. Preparation of p-nitrocinnamoylchloride. Procedure: LD, after Leclercq et al.[6]

Reaction:

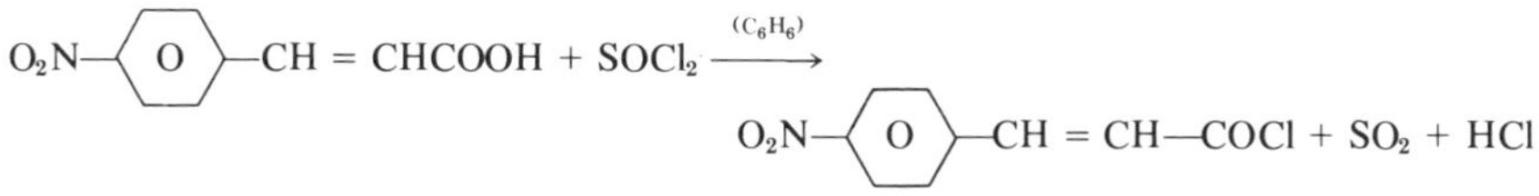

$$O_2N - \langle O \rangle - CH = CHCOOH + SOCl_2 \xrightarrow{(C_6H_6)}$$
$$O_2N - \langle O \rangle - CH = CH - COCl + SO_2 + HCl$$

[6] Leclercq, M. Billard, J., and Jacques, J., *Mol. Cryst. Liq. Cryst.* **8**, 367 (1969).

Reagents and solvents:

p-nitrocinnamic acid: O_2N—⟨O⟩—CH = CH—COOH   (c.p.)

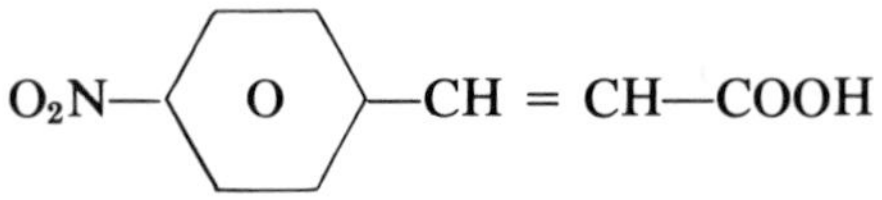

thionyl chloride:	$SOCl_2$	(c.p.)
benzene:	C_6H_6	(c.p.)
carbon tetrachloride:	CCl_4	(c.p.)

Apparatus: The apparatus for this synthesis is the same as that in Section III,1,*b,i*.

Experiment: 5.80 g (0.03 mole) of *p*-nitrocinnamic acid, 15 ml of thionyl chloride (1), and 12 ml of benzene are refluxed for 6 hr in the apparatus described above. The release of gas is verified by bubbling at the exit of the water condenser. The gas trap is useful for the safety of the experimentalist.

Purification: The solvent and excess of thionyl chloride are removed by distillation using a rotovapor, under reduced pressure. The residue is crystallized from carbon tetrachloride (150 ml). Yield: ≈60%; Mp = 150°C.

Remarks: 1. Thionyl chloride is a lachrymatory and it is essential that great care be taken in using it; *it should be kept under a hood during use.*

2. See Remark 2 in Section III,1,*b,iii*.

ii. Preparation of p-Nitrocinnamate. Procedure: LD, after Leclercq et al.[6]

Reaction:

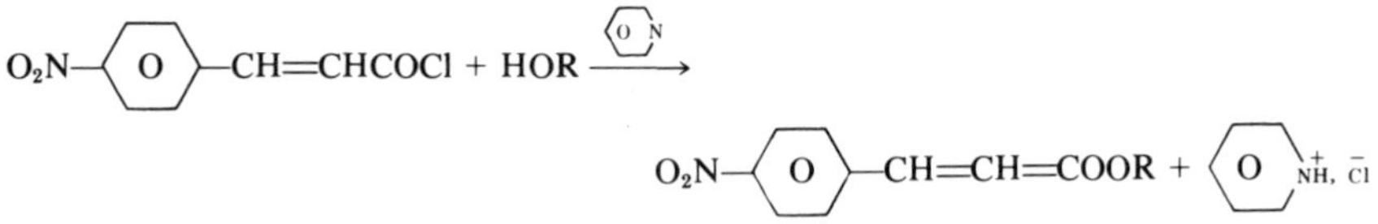

O_2N—⟨O⟩—CH=CHCOCl + HOR →

O_2N—⟨O⟩—CH=CH=COOR + ⟨O⟩NH, Cl

Reagents and solvents:

p-nitrocinnamo- O_2N—⟨O⟩—CH=CH—COCl
  ylchloride:

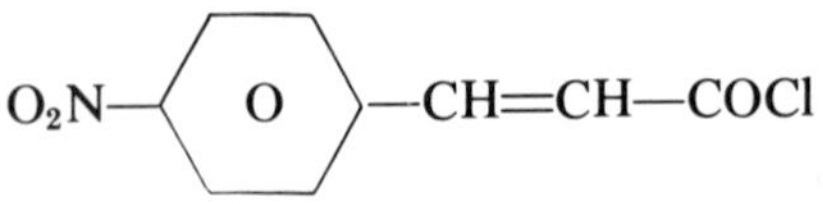

alcohol: ROH   (c.p.)

dried pyridine: ⟨O⟩N   (c.p.)

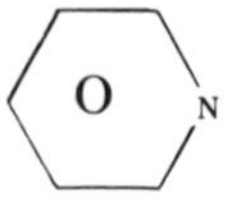

ether:	C_2H_5—O—C_2H_5	(c.p.)
sulfuric acid:	SO_4H_2	(c.p.)
sodium hydrogen carbonate:	CO_3HNa	(c.p.)
methanol:	CH_3OH	(c.p.)
anhydrous magnesium sulfate:	$MgSO_4$	(c.p.)

Apparatus: An erlenmeyer.

Experiment: A mixture of 3.8 g (0.018 mole) of *p*-nitrocinnamoyl chloride, 0.017 mole of alcohol, and 75 ml of dried pyridine (1) is allowed to stand for overnight at room temperature in an erlenmeyer (100 ml).

Purification: Water (100 ml) is added; the resulting heterogenous mixture is poured into a separatory funnel (1 l) and extracted with ether (twice 100 ml). The ether layer is always on top. The combined organic layers are washed by extraction several times with a sulfuric acid solution (2) (10% by weight), once with water, once with a sodium hydrogen carbonate solution (5% by weight), and finally with water. The ether solution is dried on anhydrous magnesium sulfate for some hours, then ether is removed by distillation using a rotovapor. The residue can be a solid or a liquid depending on the alcohol used in the reaction.

The ester obtained can be used without other purification for the last reaction. Yield: ≈50%.

Remarks: 1. Commercial pyridine must be dried by adding KOH in pellets in the bottle before use (about 30 g by liter).

2. This sulfuric acid solution is used to remove the pyridine. These washings are very important for the purity of esters.

iii. Preparation of p-Aminocinnamate. Procedure: D, after Leclercq et al.[6]

Reaction:

$$O_2N-\langle\!\!\langle\ O\ \rangle\!\!\rangle-CH{=}CH-COOR \xrightarrow[\text{[FeCl}_2+\text{NH}_3\text{]}]{\text{(C}_2\text{H}_5\text{OH)}} H_2N-\langle\!\!\langle\ O\ \rangle\!\!\rangle-CH{=}CH-COOR$$

Reagents and solvents:

p-nitrocinnamate:	$O_2N-\langle\!\!\langle\ O\ \rangle\!\!\rangle-CH{=}CH-COOR$	
anhydrous ferrous chloride:	$FeCl_2$	(c.p.)
ethanol:	C_2H_5OH	(c.p.)
ammonia (gas):	NH_3	(c.p.)

ether: $C_2H_5OC_2H_5$ (c.p.)
anhydrous magnesium $MgSO_4$ (c.p.)
    sulfate:

Apparatus: The apparatus for this synthesis is shown in Fig. 6.

Experiment: Dry ammonia (1) is passed through a heterogeneous mixture of *p*-nitrocinnamate (0.01 mole) and anhydrous ferrous chloride (20 g) in ethanol (140 ml), stirred with a magnetic stirrer. The mixture immediately turns green and becomes more viscous.

Often the magnetic stirrer loses its effectiveness and the apparatus must be shaken by hand. The reaction becomes very exothermic and ethanol can reflux. After a while the mixture turns brown and becomes more fluid, and the magnetic stirrer is again efficient.

After about 20 min, no more ammonia is absorbed and the temperature begins to decrease rapidly.

The ammonia flow is stopped; the mixture is then allowed to stand for 2 hr with continuous stirring under a hood to permit release of ammonia gas dissolved in ethanol (2).

Purification: The mixture is poured into a separatory funnel (2 l) and water is added (about 300 ml). The resulting dark-brown solution is extracted twice with 150 ml of ether. The combined organic layers are washed by extraction several times with water until they are neutral, and are then dried on anhydrous magnesium sulfate for some hours. The solvent is removed by distillation using rotovapor and the residue is crystallized from aqueous ethanol. Yield: 75–80%.

Remarks: 1. Warning: NH_3 is unpleasant to inhale. *This experiment must be carried out under a hood.*

2. If NH_3 is not completely evolved, some trouble can occur during the purification and the yield is lower.

3. Melting points of some *p*-aminocinnamates are given in Table IV.

iv. Preparation of p-Substituted Benzylidene-p'-Aminocinnamates.
Procedure: E, after Section III,1,c,ii.
Reaction:

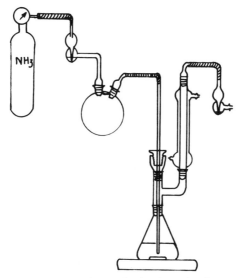

FIG. 6.   Apparatus for synthesis of p-aminocinnamate.

TABLE IV.   MELTING POINTS OF SOME p-
AMINOCINNAMATES

R	Mp (°C)
CH_3	129
C_2H_5	72
C_3H_7	93
C_4H_9	86
C_5H_{11}	68
C_6H_{13}	82.5
C_7H_{15}	80
C_8H_{17}	86
C_9H_{19}	83
$C_{10}H_{21}$	80
$C_{11}H_{23}$	89
$-CH_2-CH-C_2H_5$	80[6]
$\mid$	
CH_3	83–84[7]

[7] Coates, D., Harrisson, K. J., and Gray, G. W., *Mol. Cryst. Liq. Cryst.* **22**, 99 (1973).

Reagents:

p-substituted R_1—⟨ O ⟩—CHO   (synthesized or c.p.)
   benzaldehyde:

p'-aminocinnamate: H_2N—⟨ O ⟩—CH=CH—COOR$_2$

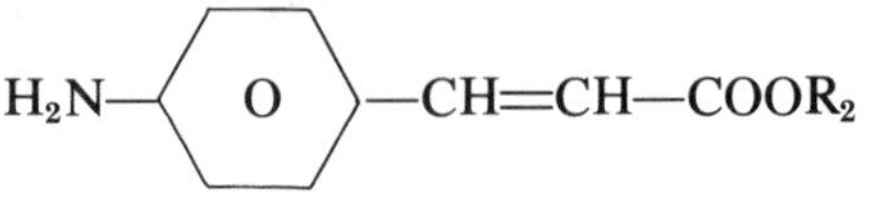

ethanol: C_2H_5OH (c.p.)
acetic acid: CH_3COOH (c.p.)

Apparatus, experiment, and purification are the same as in Section III,1,*c*,*ii*. (ethanol method), and the solvent of crystallization is generally ethanol.

4. PREPARATION OF DI SCHIFF'S BASES

Formulas:

a. 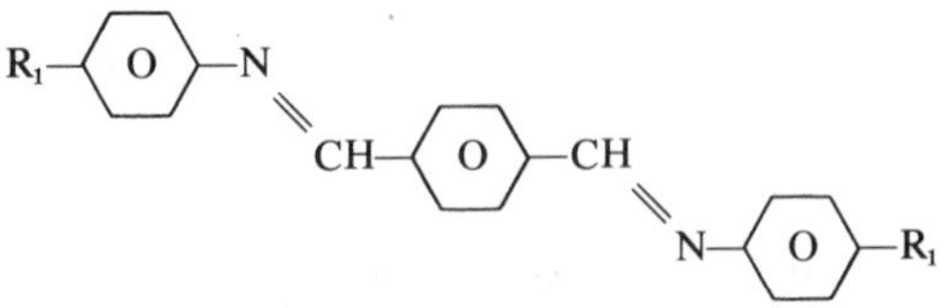 R_1—⟨ O ⟩—N
              ‖
              CH—⟨ O ⟩—CH
                          ‖
                          N—⟨ O ⟩—R_1

b. 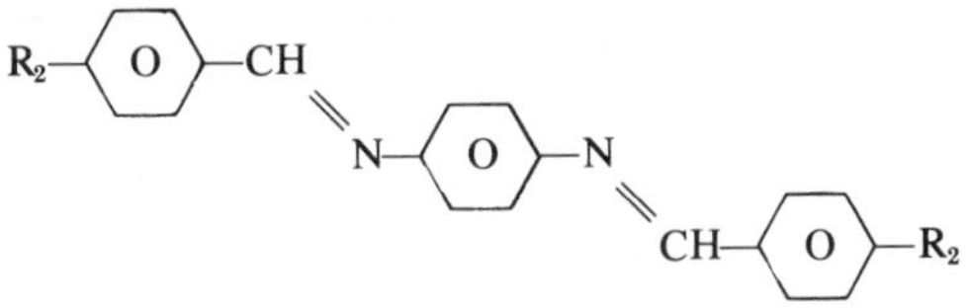 R_2—⟨ O ⟩—CH
              ‖
              N—⟨ O ⟩—N
                          ‖
                          CH—⟨ O ⟩—R_2

For the preparation of (a), p-substituted aniline (see Sections III,1,*b*, III,2,*a*, and III,3,*a*) and the terephthaldehyde, which is a commerçial product, are used.

For (b), p-substituted benzaldehyde (see Section III,1,*a*; also a c.p.) and p-phenylenediamine, which is a commercial product, are used.

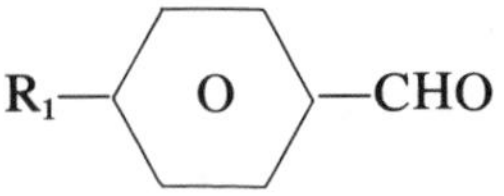

Let us describe now the synthesis of terephthalydene-bis(p-n-butylaniline), or TBBA, much studied by physicists because it possesses different mesophases:

$$C \xrightarrow{113} SmB \xrightarrow{143} SmC \xrightarrow{176} SmA \xrightarrow{201} N \xrightarrow{236} I.$$

a. Preparation of TBBA

Procedure: E, after Taylor *et al.*[8]
Reaction:

$$2C_4H_9\text{-}\langle O \rangle\text{-}NH_2 + OHC\text{-}\langle O \rangle\text{-}CHO \xrightarrow[\text{[CH}_3\text{COOH]}]{(C_2H_5OH)}$$

$$C_4H_9\text{-}\langle O \rangle\text{-}N$$
$$HC\text{-}\langle O \rangle\text{-}CH$$
$$N\text{-}\langle O \rangle\text{-}C_4H_9 + 2H_2O$$

Reagents:

p-n-butylaniline:	$C_4H_9\text{-}\langle O \rangle\text{-}NH_2$	(c.p.)
terephthaldehyde:	$OHC\text{-}\langle O \rangle\text{-}CHO$	(c.p.)
absolute ethanol:	C_2H_5OH	(c.p.)
acetic acid:	CH_3COOH	(c.p.)

Apparatus: An erlenmeyer with a magnetic stirrer.

Experiment: A mixture of 2.98 g (*0.02 mole*) of *p-n*-butylaniline (1) and 1.34 g (*0.01 mole*) of terephthaldehyde in 50 ml of ethanol with one or two drops of acetic acid is stirred for 2 hr. The products formed during the reaction precipitate.

Purification: The solid product is filtered off by suction, and washed twice with 25 ml of ethanol. Then it is recrystallized several times from ethanol until the transition temperatures remain constant (see Section VI).

Remarks: 1. The commercial *p-n*-butylaniline must be distilled before use in the synthesis.

2. TBBA is very unstable at high temperature, and the transition temperatures will decrease rapidly after heating above 200°C.

3. All di Schiff's bases like (a). and (b) can be prepared with this method. Do not forget in the procedure: "Two moles of one and one mole of another."

[8] Taylor, T. R., Arora, S. L., and Fergason, J. L., *Phys. Rev. Lett.* **25**, 722 (1970).

5. CONCLUSION

In the above section, we have seen the synthesis and purification of different series of Schiff's bases. We think that the benzene method is more instructive (one may observe the water formed during the reaction) and sometimes more successful, but benzene is a health hazard and the ethanol method is often preferable.

When only a few grams of material are needed (5 g)—which is sufficient for many physical experiments—we recommend the ethanol method.

The difficulties of the syntheses that we have described are often in the preparation of starting materials and also in the purification of Schiff's bases. Our ultimate advice is to pay great attention to the purity of the starting materials, because they determine the purity of the desired product.

Another inconvenience of the Schiff's bases is their degradation under the influence of air, light, and chemical reagents. They must be stocked in well-sealed flasks under nitrogen if possible. If there is any doubt about their purity before use, they must be recrystallized.

If great purity is required in experiments, one must find another liquid crystal; perhaps the azoxy or ester compounds, which will be studied in the following sections, will be more appropriate for these experiments.

IV. p-p'-Disubstituted Azoxybenzene

Historically, the p-p'-azoxyanisol (PAA) is the most famous nematic compound. Its uses have been numerous for many physical experiments on the nematic phase, and almost all physicists beginning work on liquid crystals have seen or employed PAA. Theoretically, the synthesis of azoxy compounds is a duplicative oxidation of aromatic amines or a duplicative reduction of nitro compounds. We can write

a. $2R\text{—}\langle O \rangle\text{—}NH_2 \xrightarrow{[O]} R\text{—}\langle O \rangle\text{—}N{=}N\text{—}\langle O \rangle\text{—}R$
$$\downarrow$$
$$O$$

and

b. $2R\text{—}\langle O \rangle\text{—}NO_2 \xrightarrow{[H]} R\text{—}\langle O \rangle\text{—}N{=}N\text{—}\langle O \rangle\text{—}R$
$$\downarrow$$
$$O$$

We will describe here the two methods.

6. DUPLICATIVE OXIDATION

The starting materials are the p-substituted anilines (see Sections III,1,b, III,2,a, and III,3,a).

a. Preparation of p,p'(di-n-alkyl) Azoxybenzene

Procedure: D, after Van der Veen et al.[3]
Reaction:

$$2C_nH_{2n+1}-\langle O \rangle-NH_2 \xrightarrow[\text{[H}_2\text{O}_2\text{][NaOH]}]{(\text{CH}_3\text{C}\equiv\text{N})(\text{C}_2\text{H}_5\text{OH})} C_nH_{2n+1}-\langle O \rangle-N=N-\langle O \rangle-C_nH_{2n+1}$$
$$\downarrow$$
$$O$$

Reagents and solvents:

p-n-alkylaniline: $C_nH_{2n+1}-\langle O \rangle-NH_2$ (see Section III,1,b)

methanol:	CH$_3$OH	(c.p.)
acetonitrile:	CH$_3$CN	(c.p.)
sodium hydroxide:	NaOH	(c.p.)
hydrogen peroxide:	H$_2$O$_2$	(c.p.)
ethanol:	C$_2$H$_5$OH	(c.p.)
ethylacetate:	CH$_3$COOC$_2$H$_5$	(c.p.)

Apparatus: The apparatus for this synthesis is shown in Fig. 7.

Experiment: A mixture of 20.5 g of acetonitrile, 200 ml of methanol, 0.09 mole of p-n-alkylaniline, and a few drops of 2 N sodium hydroxide (8 g of sodium hydroxide in 100 ml of water) is well stirred in a 500 ml beaker. To this mixture, 68 ml of 30% hydrogen peroxide is added at 35–40°C (1). The pH of the solution is regularly adjusted at 9.5 (2) with addition of 2 N sodium hydroxide through the cruet and measured with a glass electrode and a calomel electrode as reference. The resulting solution is then stirred for 5 hr at 35° and the pH maintained at 9.5 by addition of 2 N sodium hydroxide.

Purification: The mixture is poured into a separatory funnel and the organic product (3) is extracted by ether. The solvent is removed by distillation on rotovapor under reduced pressure. The resulting residue is crystallized twice from a mixture of ethanol and methanol (50/50) and once from a mixture of methanol and ethylacetate (50/50). Yield: 60–80%.

Remarks: 1. The temperature (35 to 40°C) is an important factor. In

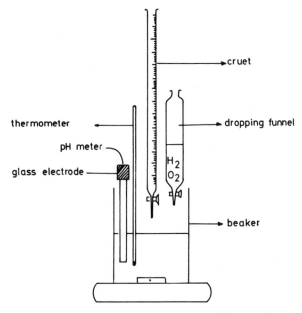

FIG. 7.   Apparatus for synthesis of p,p' (di-n-alkyl) azoxybenzene.

one of our experiments, done at about 50°C, we obtained a very low yield for a mixture of azo and azoxy compounds.

2. Careful control of the pH is essential.

3. Often the product precipitates during the reaction, and filtering will be enough to separate it.

4. Azoxy compounds can be purified by chromatography on alumina or silica gel, but our experience is that this is not necessary for obtaining good purity.

5. The following series of azoxy compounds can also be prepared by this method:

a. $C_nH_{2n+1}O$—⟨O⟩—N=N—⟨O⟩—OC_nH_{2n+1}
                            ↓
                            O

($C_nH_{2n+1}O$—⟨O⟩—NH_2: starting material) p-p'-alkyloxyazoxyben-

zene; see Section III,2,a. Purification by recrystallization in ethanol.

b. $H_{2n+1}C_nOOC$—CH=HC—⟨O⟩—N=N—⟨O⟩—CH=CH
$$\downarrow$$
$$O$$

$$—COOC_nH_{2n+1}$$

$(C_nH_{2n+1}OCOCH=CH$—⟨O⟩—NH_2: starting material) p-p' azoxy-

cinnamate; see Section III,3,a. Purification by recrystallization in ethanol.

6. We think that this preparation is one of the best of the many syntheses described in the literature.

7. DUPLICATIVE REDUCTION METHOD

The reaction steps are as follows.
a. Preparation of p-substituted nitro compounds:

$$HO—⟨O⟩—NO_2 + C_nH_{2n+1}—Br \rightarrow R_1—O—⟨O⟩—NO_2 + HBr$$

b. Duplicative reduction of nitro compounds:

$$2R_1—O—⟨O⟩—NO_2 \rightarrow R_1—O—⟨O⟩—N=N—⟨O⟩—O—R_1$$
$$\downarrow$$
$$O$$

We will take the example of p-p'-n-alkyloxy-azoxybenzene.

a. Preparation of p-n-Alkyloxynitrobenzene

Procedure: LD, after Gray and Jones[2]
Reaction:

$$2HO—⟨O⟩—NO_2 + K_2CO_3 + 2C_nH_{2n+1}Br \xrightarrow{(\langle O \rangle =O)}$$

$$C_nH_{2n+1}—O—⟨O⟩—NO_2 + CO_2 + 2KBr + H_2O$$

Reagents:

p-hydroxynitrobenzene: OH—⟨O⟩—NO_2 (c.p.)

n-alkylbromide:	$C_nH_{2n+1}Br$	(c.p.)
anhydrous potassium carbonate:	CO_3K_2	(c.p.)

cyclohexanone:

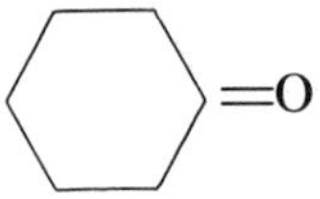

(c.p.)

hexane: C_6H_{14} (c.p.)

Apparatus: The apparatus for this synthesis is the same as that described in Section III,1,*a*.

Experiment: The same as that of Section III,1,*a*, but *p*-hydroxybenzaldehyde is substituted by *p*-hydroxynitrobenzene.

Purification: Cyclohexanone is distilled off by rotovapor (water bath at 90–100°C). The residue is crystallized several times from hexane until the melting point remains constant.

b. *Preparation of p-p'-Alkyloxyazoxybenzene*

Procedure: VD, after Dewar and Goldberg.[9]
Reaction:

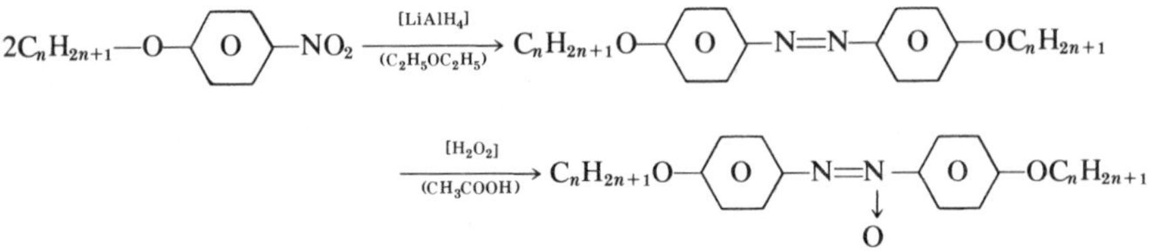

Reagents:

p-alkyloxynitrobenzene:

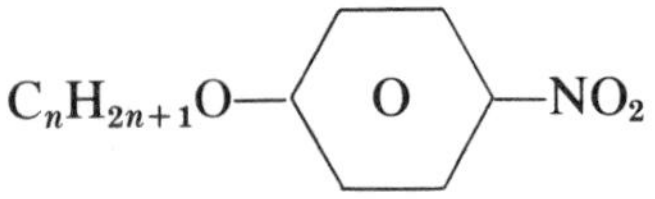

dry ether:	$C_2H_5OC_2H_5$	(c.p.)
lithium aluminum hydride:	$LiAlH_4$	(c.p.)
acetic acid:	CH_3COOH	(c.p.)
hydrogen peroxide:	H_2O_2	(c.p.)
ethanol (95%):	C_2H_5OH	(c.p.)
solid carbon dioxide:	CO_2	(c.p.)
acetone:	CH_3COCH_3	(c.p.)

[9] Dewar, M. J. S., and Goldberg, R. S., *Tetrahedron Lett.* p. 2717 (1966).

Apparatus: The apparatus for this synthesis is shown in Fig. 8.

Experiment: Under argon, 2.5 g (0.066 mole) of lithium aluminum hydride (1) is poured into 300 ml of dry ether. The temperature ($-70°C$) is maintained by a carbon dioxide/acetone bath. At $-70°C$ a solution of 0.0457 mole of *p-n*-alkyloxynitrobenzene in dry ether (150 ml) is added dropwise and slowly (the temperature should not get higher than $-60°C$).

The resulting solution is then allowed to warm up slowly to room temperature. Excess lithium aluminum hydride is decomposed by adding water (2) very cautiously because this hydrolysis is very fast and exothermic (you can observe this reaction by the reflux of ether at the water condenser and by the bubbling of hydrogen evolved during the

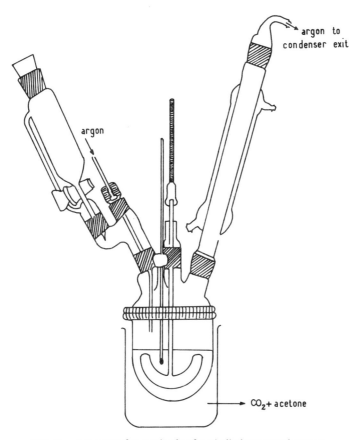

FIG. 8.   Apparatus for synthesis of *p,p'*-alkyloxyazoxybenzene.

reaction). After this hydrolysis, the ether is distilled. Thus we obtain the azo derivative and the second part of the reaction can be performed.

To the residue dissolved in 500 ml of acetic acid, 5.2 g of 30% hydrogen peroxide is added and the solution is allowed to stand 36 hr at 65°C. This operation is done in a 1 l erlenmeyer equipped with a water condenser (see Section III,1,a). Then the solution is poured into 100 ml of water. The product which precipitates (3) is filtered off.

Purification: The precipitate is recrystallized from ethanol (95%) several times until the transition temperature remains constant. Yield: ≈70–80%.

Remarks: 1. Lithium aluminum hydride, LiAlH₄, is a useful and convenient reagent for selective reduction of nitroaromatic compounds in azo compounds, but *it is a dangerous product.* It reacts violently with water and dilute mineral acids, and *all operations must be performed in a dry apparatus, with dry solvents* (the ether used is distilled once, dried for some days under CaCl₂, and stored on calcium hydride) *and under a hood.* The symbol VD takes into account these *hazards.*

2. Water for hydrolysis must be substituted by ethylacetate, which reacts more slowly than water.

3. Sometimes the product of reaction may still contain small amounts of unoxidized azo compounds. If so, the peroxide treatment can be repeated without significant loss.

8. Conclusion

The *p-p'*-substituted azoxy compounds are more stable than Schiff's bases but their transition temperatures generally are higher and they are less rich in smectic mesophases. Thus their use has been more limited in physical experiments.

At present, the physicist prefers to use the ester liquid crystals. Let us now examine this family.

V. Esters

These compounds are more stable than Schiff's bases but often they do not exhibit smectic mesophases. They can be very useful when large nematic ranges are required, and some of them are room-temperature nematics.

Generally the esters are obtained by reaction between phenols and acid chlorides:

$$R_1 \!-\!\!\bigcirc\!\!-\!O\!\!-\!OH + ClCO\!\!-\!\!\bigcirc\!\!-\!O\!\!-\!R_2 \rightarrow R_1\!\!-\!\!\bigcirc\!\!-\!O\!\!-\!OC\!\!-\!\!\bigcirc\!\!-\!O\!\!-\!R_2$$

We will describe in Sections V,1 and V,2 the following two series:

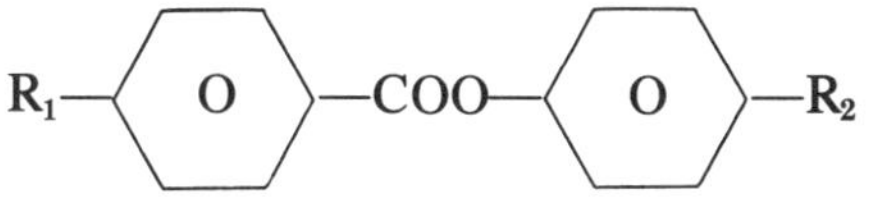

p-p'-substituted phenylbenzoates

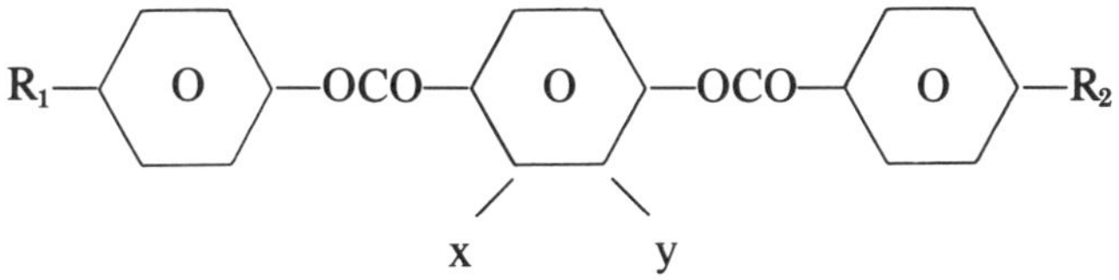

phenylbenzoyloxybenzoate derivatives.

9. PREPARATION OF *p-p'*-DISUBSTITUTED PHENYLBENZOATES

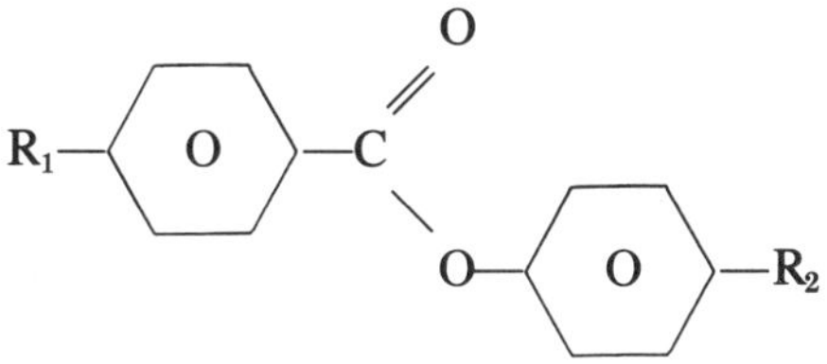

where R_1 and R_2 are

$-R_1 = -C_nH_{2n+1}$ alkyl and $R_2 = -C_nH_{2n+1}$ alkyl;

$-R_1 = -C_nH_{2n+1}$ alkyl and $R_2 = -OC_nH_{2n+1}$ alkyloxy;

$-R_1 = -OC_nH_{2n+1}$ alkyloxy and $R_2 = -C_nH_{2n+1}$ alkyl;

$-R_1 = -OC_nH_{2n+1}$ alkyloxy and $R_2 = -OC_nH_{2n+1}$ alkyloxy.

They are readily obtained by reaction between phenols and benzoylchlorides as follows:

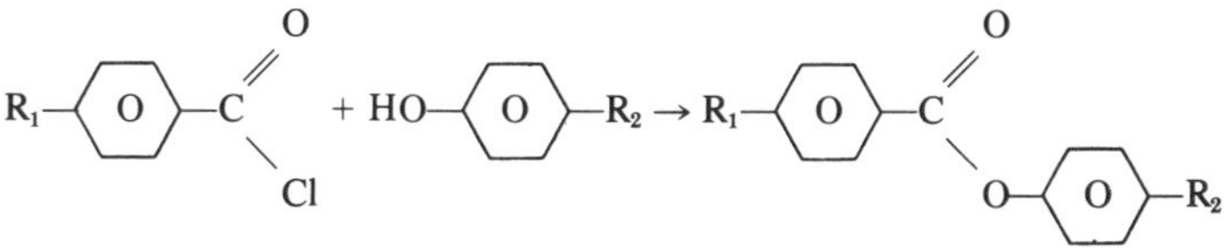

As we can see from the above, the preparation of these esters requires the synthesis of *p*-alkyl (or alkyloxy) benzoylchlorides and *p*-alkyl (or alkyloxy) phenols. We will study these syntheses successively.

a. Preparation of p-Alkylbenzoylchlorides

The steps of the reactions are as follows.
a. Friedel and Crafts reaction:

$$C_nH_{2n+1}-\!\!\langle O \rangle\!\!- + CH_3COCl \rightarrow C_nH_{2n+1}-\!\!\langle O \rangle\!\!-COCH_3 + HCl$$

b. Haloform reaction:

$$C_nH_{2n+1}-\!\!\langle O \rangle\!\!-COCH_3 + 3NaOBr \rightarrow R-\!\!\langle O \rangle\!\!-\overset{O}{\underset{ONa}{C}} + HCBr_3 + 2NaOH$$

$$C_nH_{2n+1}-\!\!\langle O \rangle\!\!-\overset{O}{\underset{ONa}{C}} + HX \rightarrow C_nH_{2n+1}-\!\!\langle O \rangle\!\!-\overset{O}{\underset{OH}{C}} + NaX$$

c. Preparation of *p-n*-alkylbenzoylchloride by thionylchloride:

$$C_nH_{2n+1}-\!\!\langle O \rangle\!\!-COOH + SOCl_2 \rightarrow C_nH_{2n+1}-\!\!\langle O \rangle\!\!-\overset{O}{\underset{Cl}{C}} + SO_2 + HCl$$

i. Preparation of p-n-Alkylacetophenone. This preparation has been described previously in Section III,1,*b,i*.

ii. Preparation of p-n-Alkylbenzoic Acid. Procedure: LD, after Johnson *et al.*[10]
Reactions:

$$C_nH_{2n+1}-\!\!\langle O \rangle\!\!-COCH_3 + 3NaOBr \xrightarrow{\text{(o \ o)}} C_nH_{2n+1}-\!\!\langle O \rangle\!\!-\overset{O}{\underset{ONa}{C}} + HCBr_3 + 2NaOH$$

$$C_nH_{2n+1}-\!\!\langle O \rangle\!\!-\overset{O}{\underset{ONa}{C}} + SO_4H_2 \rightarrow C_nH_{2n+1}-\!\!\langle O \rangle\!\!-COOH + NaHSO_4$$

[10] Johnson, W. S., Gutsche, C. D., and Offenhauer, Z. D., *J. Am. Chem. Soc.* **68**, 1648 (1946).

Reagents and solvents:

p-n-alkylacetophenone: C_nH_{2n+1}—⟨ O ⟩—$COCH_3$

sodium hydroxide:	NaOH	(c.p.)
bromine:	Br_2	(c.p.)

dioxanne: O ⟨ ⟩ O   (c.p.)

sodium bisulfite:	$NaHSO_3$	(c.p.)
sulfuric acid:	SO_4H_2	(c.p.)

Apparatus: The apparatus for this synthesis is shown in Fig. 9.

Experiment

a. Preparation of the hypobromite solution $[NaOBr/(H_2O)]$. In an erlenmeyer, 63 g of NaOH is dissolved by stirring in 300 ml of water, and the resulting solution is cooled to 0°C with an ice–salt bath. Then 22.6 g of bromine (1) is added slowly to this solution; the temperature must be kept between 0 and 3°C during the addition of bromine. The resulting sodium hypobromite solution is a clear yellow solution which is used for the following preparation.

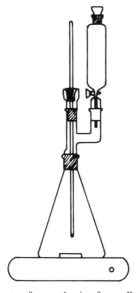

Fig. 9.   Apparatus for synthesis of *p-n*-alkylbenzoic acid.

TABLE V. MELTING POINTS OF SOME *p-n*-ALKYLBENZOIC ACIDS

$$C_nH_{2n+1}-\left\langle O \right\rangle-COOH$$

with $n =$	Mp (°C)
4	98[11]
5	88
6	97[12]
7	101.5
8	100

b. Oxidation of *p-n*-alkylacetophenone. The sodium hypobromite solution is placed in the dropping funnel and added dropwise to a stirred solution of *p-n*-alkyacetophenone (0.1 mole) in 230 ml of dioxanne as solvent. After each addition of sodium hypobromite the reaction mixture turns yellow and after a short time becomes colorless, but generally not for the last few milliliters added. During addition the temperature is allowed to rise to 35–40°C. After stirring again for 15 min, the suspension of the sparingly soluble sodium salt is treated with enough sodium bisulfite (in powder or in solution) to destroy the excess of hypobromite: sodium bisulfite must be added until discoloration of the reaction mixture takes place. Then the reaction mixture is poured into 1.5 l of water contained in a beaker (3 l).

About 300 ml of the liquid is distilled off using a rotovapor to remove the bromoform (2) formed during the reaction and a little of the dioxanne.

The acidification of the hot solution by addition of concentrated sulfuric acid gives, after cooling, the *p-n*-alkylbenzoïc acid, which precipitates. *This manipulation is dangerous and must be performed carefully.*

Purification: The precipitate is filtered off by suction, washed with water several times, and crystallized from 75/25 mixture of ethanol/water. Yield: 80–90%.

Remarks: 1. Warning: Bromine is *very, very* dangerous because its burns are very difficult to heal. To inhale it is also dangerous. *This work must be performed with goggles and gloves under a good hood.*

2. When the solution, which is turbid before removal of the bromoform, becomes clear, most of the bromoform has been removed.

3. Melting points of some *p-n*-alkylbenzoïc acids are given in Table V.

[11] Steinstrasser, R., *Z. Naturforsch., Teil B* **27**, 774 (1972).
[12] Jones, F. B., and Ratto, J. J., *J. Phys. C. Suppl.* **3**, 413 (1975).

iii. Preparation of p-n-Alkylbenzoylchlorides. Procedure: E, after Vögel.[13]

Reaction:

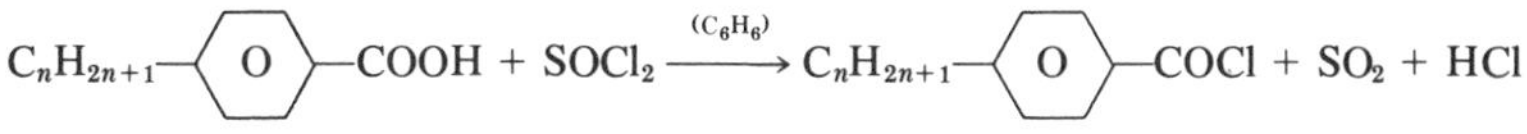

$$C_nH_{2n+1}\text{—}\langle O \rangle\text{—COOH} + SOCl_2 \xrightarrow{(C_6H_6)} C_nH_{2n+1}\text{—}\langle O \rangle\text{—COCl} + SO_2 + HCl$$

Reagents and solvents:

p-n-alkylbenzoic acid: $C_nH_{2n+1}\text{—}\langle O \rangle\text{—COOH}$

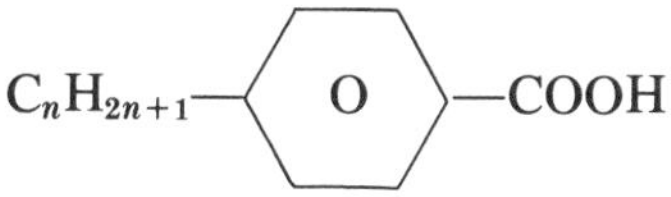

thionylchloride: $SOCl_2$ (c.p.)

benzene: $\langle O \rangle$ (c.p.)

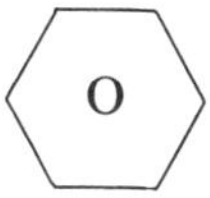

Apparatus: The apparatus for this synthesis is the same as that in Section III,3,*a,i*.

Experiment: Thionylchloride (1) (0.15 mole) and *p-n*-alkylbenzoîc acid (0.1 mole) are combined with 75 ml of benzene (2). The resulting solution is refluxed until the evolution of HCl and SO_2 ceases.

Purification: The excess of thionylchloride is removed by distillation with a rotovapor and the residue is distilled in vacuum (3) (see Section VI). Yield: nearly 100% in most cases.

Remarks: 1. Thionylchloride is a lachrymatory and it is most important to take good precautions for its use and to *work only under a hood*.

2. The same as Remark 2 in Section III,1,*b,iii*.

3. The acid chlorides are very sensitive to moisture and must be used immediately after synthesis.

b. Preparation of p-n-Alkyloxybenzoylchlorides

The step reactions are as follows.

a. Alkylation of *p*-hydroxyacetophenone:

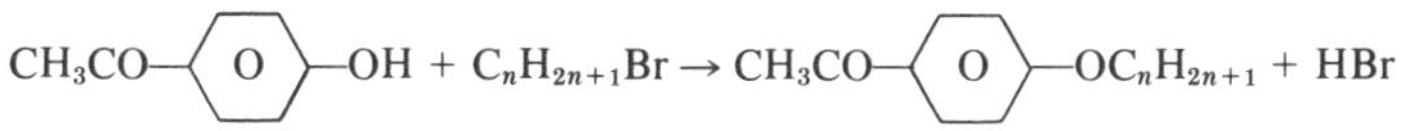

$$CH_3CO\text{—}\langle O \rangle\text{—OH} + C_nH_{2n+1}Br \rightarrow CH_3CO\text{—}\langle O \rangle\text{—OC}_nH_{2n+1} + HBr$$

---

[13] Vögel, A. I., "Textbook of Practical Organic Chemistry," p. 792. Longmans, Green, New York, 1956.

TABLE VI.   MELTING POINTS OF SOME *p-n*-ALKYLOXYBENZOIC
ACIDS[a]

| | Mp (°C) | | | | |
n	C→Sc	C→N	Sc→N	Sc→I	N→I
3		145			154
4		147			160
5		124			151
6		106			154
7	92		98		146
8	101		108		147
9	94		117		143
10	97		122		142
12	95		129		137
16	85			132.5	
18	102			131	

a After Gray and Jones.[14]

b. Haloform reaction:

$$CH_3CO\text{—}\langle O \rangle\text{—}OC_nH_{2n+1} + 3NaOBr \rightarrow C_nH_{2n+1}O\text{—}\langle O \rangle\text{—}C\overset{O}{\underset{ONa}{\diagdown}} + HCBr_3 + 2NaOH$$

$$C_nH_{2n+1}O\text{—}\langle O \rangle\text{—}COONa + H_2SO_4 \rightarrow C_nH_{2n+1}O\text{—}\langle O \rangle\text{—}COOH + NaHSO_4$$

c. Preparation of *p-n*-alkyloxybenzoylchloride by thionylchloride:

$$C_nH_{2n+1}\text{—}O\text{—}\langle O \rangle\text{—}COOH + SOCl_2 \rightarrow C_nH_{2n+1}O\text{—}\langle O \rangle\text{—}COCl + SO_2 + HCl$$

Reactions b and c are similar to those in Sections V,1,*a,ii* and
V,1,*a,iii*; *p-n*-alkylacetophenone must be replaced by *p-n*-alkyloxyaceto-
phenone. This last compound is prepared as described in Section III,1,*a*,
and *p*-hydroxybenzaldehyde must be replaced by *p*-hydroxyacetophen-
one (c.p.).
The   melting   points   of   some   *p-n*-alkyloxybenzoïc   acids

$$(C_nH_{2n+1}O\text{—}\langle O \rangle\text{—}COOH)$$ are given in Table VI.

[14] Gray, G. W., and Jones, B., *J. Chem. Soc.* p. 4197 (1953).

c. Preparation of p-n-Alkylphenols

p-n-alkylphenols are conveniently prepared in two steps as follows.
a. Alkoylation of phenol by a Friedel and Crafts reaction:

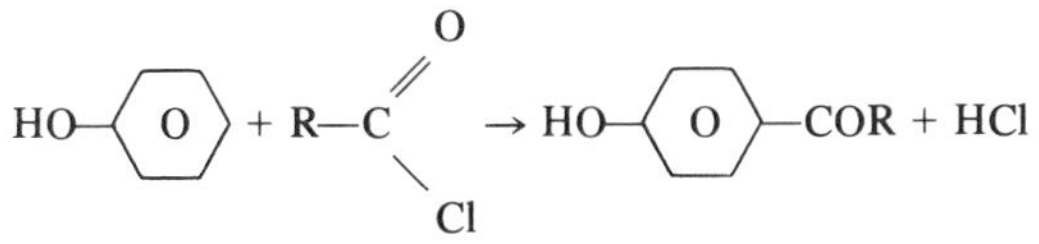

$$HO-\langle O \rangle + R-C{\overset{O}{\underset{Cl}{\lessgtr}}} \rightarrow HO-\langle O \rangle -COR + HCl$$

b. Reduction of ketone by a Clemmensen reaction:

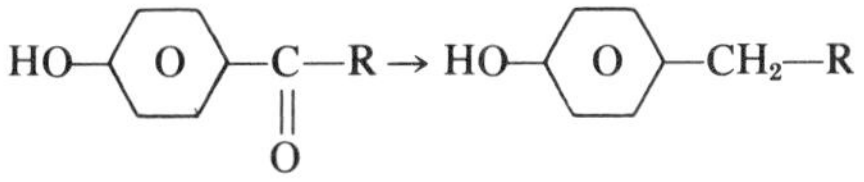

$$HO-\langle O \rangle -\underset{\underset{O}{\|}}{C}-R \rightarrow HO-\langle O \rangle -CH_2-R$$

i. Preparation of p-Hydroxyphenyl-n-Alkylketone. Procedure: D, after Close et al.[15]

Reaction:

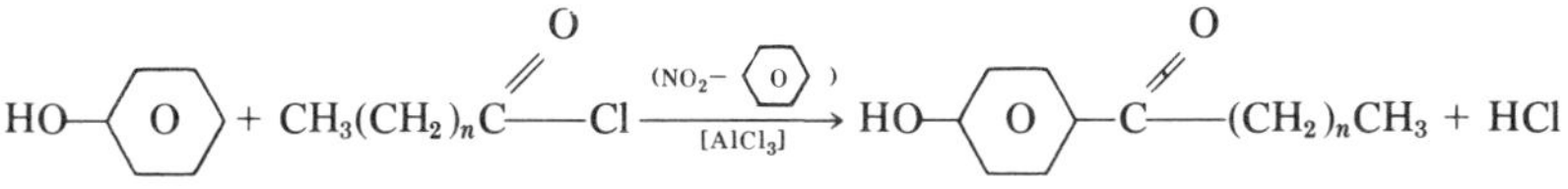

$$HO-\langle O \rangle + CH_3(CH_2)_n\overset{O}{\overset{\|}{C}}-Cl \xrightarrow[\text{[AlCl}_3]{(NO_2-\langle O \rangle)}} HO-\langle O \rangle -\overset{O}{\overset{\|}{C}}-(CH_2)_nCH_3 + HCl$$

Reagents and solvents:

phenol:	$HO-\langle O \rangle$	(c.p.)
n-alkyloylchloride:	$CH_3(CH_2)_n-\underset{\underset{O}{\|}}{\overset{\overset{O}{\|}}{C}}-Cl$	(c.p.)
dried aluminum chloride:	$AlCl_3$	(c.p.)
nitrobenzene:	$O_2N-\langle O \rangle$	(c.p.)
sodium hydroxide:	$NaOH$	(c.p.)

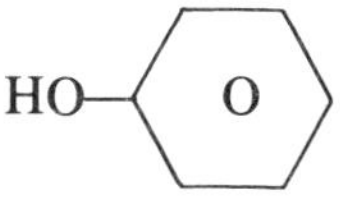

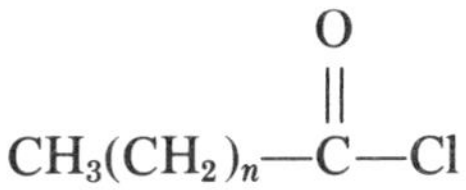

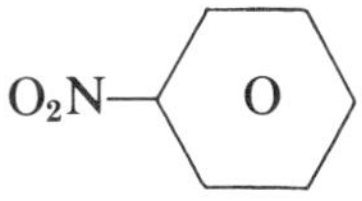

[15] Close, W. J., Tiffany, B. D., and Spielman, M. A., *J. Am. Chem. Soc.* **71**, 1265 (1949).

hydrochloric acid: HCl   (c.p.)
anhydrous magnesium sulfate: $MgSO_4$ (c.p.)

Apparatus: The apparatus for this synthesis is the same as that in Section III,1,*b,i.*

Experiment: To a stirred mixture of 205 g (1.5 mole) of $AlCl_3$ in 400 ml of nitrobenzene (as solvent) cooled with an ice bath, phenol (72 g, 0.76 mole) dissolved in 144 ml of nitrobenzene is added dropwise through a dropping funnel at 10–12°C. The resulting mixture is then cooled at 5–10°C and *n*-alkyloylchloride (0.62 mole) is slowly added dropwise. After standing overnight, the mixture is poured *carefully* into a beaker, into a mixture of 300 ml of concentrated hydrochloric acid and 600 g of ice.

Purification: The resulting heterogeneous solution is poured into a separatory funnel (2 l) for extraction (see Section VI). Two layers are formed: The lower is here the organic layer (organic products in nitrobenzene) and the higher is the aqueous layer. The nitrobenzene is kept and the aqueous layer is discarded.

The organic layer is then extracted several times with dilute sodium hydroxide. In this operation, phenolic derivatives which were previously in the nitrobenzene layer are now in the aqueous layers, because sodium hydroxide converts them into a sodium salt of phenol, soluble in water:

$$RC{-}\langle O \rangle{-}OH + NaOH \longrightarrow RC{-}\langle O \rangle{-}O^-, Na^+ + H_2O$$
$$\;\;\;\;\;\| \qquad\qquad\qquad\qquad\qquad\quad \|$$
$$\;\;\;\;\;O \qquad\qquad\qquad\qquad\qquad\quad O$$

  (nitrobenzene layer)   (aqueous layer)

The aqueous layer *must be kept,* whereas the nitrobenzene is discarded. The combined aqueous layers are neutralized by adding hydrochloric acid; the oil obtained is extracted from water with ether in a separatory funnel: The ether layer is the higher.

The ether solution is dried on magnesium sulfate; then the solvent is removed by distillation with a rotovapor and the residue is distilled in vacuum (see Section VI). Yield: 80–90%.

ii. Preparation of p-n-Alkylphenols. Procedure: D, after Read and Wood.[16]

Reaction:

$$HO{-}\langle O \rangle{-}\overset{\overset{O}{\|}}{C}{-}R + Zn/HgCl_2 + HCl \xrightarrow{(C_2H_5OH)} HO{-}\langle O \rangle{-}CH_2{-}R$$

[16] Read, R. R. and Wood, J., *Org. Synth. Collect.* **3**, 444 (1955).

Reagents and solvents:

p-hydroxyphenyl-*n*-alkylketone:

$$HO\!-\!\langle\ O\ \rangle\!-\!\underset{\underset{O}{\|}}{C}\!-\!R$$

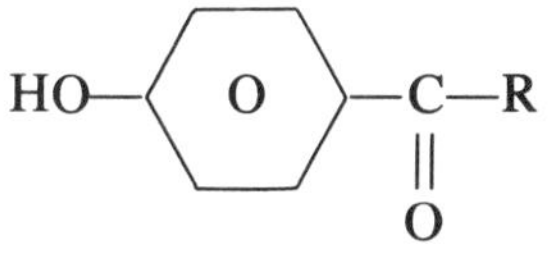

zinc:	Zn	(c.p.)
mercuric chloride:	HgCl$_2$	(c.p.)
ethanol:	C$_2$H$_5$OH	(c.p.)
concentrated hydrochloric acid:	HCl	(c.p.)

toluene: $\langle\ O\ \rangle\!-\!CH_3$ (c.p.)

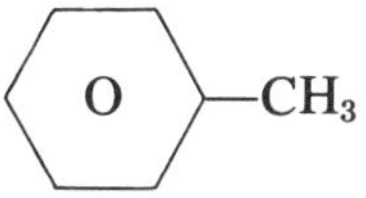

Apparatus: The apparatus for this synthesis is the same as that in Section III,1,*b,i* (Fig. 2), without the thermometer.

Experiment: The reaction requires amalgamated mossy zinc, which can be made as follows.

a. Preparation of the amalgamated zinc: The zinc (100 g) is amalgamated in a beaker by covering it with a solution of 2 g of mercuric chloride in 150 ml of water. The mixture is occasionally shaken during one-half hour, then the solution is poured off. The zinc is washed once with water and the water poured off also. The resulting amalgamated zinc is ready for use in the following reaction.

b. Reduction of *p*-hydroxyphenyl-*n*-alkylketone: The amalgamated zinc is placed in a reactor (1 l) fitted with an efficient mechanical stirrer and a water condenser. A mixture of 100 ml of water and 100 ml of concentrated hydrochloric acid (1) is added in the reactor and then a solution of *p*-hydroxyphenyl-*n*-alkylketone (0.15 mole) in 50 ml of ethanol. The mixture is stirred vigorously and refluxed for 8 to 10 hr. Heating is then stopped and 80 ml of toluene is added to the mixture and the stirring is continued for a few minutes.

Purification: The reaction mixture is poured into a separatory funnel (see Section VI,13). Two layers are formed: the organic layer (the higher one) is separated from the aqueous solution (the lower one) and washed three times with water (at each time the organic layer is the higher one).

The toluene layer is filtered from suspended matter with a paper filter, then removed by distillation with a rotovapor. The residue is then distilled in vacuum (see Section VI).

TABLE VII. BOILING POINTS OF SOME *p-n-*

ALKYLPHENOLS (C_nH_{2n+1}—⟨ O ⟩—OH)

n	Bp (°C)	Pressure (mmHg)
3	79–80	0.75
4	86	0.1
5	134	0.1
6	112	1
7	123	1
8	134	1

Remarks: 1. Warning: *Concentrated hydrochloric acid must be poured into water and not the reverse.*
2. Boiling points of some *p-n*-alkylphenols are given in Table VII.

d. Preparation of p-n-Alkyloxyphenols

Procedure: LD, after Klarman *et al.*[17]
Reaction:

$$HO—⟨O⟩—OH + C_nH_{2n+1}Br \xrightarrow[\text{[KOH]}]{\text{(C}_2\text{H}_5\text{OH)}} HO—⟨O⟩—OC_nH_{2n+1} + KBr + H_2O$$

Reagents and solvents:

hydroquinone:	HO—⟨ O ⟩—OH	(c.p.)
n-alkylbromide:	$C_nH_{2n+1}Br$	(c.p.)
ethanol:	C_2H_5OH	(c.p.)
potassium hydroxide:	KOH	(c.p.)
hydrochloric acid:	HCl	(c.p.)
ether:	$C_2H_5OC_2H_5$	(c.p.)
heptane:	C_7H_{16}	(c.p.)

Apparatus: The apparatus for this synthesis is the same as that in Section III,2,*a,i.*

Experiment: In an erlenmeyer (250 ml), hydroquinone (33 g, 0.3 mole) is dissolved in 36 ml of ethanol. To this solution, *n*-alkylbromide (0.29

[17] Klarman, E., Gatyas, L. W., and Shternov, V. A., *J. Am. Chem. Soc.* **54**, 298 (1932).

mole) is added. The resulting solution is refluxed and then a solution of 17.8 g (0.31 mole) of potassium hydroxide in 54 ml of water (1) is added dropwise in the course of 1 hr. Boiling is continued for 3 more hours.

Purification: The mixture is allowed to cool and then is poured into a separatory funnel. The reaction mixture is extracted (see Section VI) twice with 50 ml of ether (each time, the ether solution is the higher).

The combined organic layers are extracted several times with a 10% solution of potassium hydroxide (5.4 g of KOH in 100 ml of water). In this operation phenolic derivatives which were previously in the ether layer are now in the aqueous extracts *which must be kept* because potassium hydroxide converts them into the potassium salt of phenol soluble in water:

$$R-O-\left\langle O \right\rangle-OH + KOH \longrightarrow R-O-\left\langle O \right\rangle-O^-K^+ + H_2O$$

(ether layer)   (aqueous layer)

The combined aqueous layers are acidified with dilute hydrochloric acid. A crystalline precipitate of *p-n*-alkyloxyphenol is formed. It is filtered off and purified by crystallization from a suitable solvent.   The lower aliphatic ethers, which are solid at room temperature, are crystallized from water, the higher ones (from the pentyl ether upward) from heptane. Yield: 70–80%.

Remarks: 1. The same as Remark 1 of Section III,1,*a*.

2. Melting points of some monoethers of hydroquinone are given in Table VIII.

TABLE VIII. MELTING POINTS OF
SOME MONOETHERS OF
HYDROQUINONE

$$(C_nH_{2n+1}O-\left\langle O \right\rangle-O-OH)$$

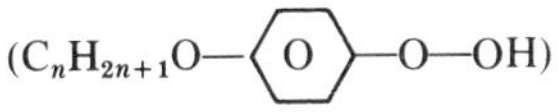

n	Mp (°C)
2	65–66
3	56–57
4	64–65
5	49–50
6	48
7	60
8	60–61
9	68.5
sec pentyl	48–49

e. Preparation of p-p'-Disubstituted Phenylbenzoates

Procedure: E.
Reaction:

$$R_1 \!-\!\!\langle O \rangle\!-\!OH + R_2 \!-\!\!\langle O \rangle\!-\!COCl \xrightarrow{\;(\langle O \rangle N)\;} R_1 \!-\!\!\langle O \rangle\!-\!O\!-\!\overset{\displaystyle O}{\underset{\displaystyle \|}{C}}\!-\!\langle O \rangle\!-\!R_2 + HCl,$$

where R_1 and R_2 are as described in Section V,9.
Reagents and solvents:

p-substituted phenol: $R_1 \!-\!\!\langle O \rangle\!-\!OH$

p-substituted benzoylchloride:

$$R_2 \!-\!\!\langle O \rangle\!-\!C \overset{\displaystyle O}{\big\|}\!-\!Cl$$

dried pyridine: $\langle O \; N \rangle$ (c.p.)

hydrochloric acid: HCl   (c.p.)
ethanol: C_2H_5OH (c.p.)
methanol: CH_3OH (c.p.)

Apparatus: An erlenmeyer of 250 ml and a magnetic stirrer.

Experiment: To a solution of *p*-substituted phenol (0.0115 mole) in dried pyridine (1) (50 ml) is added *p*-substituted benzoylchloride (0.0118 mole). The mixture is stirred at room temperature for 24 hr and is then acidified with dilute hydrochloric acid.

Purification: The resulting precipitate of ester is filtered off, washed with cold ethanol to remove pyridine, and then crystallized several times from methanol until the transition temperatures remain constant.

Remarks: 1. Commercial pyridine must be dried by adding KOH in pellets in the bottle before use (about 30 g per liter).

10. PREPARATION OF PHENYL-4-BENZOYLOXYBENZOATE DERIVATIVES

These esters are conveniently prepared by a two-step procedure as follows:

a. 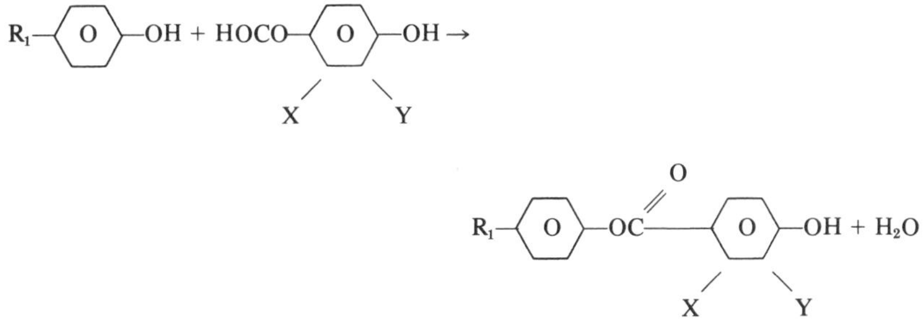 R_1—⟨O⟩—OH + HOCO—⟨O⟩—OH →
         X   Y

R_1—⟨O⟩—OC—⟨O⟩—OH + H_2O
           ‖O   X Y

b. 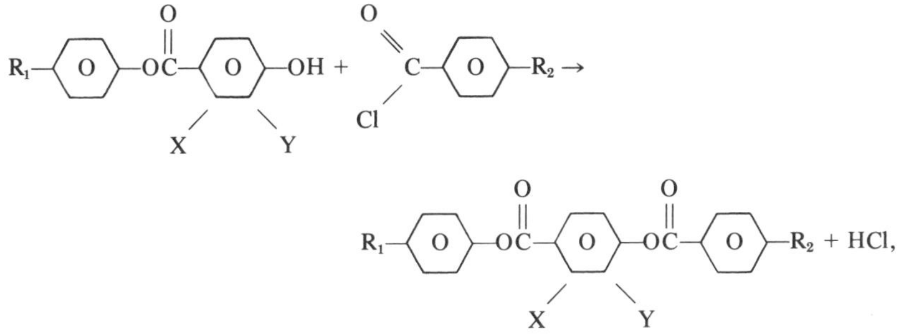 R_1—⟨O⟩—OC—⟨O⟩—OH + C—⟨O⟩—R_2 →
           ‖O   X Y   ‖O Cl

R_1—⟨O⟩—OC—⟨O⟩—OC—⟨O⟩—R_2 + HCl,
           ‖O   ‖O   X Y

where R_1 = alkyl or alkyloxy (C_nH_{2n+1}; —OC_nH_{2n+1}); X = H and Y = H; X = Cl and Y = H; X = H and Y = Cl; R_2 = alkyl or alkyloxy.

Let us examine the a and b reactions successively.

a. Preparation of 4'-Substituted Phenyl-4-Hydroxybenzoate Derivatives

Procedure: D, after Van Meter and Seidel[18] and Lowrance.[19]

Reaction:

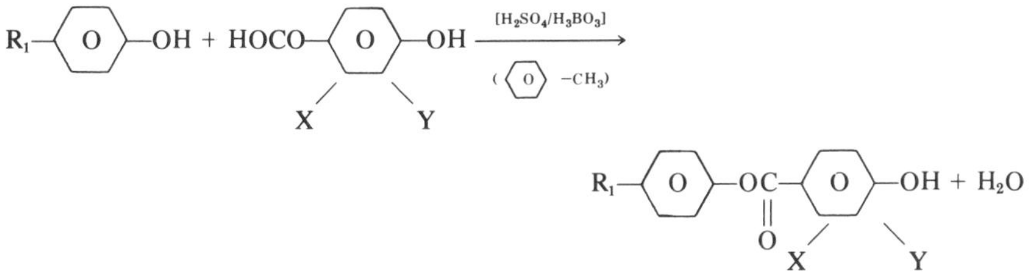

R_1—⟨O⟩—OH + HOCO—⟨O⟩—OH $\xrightarrow[(\langle O \rangle -CH_3)]{[H_2SO_4/H_3BO_3]}$
         X Y

R_1—⟨O⟩—OC—⟨O⟩—OH + H_2O
           ‖O   X Y

---

[18] Van Meter, J. P., and Seidel, A. K., *J. Org. Chem.* **40**, 2998 (1975).
[19] Lowrance, W. W., *Tetrahedron Lett.* p. 3453 (1971).

Reagents and solvents:

p-substituted   R₁—⟨ O ⟩—OH   (see Sections V,1,*c*
    phenol: and V,1,*d*)

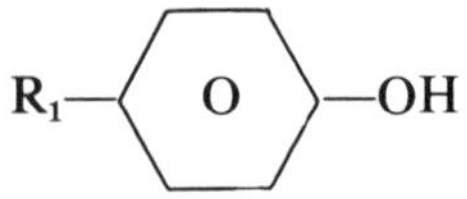

p-hydroxybenzoic
    acid derivatives: H—O—C—⟨ O ⟩—OH (c.p.)
                              ‖
                              O   X   Y

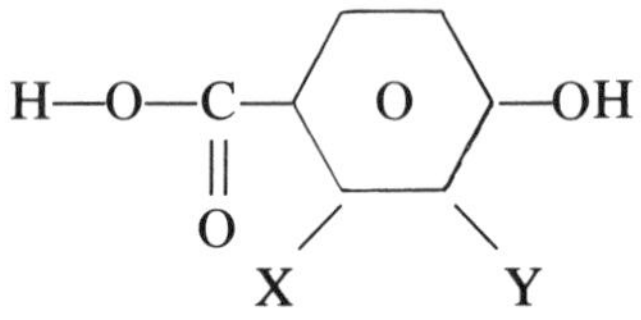

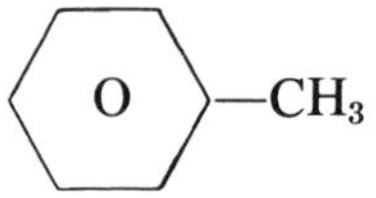

toluene: ⟨ O ⟩—CH₃   (c.p.)

concentrated H_2SO_4 (c.p.)
    sulfuric acid:
boric acid: H_3BO_3 (c.p.)
ethanol: C_2H_5OH (c.p.)

Apparatus: The apparatus for this synthesis is the same as in Section III,1,*c,i*.

Experiment: In a 2 l erlenmeyer, p-substituted phenol (0.22 mole) and the p-hydroxyacid derivative (0.2 mole) in toluene (1 l) containing concentrated sulfuric acid (1 g, 0.010 mole) and boric acid (0.6 g, 0.010 mole) as catalyst, are refluxed for 6 to 7 hr. The water formed during the reaction is removed azeotropically with a Dean–Stark trap (1).

Purification: The solution is cooled and the solvent removed by distillation with a rotovapor. The resulting solid residue is crystallized twice from ethanol–water (50/50). Yield: 75–85%.

Remark: 1. See Section III,1,*c,i* for the apparatus and explanations of azeotropic distillation.

b. Preparation of Phenyl-4-Benzoyloxybenzoate Derivatives

Procedure: E, after Van Meter and Seidel.[18]
Reaction:

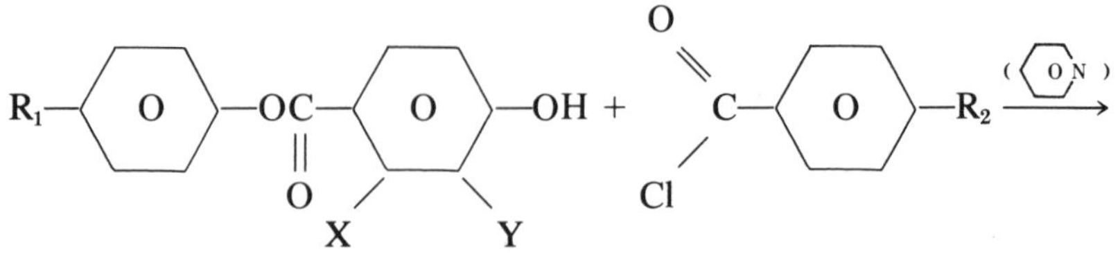

R₁—⟨ O ⟩—OC—⟨ O ⟩—OH + C—⟨ O ⟩—R₂ ⟶ (⟨ O N ⟩)
                ‖   ‖
                O   X   Y   O   Cl

Reagents and solvents:

4'-substituted phenyl-4-hydroxybenzoate derivatives:		
p-substituted benzoylchlorides:		(see Section V,1,*a*)
dried pyridine:		(c.p.)
ethanol:	C_2H_5OH	(c.p.)

Apparatus: The apparatus for this synthesis is the same as in Section V,9,*e*.

Experiment: In a 100 ml erlenmeyer, to a solution of 4'-substituted phenyl-4-hydroxybenzoate derivative (0.0094 mole) in dry pyridine (see remarks for Section V,9,*e*) (50 ml), *p*-substituted benzoylchloride (0.01 mole) is added. The mixture is stirred at room temperature for 18 hr.

Purification: The resulting mixture is poured into an ice–water mixture placed in a beaker. The precipitate is isolated by filtration, then crystallized several times from ethanol until the transition temperatures remain constant. Yield: 70–80%.

11. CONCLUSION

This family of esters is very important in number, but also for the stability of its members to external agents and temperature. The white color of these products offers advantages for some applications and the chemical research on new compounds of this kind is well developed.

VI. Purification

The product obtained after a synthesis is often unsuitable for physical applications, as it is impure. A purification stage is thus necessary.

12. CRYSTALLIZATIONS

This purification process is required for the organic substances that are solid at room temperature. After organic reactions the products are often contaminated with impurities, secondary compounds of reactions, or raw materials which have not reacted. The crystallization is based upon differences in the solubility of these impurities and of the desired product in a solvent or a mixture of solvents.

For the crystallization, you must:

1. Find the appropriate solvent for crystallization among the following: water, diethylether, acetone, ethanol, methanol, methylene chloride, chloroform, carbon tetrachloride, ethylacetate, benzene, toluene, and light petroleum (Bp = 40–60°; 60–80°). This means that you must take a small amount of the residue (about 10 mg) and try the solvent (1 or 2 cm^3) at or near its boiling point in a test tube. If the product is not soluble when cold but soluble when warm, the solvent may be used.

In the syntheses we have described, the solvent for each crystallization has been indicated.

2. Pour the product of reaction in the solvent (about 1 g for 10 g of solvent) into an appropriate flask equipped with an air condenser. Heat until the product is dissolved (sometimes some solvent must be added for complete dissolution). Filter off the hot solution to remove insoluble materials and dust: This is usually done through a fluted filter paper supported in a glass funnel previously warmed externally by an electrical jacket.

Allow the hot solution to cool: The product crystallizes out. Filter off the crystals by suction on a sintered glass funnel. Then dry in the apparatus shown in Fig. 10, which is used for small quantities (about 5 g).

Take the melting point of the product. (For large quantities, a vacuum desiccator may be used.)

You must repeat all these operations until the melting point remains constant; therefore the time of preparation is always approximate.

Remarks: 1. Sometimes it is very difficult to remove traces of coloring or resinous matter, and a few grams of decolorizing carbon in the mixture of solvent and product may be added. In this case, it is necessary to filter off the solution while hot.

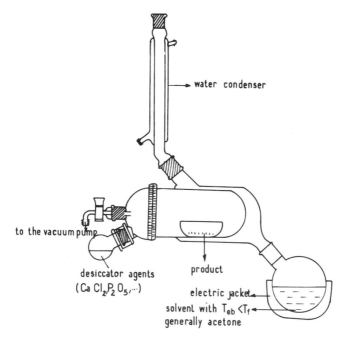

water condenser

to the vacuum pump

desiccator agents
($Ca\ Cl_2, P_2\ O_5, \cdots$)

product

electric jacket
solvent with $T_{eb} < T_f$
generally acetone

FIG. 10.   Apparatus for drying small quantities of crystals.

2. Warning: *Almost all common solvents used in the crystallization techniques are flammable.*

13.   EXTRACTION TECHNIQUE

We have often used this technique for the first step of the purification. As we have previously mentioned, if the reader wants to know the theory of this procedure, he may look it up in a classical book of organic chemistry. He must merely know that this operation is the separation of an organic compound from a solution or suspension in a liquid by shaking with a second solvent in which the organic compound is soluble and which is immiscible or hardly miscible with the liquid containing the substance. The liquid is often water and the solvent is selected from among those common solvents immiscible in water: diethylether ($C_2H_5OC_2H_5$), petroleum ether (Bp = 40–60°; 60–80°), benzene, chloroform, methylene chloride, and carbon tetrachloride. In the purification we have described, the extraction solvent has always been mentioned.

Apparatus: The extraction apparatus is shown in Fig. 11.

Experiment: Pour the mixture of water and organic compound into the separatory funnel (its volume must be twice the volume of the mixture).

Add the selected organic solvent (about one-third of the volume of the solution). Take the separatory funnel by its top with your left hand and near the stopcock with your right hand. Close the top of the funnel and gently tip it upside down; open the stopcock in order to relieve the excess pressure of solvent. Close the stopcock, shake gently and repeat the above operation (shake, open, . . .). When the funnel is saturated with solvent vapor, further shaking does not develop additional pressure. At this stage, shake vigorously and then open several times, and the organic compound is transferred to the organic solvent layer.

Now allow the funnel to rest on the ring stand. Two layers form and the aqueous layer is separated from the organic layer. When the organic solvent is denser than water, the organic layer is removed through the stopcock and the aqueous layer is again extracted twice in the same manner. When the organic solvent is less dense than water, the water layer is removed through the stopcock and the organic layer through the upper stopper. Then the water layer is poured into the separatory funnel and the whole operation is repeated twice.

The different washings are carried out in the same conditions and the organic layer is dried as indicated in the purification section.

Remarks: 1. Almost all the common solvents used are very flammable and the extractions must be carried out without flames in the vicinity.

2. Sometimes emulsions are formed in this operation and the separation is not possible. To break down these emulsions, you may successively: (a) allow the emulsion to stand for some time; one night is often

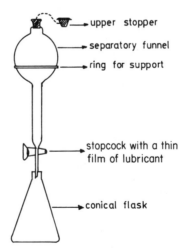

FIG. 11.   Extraction apparatus.

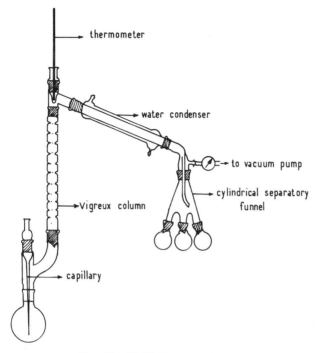

thermometer

water condenser

to vacuum pump

cylindrical separatory funnel

Vigreux column

capillary

FIG. 12. Distillation apparatus.

very satisfactory; (b) saturate the aqueous phase with sodium chloride; (c) centrifuge the mixture.

14. DISTILLATION

This technique is used to purify liquids. We will describe distillation under reduced pressure.

This method is required when the boiling point of the product is too high and sometimes so high that decomposition occurs. In general, the product is liquid, but it can be solid with a low melting point. When the pressure is reduced, the boiling point is lowered.

Apparatus: The distillation apparatus is shown in Fig. 12.

Experiment: The residue of the reaction is progressively heated with a Bunsen burner up to its boiling point. Bubbling of air through a capillary prevents bumping of the liquid and regulates the boiling. The Vigreux column permits a separation of different liquids with different boiling points. When the temperature read by the thermometer does not increase, the product with a determined point at a given pressure is retrieved in a first flask by means of a cylindrical separatory funnel.

Often, the first distillate is composed of the solvent residue of the reaction and the good product which distills at a higher temperature. If a very pure compound is desired, one must repeat the operations with the distillation core. Thus the temperature is best defined and the purity is assured.

Remarks: 1. Warning: This operation is always dangerous because the apparatus may implode or explode under reduced pressure. *You must always distill with safety glasses or, preferably, with goggles.*

2. It is not easy to heat the flask uniformly and we suggest continuously moving the burner under it by hand. At the beginning, one must heat very slowly and regularly while observing the reactions of the liquid.

3. If the product recrystallizes ahead of the junction leading to the pump, the apparatus can explode. Therefore, if this happens one must stop heating, stop running the water in the water condenser, stop the pump, and melt the product. After that, one can distill without water in the water condenser.

4. The pump may be a water pump for the product of low boiling point (typically Bp < 200°C at 760 mm) or a pump with a higher vacuum (~2–3 mm) for the products of higher boiling points.

15. CONCLUSION

As a reminder, we note that other purification methods also exist, such as steam distillation, chromatography on alumina or silica gel, and zone refining. If the reader wants to make other syntheses requiring these purifications, he will always be able to find them in a classical book of organic chemistry.

After purification of organic compounds, the purity must be checked. Let us consider now the means available for this verification.

16. MELTING POINTS

The melting point of a crystalline solid is the temperature at which the solid begins to change into a liquid. For the liquid crystals, the different transition temperatures are related to the different existing mesophases of a determined compound. Generally, these transition temperatures are reversible and permit an identification. Moreover, they are considerably influenced by the presence of other substances. They are an important criterion of purity: If the determined temperature is lower by 1°C than the temperature indicated in the literature, the product is not pure enough.

The Kofler hot bench may be used for all organic compounds melting between 50 and 260°C. It must be calibrated before use by standard organic substances.

The best apparatus for the characterization of liquid crystals is a microscope equipped with an electrically heated hot-stage. It is particularly valuable when the melting point of a very small amount of liquid crystal has to be determined. The use of crossed Polaroids is very helpful to determine the exact moment of the liquid crystals' isotropic transition.

Sharp melting or transition points are usually indicative of the high purity of a substance. If these points are not sharp, the recrystallizations must be repeated until they remain constant.

17. OTHER PURITY DETERMINATION METHODS

Among the different methods used by the organic chemist for the determination of purity of organic substances we can also mention:

Elementary analysis: We think it is difficult for a physicist to perform this analysis. However, if a company or a chemical group exists near by, the product may be sent for analysis by a specialist.

Refractive index: Using an Abbé refractometer, one can very rapidly measure the purity of the organic liquid. A discrepancy in the third decimal of the refractive index value, measured in general at 20°C for the D sodium line, reveals an impure product.

Differential thermal analysis: The purity test is excellent with this method. To every transition there corresponds a peak, the symmetry and sharpness of which give an indication of the purity of the product. It is also possible to calculate the enthalpy of every transition and determine if the transition is of the first or almost of the second order.

Infrared, nuclear magnetic resonance, UV spectroscopy: The interested reader can find all the details of these methods elsewhere.

VII. Conclusion

There are some final remarks to be made. When we began this work, it seemed easy for a specialist in organic chemistry to describe a synthesis. We have had to keep in mind that it was intended for physicists, and the enterprise proved to be difficult because many years of chemistry often transform a scientist into a faddist for esoteric formulae and complicated reactions. Thus, we want to apologize if we have not always been clear enough. On the other hand, we have learned personally many things during the write-up because achieving simplicity requires a deepening of knowledge.

Among the many families of liquid crystals we have selected only three. As this choice was based on our continuous experience and discussions with the physicists of our laboratory, it may seem arbitrary to other chemists. We think that with the liquid crystals we have chosen, many physical problems can be treated, particularly those concerning the relations between molecular and macroscopic structure.

In our introduction, we mentioned mountain hikes. We wish the reader fine weather for his experiment, and no rain, avalanche, nor accident.

ACKNOWLEDGMENTS

We are very grateful to our friend Maryvonne Guyon for his considerable help in preparing the English version of the article. We want also thank very much P. G. de Gennes, E. Guyon, and J. Charvolin for their critical reading.

Elasticity of Nematic Liquid Crystals

H. J. DEULING

Gesamthochschule Kassel, Kassel, West Germany

I. Introduction

A liquid crystal in its nematic state is a liquid consisting of rod-like molecules, which on average line up parallel to a preferred direction. Enclosed in a sandwich cell between two glass plates, a nematic slab will usually be polycrystalline. It will be made up of many domains having different directions of alignment. Nematic single crystals can be prepared by proper treatment of the glass surfaces. Coatings with a thin layer of lecithin produces the homeotropic structure, where the preferred direction of alignment is normal to the bounding plates.[1-3] A

[1] Haas, W., Adams, J., and Flannery, J., *Phys. Rev. Lett.* **25**, 1326 (1970).
[2] Creagh, L. T., and Kmetz, A. R., *SID Symp. Dig. Tech. Pap., 1972* pp. 90-91 (1972).
[3] Proust, J. E., Ter-Minassian, L., and Guyon, E., *Solid State Commun.* **11**, 1227 (1972).

Splay   Twist   Bend

FIG. 1.   Three types of distortion of a director field $\mathbf{n}(x)$. The splay mode gives div $\mathbf{n} \neq$ 0, for the twist mode $\mathbf{n} \cdot \mathbf{curl}\ \mathbf{n} \neq 0$, and the bending mode gives $\mathbf{n} \times \mathbf{curl}\ \mathbf{n} \neq 0$. The figure gives only a lateral view of the director field.

planar structure in which the direction of alignment is parallel to the cell is obtained by rubbing the glass plates in one direction[4,5] or by oblique deposition of silicon oxide.[6,7] In a distorted nematic layer the preferred direction of alignment, which we denote by a unit vector $\mathbf{n}$, is no longer uniform throughout the sample but is gradually changing from point to point. The distortion is thus described by the vector field $\mathbf{n}(x)$. There are three different types of distortion (see Fig. 1). The splay mode is characterized by a nonvanishing divergence of the vector field $\mathbf{n}$ ($\nabla \cdot \mathbf{n} \neq$ 0). The bending mode and the twist mode have $\mathbf{curl}\ \mathbf{n} \neq 0$. The degree of bending is given by the component of $\mathbf{curl}\ \mathbf{n}$ perpendicular to $\mathbf{n}$, whereas the amount of twist is given by the component of $\mathbf{curl}\ \mathbf{n}$ parallel to $\mathbf{n}$. The amount of elastic energy per unit volume g_{d} which is stored in the distortion is found by analogy to Hooke's law to be

$$g_{\mathrm{d}} = \tfrac{1}{2}k_1(\nabla \cdot \mathbf{n})^2 + \tfrac{1}{2}k_2(\mathbf{n} \cdot \nabla \times \mathbf{n})^2 + \tfrac{1}{2}k_3(\mathbf{n} \times \nabla \times \mathbf{n})^2, \qquad (1)$$

where k_1, k_2, and k_3 are elastic constants. They have the dimensions of energy/length and are of the order of 10^{-6} dynes. A detailed derivation and discussion of this expression for the elastic energy density is found in the book by de Gennes[8] and in the papers by Zocher,[9] Oseen,[10] and Frank.[11]

Besides these bulk energies, we may have surface energy terms due to elastic anchoring of the liquid crystal at the bounding surfaces. For a slab with homeotropic alignment the surface energy per unit area may be

[4] Chatelain, P., *Bull. Soc. Fr. Mineral.* **66**, 105 (1943).
[5] Berreman, D. W., *Phys. Rev. Lett.* **28**, 1683 (1972).
[6] Janning, J. L., *Appl. Phys. Lett.* **21**, 173 (1973).
[7] Urbach, W., Boix, M., and Guyon, E., *Appl. Phys. Lett.* **25**, 479 (1974).
[8] de Gennes, P., "The Physics of Liquid Crystals." Oxford Univ. Press (Clarendon), London and New York, 1974.
[9] Zocher, H., *Z. Phys.* **28**, 790 (1927).
[10] Oseen, C. W., *Trans. Faraday Soc.* **29**, 883 (1933).
[11] Frank, F. C., *Discuss. Faraday Soc.* **25**, 19 (1958).

assumed to be of the form

$$w_s = \tfrac{1}{2}C \sin^2\phi_s,$$

where ϕ_s is the angle between the surface normal and the director at the surface. The effect of such a surface energy term on the elastic response of a nematic slab has been considered by Rapini and Papoular[12] and, more recently, by Nehring et al.[13] Throughout the present paper we shall assume rigid anchoring. More references on anchoring properties can be found in a recent review article by Guyon and Urbach.[14]

II. Freedericksz Transition in Nematics

1. Distortion by Magnetic Fields

The distortion of a nematic slab in a magnetic field has been considered by several authors.[15–18] Here we will follow closely the treatment by Saupe.[16]

The susceptibility χ of a nematic liquid depends on the direction of the magnetic field $\mathbf{H}$ relative to $\mathbf{n}$. For $\mathbf{H} \parallel \mathbf{n}$ one finds the susceptibility $\chi_\parallel$ to be larger than the value $\chi_\perp$ which is measured when $\mathbf{H} \perp \mathbf{n}$. This anisotropy $\chi_a = \chi_\parallel - \chi_\perp$ produces an excess magnetization $\Delta M = x_a(\mathbf{n} \cdot \mathbf{H})$ in the direction of $\mathbf{n}$. This excess magnetization produces a magnetic torque per volume $\Gamma_m = \chi_a(\mathbf{n} \cdot \mathbf{H})\mathbf{n} \times \mathbf{H}$. By this torque per volume the magnetic field distorts the nematic liquid. An equilibrium is reached when the elastic torque balances the magnetic torque. The balance of torques can be found by minimizing the total free energy. We find the magnetic contribution to the total free energy per volume by integrating the excess magnetization ΔM over the component $H_\parallel = \mathbf{n} \cdot \mathbf{H}$ of the magnetic field $\mathbf{H}$:

$$g_m = -\tfrac{1}{2}\chi_a(\mathbf{n} \cdot \mathbf{H})^2. \tag{2}$$

Let us consider a slab of thickness d and let us take the z-axis of a Cartesian coordinate system perpendicular to the slab. The total free

[12] Rapini, A., and Papoular, J., J. Phys. (Paris) 30-C4, 54 (1969).

[13] Nehring, J., Kmetz, A. R., and Scheffer, T. J., J. Appl. Phys. 47, 850 (1976).

[14] Guyon, E., and Urbach, W., in "Nonemissive Electrooptic Displays" (A. R. Kmetz and F. K. von Willisen, eds.), p. 121. Plenum, New York, 1976.

[15] Freedericksz, V., and Zolina, V. Z Kristallogr., Kristallgeom., Kristallphys., Kristallchem. 79, 225 (1931)..

[16] Saupe, A., Z. Naturforsch., Teil A 15 815 (1960).

[17] Gruler, H., Scheffer, T·., and Meier, G., Z. Naturforsch., Teil A 27, 966 (1972).

[18] Pincus, P., J. Appl. Phys. 41, 974 (1970).

energy per unit area of the slab is then

$$G = \frac{1}{2} \int_0^d \{k_1(\nabla \cdot \mathbf{n})^2 + k_2(\mathbf{n} \cdot \nabla \times \mathbf{n})^2$$

$$+ k_3(\mathbf{n} \times \nabla \times \mathbf{n})^2 - \chi_a(\mathbf{n} \cdot \mathbf{H})^2\}dz. \tag{3}$$

a. Twist Mode

We consider a sandwich cell with planar orientation in a magnetic field **H** parallel to the cell. We take **n** $\parallel$ y-axis and **H** $\parallel$ x-axis. If **H** is strong enough it will turn the director **n** by an angle ω. At the boundaries we have $\omega = 0$; for $z = d/2$ the twist angle ω takes on its maximum value ω_m. For the total free energy G per unit area we have

$$G = \frac{1}{2} \int_0^d \{k_2(d\omega/dz)^2 - \chi_a H^2 \sin^2\omega\}dz. \tag{4}$$

G does not depend on coordinate z explicitly. Therefore we find it convenient to introduce ω as the independent variable:

$$G = \int_0^{\omega_m} \{k_2(dz/d\omega)^{-1} - \chi_a H^2 \sin^2\omega(dz/d\omega)\}d\omega.$$

Minimizing the energy G with respect to the function $z(\omega)$ gives directly

$$k_2(dz/d\omega)^{-2} + \chi_a H^2 \sin^2\omega = \text{const.} \tag{5}$$

The constant is determined by the condition $z = d/2$ for $\omega = \omega_m$. The final result is

$$(2/\pi) \int_0^{\pi/2} (1 - \sin^2\omega_m \sin^2\psi)^{-1/2}d\psi = (d/\pi)(\chi_a/k_2)^{1/2}H. \tag{6}$$

This equation relates the maximum twist angle ω_m to the magnetic field strength. The left-hand side of this equation is always ≥ 1, so Eq. (6) can have a solution only if H exceeds a critical value $H_{c2} = (\pi/d)(k_2/\chi_a)^{1/2}$. This critical field can be observed optically, as will be discussed below.

b. Splay and Bend Mode

To induce a splay distortion we take the magnetic field **H** along the z-direction. The field will now tilt the director by an angle φ. At $z = 0$ and at $z = d$ we have $\varphi = 0$. In the middle of the layer φ will be at its maximum value φ_m. Again we introduce the angle as the independent variable instead of z. The total free energy per unit area is then given by the expression

$$G = \int_0^{\varphi_m} \{(k_1 \cos^2\varphi + k_3 \sin^2\varphi)(dz/d\varphi)^{-1}$$

$$- \chi_a H^2 \sin^2\varphi(dz/d\varphi)\}d\varphi. \tag{7}$$

The energy has a minimum value when

$$(k_1 \cos^2\varphi + k_3 \sin^2\varphi)(dz/d\varphi)^{-2} + \chi_a H^2 \sin^2\varphi = \text{const.} \tag{8}$$

The constant is determined by the boundary condition $\varphi = \varphi_m$ at $z = d/2$.

A straightforward calculation gives

$$H(d/\pi)(\chi_a/k_1)^{1/2}$$

$$= (2/\pi) \int_0^{\pi/2} \{(1 + \kappa\eta^2 \sin^2\psi)/(1 - \eta^2 \sin^2\psi)\}^{1/2}d\psi, \tag{9}$$

where $\eta = \sin\varphi_m$ and $\kappa = (k_3 - k_1)/k_1$. This equation relates the maximum tilt angle φ_m to the magnetic field H. The right-hand side of Eq. (9) is always ≥ 1. Therefore a solution exists only if H exceeds a critical field $H_{c1} = (\pi/d)(k_1/\chi_a)^{1/2}$. The d^{-1} dependence of the critical field was first observed by Freedericksz and Zolina.[15] We can deform a slab with homeotropic orientation if we take **H** parallel to the bounding plates. In Eqs. (7)–(9) we have to replace φ by $\pi/2 - \varphi$. This is equivalent to replacing k_1 by k_3 and vice-versa. For the homeotropic slab we get therefore the critical field $H_{c3} = (\pi/d)(k_3/\chi_a)^{1/2}$ of the bending mode. A critical field is obtained only if the magnetic field is initially perpendicular to **n**.[19-21] The transition of the nematic slab at the critical field from the undistorted state to the distorted state resembles a second-order phase transition and is called the Freedericksz transition. The critical fields can be obtained from a Landau expansion of the free energy G in terms of the order parameter $|\varphi_m|$ as was pointed out by Brochard et al.[22]

c. Freedericksz Transition in an Oblique Field

The splay critical field H_{c1} and the twist critical field H_{c2} are only limiting forms of the more general case in which the magnetic field is in

[19] Rapini, A., Papoular, M., and Pincus, P., *C. R. Hebd. Seances Acad. Sci., Ser. B* **267**, 1230 (1968).
[20] See ref. 12, p. 55.
[21] Schneider, F., *Z. Naturforsch., Teil A* **28** 1660 (1973).
[22] Brochard, F., Pieranski, P., and Guyon, E., *Phys. Rev. Lett.* **28**, 1681 (1972).

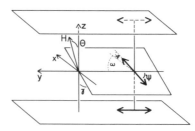

FIG. 2. Geometry of the distortion of a planar nematic slab by an oblique magnetic field. For small distortions the director stays in a plane through the y-axis at an angle γ with the z-axis. The angle γ is independent of z. The distortion can then be described as a rotation by an angle ψ in this plane.

the xz-plane at an angle θ with the z-axis.[23] Now all three types of deformation are present. The director is turned by an angle ω around the z-axis and is tilted by an angle φ with respect to the xy-plane (see Fig. 2). In Cartesian coordinates we have $\mathbf{H} = (H \sin \theta, 0, H \cos \theta)$ and $\mathbf{n} = (\cos \varphi \sin \omega, \cos \varphi \cos \omega, \sin \varphi)$. The energy now becomes a rather complicated expression:

$$G = \frac{1}{2} \int_0^d \{ (k_1 \cos^2\varphi + k_3 \sin^2\varphi)(d\varphi/dz)^2$$
$$+ (k_2 \cos^2\varphi + k_3 \sin^2\varphi) \cos^2\varphi (d\omega/dz)^2 \tag{10}$$
$$- \chi_a H^2 (\sin \theta \cos \varphi \sin \omega + \cos \theta \sin \varphi)^2 \} \, dz.$$

By variational calculus we can find from this expression the equations which express the balance of torques per volume element. These equations are complicated and cannot be solved analytically. In the limit of small distortions they read

$$d^2\varphi/dz^2 = -(\pi/d)^2(H/H_{c1})^2 \cos \theta(\omega \sin \theta + \varphi \cos \theta);$$
$$d^2\omega/dz^2 = -(\pi/d)^2(H/H_{c2})^2 \sin \theta(\omega \sin \theta + \varphi \cos \theta). \tag{11}$$

For small distortions the director will stay in a plane containing the y-axis. Let this plane be at an angle γ with the z-axis and let the director be turned by an angle ψ in this plane. This assumption gives us an ansatz

$$\varphi(z) = \psi(z) \cos \gamma; \qquad \omega(z) = \psi(z) \sin \gamma.$$

Inserting into Eq. (11) gives

$$\tan \gamma = (k_1/k_2) \tan \theta$$

and

$$d^2\psi/dz^2 = -(\pi/d)^2 H^2 [H_{c1}^{-2} \cos^2 \theta + H_{c2}^{-2} \sin^2 \theta]\psi(z).$$

[23] Deuling, H., Gabay, M., Guyon, E., and Pieranski, P., *J. Phys.* **36,** 689 (1975).

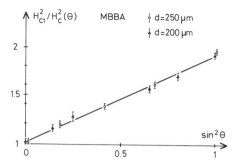

FIG. 3. Plot of $[H_{c1}/H_c(\Theta)]^2$ versus $\sin^2 \Theta$. The data obtained with two samples of MBBA are well represented by a straight line. The slope gives $\beta = (k_1 - k_2)/k_2 = 0.86$. From Deuling et al.[23] (Copyright by Commission des Publications de Françaises de Physique. Used with permission.)

The solution is of the form $\psi = \psi_m \sin(\pi z/L)$ and the expression in brackets is just the inverse square of the critical field $H_c(\theta)$:

$$H_c^{-2}(\theta) = H_{c1}^{-2} \cos^2 \theta + H_{c2}^{-2} \sin^2 \theta. \tag{12}$$

This is the equation of an ellipse with semiaxes H_{c1} and H_{c2}. Plotting $[H_{c1}/H_c(\theta)]^2$ versus $\sin^2 \theta$, we obtain a linear plot with slope $(k_1 - k_2)/k_2$. Such a plot is shown for MBBA in Fig. 3. Gabay[24] has obtained the critical field $H_c(\theta)$ and the relation $\tan \gamma = (k_1/k_2) \tan \theta$ from a Landau expansion of G in terms of the order parameter $|\psi_m|$.

d. Freedericksz Transition of a Twisted Nematic

In a nematic slab with planar orientation we can induce a twist distortion simply by turning one of the bounding plates by an angle Φ. The Freedericksz transition of such a twisted nematic in a magnetic field was first studied theoretically by Leslie.[25] We obtain the energy of a twisted nematic slab in a magnetic field perpendicular to the bounding plates putting $\theta = 0$ in expression (10). The distortion pattern will be symmetric, with a maximum tilt angle φ_m at $z = d/2$. Introducing φ again as the independent variable we get

$$G = \int_0^{\varphi_m} \{[k_1 \cos^2 \varphi + k_3 \sin^2 \varphi + \cos^2 \varphi(k_2 \cos^2 \varphi$$

$$+ k_3 \sin^2 \varphi)(d\omega/d\varphi)^2](dz/d\varphi)^{-1}$$

$$- \chi_a H^2 \sin^2 \varphi(dz/d\varphi)\}d\varphi.$$

[24] Gabay, M., Appendix A of Deuling et al.[23]
[25] Leslie, F. M., Mol. Cryst. Liq. Cryst. **12**, 57 (1970).

By variational calculus we find the equations

$$(k_2 \cos^2 \varphi + k_3 \sin^2 \varphi) \cos^2 \varphi (d\omega/dz) = (k_2 \chi_a)^{1/2} H\mu$$

$$(k_1 \cos^2 \varphi + k_3 \sin^2 \varphi)(d\varphi/dz)^2$$

$$+ \chi_a H^2 \{\sin^2 \varphi + \mu^2 [\cos^2 \varphi (\cos^2 \varphi + (k_3/k_2) \sin^2 \varphi)]^{-1}\} = \text{const}, \quad (13)$$

where μ is a constant of integration. A straightforward calculation gives

$$H/H_{c1} = (2/\pi) \int_0^{\varphi_m} (1 + \kappa \sin^2\varphi)^{1/2} \{g(\varphi_m) - g(\varphi)\}^{-1/2} d\varphi; \quad (14)$$

$$\Phi/\pi = \mu(k_1/k_2)^{1/2}(2/\pi) \int_0^{\varphi_m} [\cos^2\varphi (1 + \alpha \sin^2\varphi)]^{-1}$$

$$\times (1 + \kappa \sin^2\varphi)^{1/2} \{g(\varphi_m) - g(\varphi)\}^{-1/2} d\varphi, \quad (15)$$

where

$$g(\varphi) = \sin^2\varphi + \mu^2 \{\cos^2\varphi(1 + \alpha \sin^2\varphi)\}^{-1}$$

and $\alpha = (k_3 - k_2)/k_2$. For small distortions we find

$$g(\varphi) = \mu^2 + \{1 - \mu^2(k_3 - 2k_2)/k_1\}\varphi^2 + \cdots.$$

In the limit $\varphi_m \to 0$ we have, therefore,

$$H \to H_T = H_{c1}\{1 + (\Phi/\pi)^2(k_3 - 2k_2)/k_1\}^{1/2}; \quad (16)$$

$$\mu \to \{(k_3 - 2k_2)/k_2 + (k_1/k_2)(\pi/\Phi)^2\}^{-1/2}. \quad (17)$$

The critical field (16) has again the d^{-1} dependence of H_{c1}. For the twisted nematic with $\Phi = \pi/2$ this dependence on thickness of the critical field was found experimentally by Gerritsma *et al.*[26] For $2k_2 > k_3$ there exists a critical twist angle $\Phi_c = \pi[k_1/(2k_2 - k_3)]^{1/2}$ at which $H_T = 0$. For twist angles above Φ_c the sample can relax spontaneously by a splay and bend distortion. This relaxation has been observed by Turner and Faber.[27]

Figure 4 shows a solution of the Leslie equations (14) and (15) for MBBA. We plot the maximum tilt angle φ_m as a function of Φ for fixed values of $H > H_{c1}$. With increasing Φ the maximum tilt angle φ_m increases. At a critical value $\Phi_c < \pi$ the slope $d\varphi_m/d\Phi$ becomes infinite. For $\Phi_c < \Phi < \pi$ Eqs. (14) and (15) do not have a solution.

[26] Gerritsma, C. J., de Jeu, W. H., and van Zanten, P., *Phys. Lett A* **36**, 389 (1971).
[27] Turner, R., and Faber, T. E., *Phys. Lett A* **49** 423 (1974).

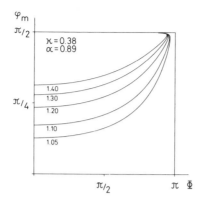

FIG. 4. Plot of the maximum tilt angle φ_m of a twisted nematic layer in a magnetic field perpendicular to the slab. Φ is the angle between the preferred directions of alignment at the bounding plates. The parameters have been chosen so as to correspond to MBBA. The parameter indexing the different curves is the magnetic field H in units of the splay critical field H_{c1}.

2. DISTORTION BY ELECTRIC FIELDS

The distortion of nematic layers by electric fields is complicated by the fact that the electric field is in general not uniform across the sample. Only the voltage remains constant. We consider a nematic slab with planar orientation and with positive dielectric anisotropy. We apply a voltage U to the slab. The resulting electric field E will have a component only in the z-direction. E is not uniform across the layer but depends on the coordinate z. The dielectric anisotropy ϵ_a produces an excess polarization $\Delta P = (4\pi)^{-1}\epsilon_a E \sin \varphi$ in the direction of $\mathbf{n}$. φ is the angle of $\mathbf{n}$ with the bounding plates. Through this polarization the field exerts a torque Γ_e per volume

$$\Gamma_e = (4\pi)^{-1}\epsilon_a E^2 \sin \varphi \cos \varphi.$$

The elastic torque per volume is obtained easily from the elastic energy density. The balance of torques requires

$$(k_1 \cos^2\varphi + k_3 \sin^2\varphi)d^2\varphi/dz^2 + (k_3 - k_1) \sin \varphi \cos \varphi \ (d\varphi/dz)^2 \qquad (18)$$

$$+ (4\pi)^{-1}\epsilon_a E^2(z) \sin \varphi \cos \varphi = 0.$$

In the limit of small distortions, i.e., near the critical field, $E(z)$ is nearly constant. The problem is then analogous to the Freedericksz transition in a magnetic field. We obtain a critical field $E_{c1} = (2\pi^{3/2}/d)(k_1/\epsilon_a)^{1/2}$. This corresponds to a critical voltage $U_{c1} = 2\pi^{3/2}(k_1/\epsilon_a)^{1/2}$. Treating the distortion above threshold, we have to distinguish between a conducting nematic and an insulating nematic. The latter has been considered by Gruler et al.,[17,28] by Deuling,[29] and by Cheung.[30] The former case has been considered by Deuling and Helfrich[31] and by Cheung.[30]

a. Freedericksz Transition of an Insulating Nematic Slab

In an insulating liquid the condition div $D = 0$ requires D_z to be constant:

$$D_z = (\epsilon_\perp \cos^2\varphi + \epsilon_\parallel \sin^2\varphi)E(z) = \text{const.} \qquad (19)$$

With this relation we can express E as a function of φ. The balance of torques equation is then solved in the same way as for the magnetic case. (Multiply the equation by $2d\varphi/dz$ and integrate once.) Introducing an anisotropy constant $\gamma = (\epsilon_\parallel - \epsilon_\perp)/\epsilon_\perp$ and using again the notation $\eta = \sin\varphi_m$, we find after some algebra

$$D_z/D_z^0 = (1 + \gamma\eta^2)^{1/2}(2/\pi) \int_0^{\pi/2} \{(1 + \gamma\eta^2 \sin^2\psi)$$

$$\times (1 + \kappa\eta^2 \sin^2\psi)/(1 - \eta^2 \sin^2\psi)\}^{1/2}d\psi;$$

$$U/U_{c1} = (2/\pi) \int_0^{\pi/2} \{(1 + \gamma\eta^2)/(1 + \gamma\eta^2 \sin^2\psi)\}^{1/2}$$

$$\times \{(1 + \kappa\eta^2 \sin^2\psi)/(1 - \eta^2 \sin^2\psi)\}^{1/2}d\psi,$$

$$(20)$$

where D_z^0 is the dielectric displacement D_z at threshold. The expression for U/U_{c1} differs from the corresponding equation for the magnetic field [Eq. (9)] only by the first term in the integral. This term is always ≥ 1. To produce the same amount of distortion as with a magnetic field H/H_{c1} we have to use a higher voltage, i.e., $U/U_{c1} \geq H/H_{c1}$. This has been demonstrated nicely by Gruler et al.[17] For a cell with homeotropic alignment we have to replace φ by $\pi/2 - \varphi$ which is equivalent to replacing k_1 and $\epsilon_\perp$ by k_3 and $\epsilon_\parallel$ and vice-versa. So we obtain a threshold voltage $U_{c3} = 2\pi^{3/2}[k_3/(-\epsilon_a)]^{1/2}$ for the bending mode. The dependence of U_{c3} on $(-\epsilon_a)^{-1/2}$ has been verified by Michel and Smith.[32] They

[28] Scheffer, T., and Gruler, H., in "Molecular Electro-Optics and the Electrical Properties of Macromolecules"(C. T. O'Konski, ed.). Dekker, New York.

[29] Deuling, H., Mol. Cryst. Liq. Cryst. **19**, 123, (1972).

[30] Cheung, L., Thesis, Division of Engineering and Applied Physics, Harvard University, Cambridge, Massachusetts, 1973.

[31] Deuling, H., and Helfrich, W., Appl. Phys. Lett. **25**, 129, (1974)

[32] Michel, R. E., and Smith, G. W., J. Appl. Phys. **45**, 3234 (1974).

measured U_{c3} for binary mixtures of PEBAB in MBBA with various concentrations of PEBAB. A homeotropic cell is uniaxial in the undistorted state and becomes birefringent above threshold. This electrically controllable birefringence is interesting for device applications and has been studied extensively by several workers.[33-35] Decomposing the director **n** in the distorted homeotropic sample into components parallel and perpendicular to the slab we may write for the parallel component

$$\mathbf{n}_\parallel = S(z) \cdot \mathbf{C}(x, y),$$

where **C** is a unit vector in the xy-plane. If the distortion is produced by a magnetic field applied parallel to the sample the vector $\mathbf{C}(x, y)$ clearly has to be parallel to H. For a material with $\epsilon_a < 0$, however, we produce the distortion by an electric field perpendicular to the sample. The vector $\mathbf{C}(x, y)$ may now have any direction. Of particular interest are point singularities in the vector field $\mathbf{C}(x, y)$, so-called "umbilics," which have been studied by Rapini et al.[36,37]

b. Freedericksz Transition of a Conducting Nematic Slab

In a conducting medium the condition div **j** = 0 requires the z-component of the current density **j** to be constant:

$$j_z = (\sigma_\perp \cos^2\varphi + \sigma_\parallel \sin^2\varphi)E(z) = \text{const}, \tag{21}$$

where $\sigma_\parallel$ and $\sigma_\perp$ are the conductivities along directions parallel and perpendicular to **n**, respectively. We take $\sigma_\parallel$ and $\sigma_\perp$ to be independent of the field. The solution is now exactly as for the insulating nematic slab, if we replace $\gamma = (\epsilon_\parallel - \epsilon_\perp)/\epsilon_\perp$ by $(\sigma_\parallel - \sigma_\perp)/\sigma_\perp$. For a material with strong conductive anisotropy and a dielectric anisotropy of opposite sign one can obtain hysteresis.[31] Let us assume $\epsilon_a < 0$ and $\sigma_\parallel > \sigma_\perp$. To induce a deformation we have to use a slab with homeotropic orientation. Our derivation holds for the homeotropic case if we replace φ by $\pi/2 - \varphi$, which is equivalent to replacing $k_1, \sigma_\perp, \epsilon_\perp$ by $k_3, \sigma_\parallel, \epsilon_\parallel$ and vice-versa. We get the result

$$U/U_{c3} = (2/\pi) \int_0^{\pi/2} \{(1 - \tilde{\gamma}\eta^2)/(1 - \tilde{\gamma}\eta^2 \sin^2\psi)\}^{1/2}$$

$$\times \{(1 - \tilde{\kappa}\eta^2 \sin^2\psi)/(1 - \eta^2 \sin^2\psi)\}^{1/2}d\psi. \tag{22}$$

[33] Heilmeier, G. H., Castellano, J. A., and Zanoni, L. A., Mol. Cryst. Liq. Cryst. **8**, 293 (1969).

[34] Mailer, H., Likins, K. L., Taylor, T. R., and Fergason, J. L., Appl. Phys. Lett. **18**, 105 (1971).

[35] Soref, R. A., and Rafuse, M. J., J. Appl. Phys. **43**, 1029 (1972).

[36] Rapini, A., J. Phys. (Paris) **34**, 629 (1973).

[37] Rapini, A., Léger, L., and Martinet, A., J. Phys. (Paris) **36-C1**, 189 (1975).

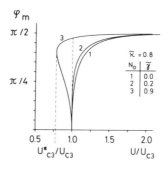

FIG. 5. Plot of maximum tilt angle φ_m versus voltage U for a conducting nematic slab with homeotropic alignment. For sufficiently strong anisotropy of conductivity (curve 3) the curve φ_m versus U contains an unstable part where $d\varphi_m/dU < 0$.

The parameters η, $\tilde{\kappa}$, and $\tilde{\gamma}$ are

$$\eta = \sin \varphi_m, \quad \tilde{\kappa} = (k_3 - k_1)/k_3, \quad \tilde{\gamma} = (\sigma_\parallel - \sigma_\perp)/\sigma_\perp.$$

For small distortions we have

$$U/U_{c3} = 1 + \tfrac{1}{4}(1 - \tilde{\kappa} - \tilde{\gamma})\varphi_m{}^2.$$

For $1 - \tilde{\kappa} - \tilde{\gamma} < 0$ the curve $\varphi_m{}^2$ versus U has negative slope at $U = U_{c3}$, i.e., small distortions are not stable. The Freedericksz transition is now of first order and we have hysteresis. This is shown in Fig. 5 for an hypothetical case with $\tilde{\kappa} = 0.8$ and $\tilde{\gamma} = 0.9$. With increasing voltage the layer switches at the critical field from the undistorted state to a state with a finite amount of distortion (φ_m). With decreasing voltage the layer switches back to the undistorted state at a lower critical voltage U_{c3}^*, which can be determined by the condition $dU/d\varphi_m = 0$.

c. Distortion of Twist Cells by an Electric Field (Schadt–Helfrich Cell)

The calculation of the distortion of a twist cell by an electric field is analogous to the calculation for the planar cell.[38] To the elastic torque in Eq. (18) we have to add a new twist-dependent term:

$$-\{(k_3 - k_2) \sin \varphi \cos^3\varphi - (k_2 \cos^2\varphi + k_3 \sin^2\varphi) \sin \varphi \cos \varphi\}(d\omega/dz)^2.$$

The torque around the z-axis is zero. This gives an equation for the twist angle $\omega(z)$:

$$(d/dz)\{\cos^2\varphi(k_2 \cos^2\varphi + k_3 \sin^2\varphi)(d\omega/dz)\} = 0.$$

As before we obtain the z-dependence of E from the condition $D_z =$

[38] Deuling, H., Mol. Cryst. Liq. Cryst. 27, 81 (1975).

const (Eq. 19). A straightforward calculation gives expressions similar to those obtained for the twist cell in a magnetic field:

$$U/U_{c1} = (2/\pi) \int_0^{\varphi_m} (1 + \kappa \sin^2\varphi)^{1/2}$$

$$\times (1 + \gamma \sin^2\varphi)^{-1}\{h(\varphi_m) - h(\varphi)\}^{-1/2} d\varphi; \qquad (24)$$

$$\Phi/\pi = \mu(k_1/k_2)^{1/2}(2/\pi) \int_0^{\varphi_m} [\cos^2\varphi(1 + \alpha \sin^2\varphi)]^{-1}$$

$$\times (1 + \kappa \sin^2\varphi)^{1/2}\{h(\varphi_m) - h(\varphi)\}^{-1/2} d\varphi. \qquad (25)$$

The function $h(\varphi)$ is defined by

$$h(\varphi) = \mu^2\{\cos^2\varphi(1 + \alpha \sin^2\varphi)\}^{-1} - \gamma^{-1}(1 + \gamma \sin^2\varphi)^{-1}. \qquad (26)$$

Note that in the limit $\gamma \to 0$ $h(\varphi_m) - h(\varphi) \to g(\varphi_m) - g(\varphi)$ and we recover all the results for the magnetic field solution. As before we get in the limit $\varphi_m \to 0$

$$U \to U_T = U_{c1}\{1 + (\Phi/\pi)^2(k_3 - 2k_2)/k_1\}^{1/2}. \qquad (27)$$

The threshold voltage U_T has the same dependence on ϵ_a as has the threshold U_{c1} of the planar cell. For twist cells filled with nematic mixtures having different values of $\epsilon_a > 0$. Alder and Raynes[39] found an $\epsilon_a^{-1/2}$ dependence of the threshold voltage.

Twist cells have unusual optical properties, which we will touch upon only briefly in the present paper. The pitch of the usual quarter-turn cell is large compared to the wavelength of visible light. At vertical incidence the polarization plane of linearly polarized light is turned along with the director as the light passes through the sample. A quarter-turn cell placed between parallel polarizers transmits no light. In a field (electric or magnetic) sufficiently far above threshold the sample is nearly homeotropic in the middle, and almost all of the elastic energy is located in two thin layers near the bounding plates. All the twist occurs in a thin layer in the middle. The nematic layer therefore does not turn the polarization plane and the light goes through nearly unchanged. This switch effect was first demonstrated by Schadt and Helfrich.[40] Clearly this optical threshold is higher than the threshold for distortion. This was demonstrated by Gerritsma et al.,[26] who obtained the distortion threshold in a magnetic field from capacitance measurements and found it to be lower than the optical threshold. Extending earlier work by de

[39] Alder, C. J., and Raynes, E. P., *J. Phys. D.* 6, L33 (1973).
[40] Schadt, M., and Helfrich, W., *Appl. Phys. Lett.* 18, 127 (1971).

Vries,[41] van Doorn[42] calculated the transmission of light at vertical incidence and the capacitance change of a twist cell as a function of magnetic field and found satisfactory agreement with the data of Gerritsma *et al.*[26] On the basis of his earlier work,[43] Berreman[44] formulated a powerful method for computing the propagation of light in twisted nematics and calculated the transmission of a twist cell in an electric field for oblique angles of incidence. He took the electric field to be uniform throughout the sample, i.e., assumed small dielectric anisotropy. Independently van Doorn and Heldens[45] calculated the angular dependence of the transmission of a twist cell and, comparing their results with experimental data, obtained good agreement. Baur and Meier[46] also measured the angular dependence of the transmission of light through a twist cell.

3. FREEDERICKSZ TRANSITION IN CROSSED ELECTRIC AND MAGNETIC FIELDS

In the previous sections we have seen that twisting a planar nematic slab changes the critical field. Instead of twisting the nematic mechanically as is done for the usual twist cell we can also twist the nematic by applying a magnetic field parallel to the bounding plates and perpendicular to the initial direction of alignment. For $H < H_{c2}$ the nematic layer is in a twisted state. If we then apply a voltage across the slab there will be a critical voltage U_s above which a splay distortion is induced. This critical voltage U_s depends on the strength of the magnetic field H and is always larger than U_{c1}. Calculation of $U_s(H)$ requires solving a nonlinear eigenvalue equation and can only be done numerically.[47] If we apply a sufficiently strong voltage $U > U_{c1}$ in zero magnetic field and then increase the field, we start with a splay and bend distortion but have no twist. We increase the field H only so far above H_{c2} that we still have $U \gg U_s(H)$. If we now reduce the voltage we will reach a critical voltage $U_t(H)$ at which the magnetic field is strong enough to induce a twist distortion. These two critical voltages are different in general: $U_s(H) \neq U_t(H)$. Figure 6 shows the dependence of U_s and U_t on magnetic field H.

[41] de Vries, H. *Acta Crystallogr.* **4**, 219 (1951).
[42] van Doorn, C. Z., *Phys. Lett. A*. **42**, 537 (1973).
[43] Berreman, D. W., *J. Opt. Soc. Am.* **62**, 502 (1972).
[44] Berreman, D. W., *J. Opt. Soc. Am.* **63**, 1374 (1973).
[45] van Doorn, C. Z., and Heldens, J. L., *Phys. Lett. A*. **47**, 135 (1974).
[46] Baur, G., and Meier, G., *Phys. Lett. A*. **50**, 149 (1974).
[47] Deuling, H., Guyon, E., and Pieranski, P., *Solid State Commun.* **15**, 277 (1974).

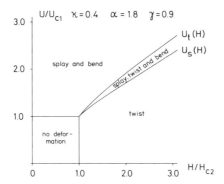

FIG. 6. "Phase diagram" of a planar nematic layer in crossed electric and magnetic fields, giving the regions with different types of distortion.

The behavior of a nematic slab in crossed electric and magnetic fields is thus strikingly different from that in an oblique magnetic field (see Section II,1,c), where we found one critical field above which both splay and twist were induced. The difference can be understood as follows. The torques exerted by the two fields are proportional to H^2 and E^2, respectively. In an oblique magnetic field, however, the torque is proportional to $H_{\parallel}^2 + H_{\perp}^2 + 2H_{\parallel}H_{\perp}$. This last cross-term of the two field components explains the difference. Figure 7 shows the capacitance increase ΔC as a function of voltage for various values of H/H_{c2}.

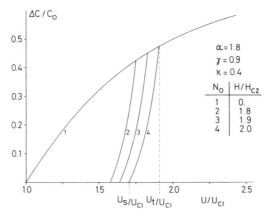

FIG. 7. Capacitance increase ΔC as a function of voltage U for a planar nematic slab in crossed electric and magnetic fields.

III. Flexo-electric Effects

Meyer[48] has introduced the concept of curvature electricity into the field of liquid crystals. Curvature electricity, or flexoelectricity as it is also called, is the liquid-crystal analog of piezoelectricity in solids. A mechanical distortion of a nematic slab can induce an electric polarization. According to Meyer this can be visualized as follows. In a bent nematic consisting of banana-shaped molecules these molecules will prefer a configuration in which their curved shape fits in with the curvature of the director field $\mathbf{n}(\mathbf{x})$ (see Fig. 8). If the molecules carry an electric dipole moment perpendicular to their long axis, as indicated by the arrow in Fig. 8, an electric polarization will result, whose magnitude is proportional to the degree of bending and which is perpendicular to $\mathbf{n}$. A similar argument can be given for gumdrop-shaped molecules in a splayed nematic leading to an electric polarization induced by a splay distortion. The net polarization can be written as

$$\mathbf{P} = e_1\mathbf{n}\nabla \cdot \mathbf{n} - e_3\mathbf{n} \times \nabla \times \mathbf{n}, \qquad (28)$$

where e_1 and e_3 are flexo-electric constants. Helfrich[49] has estimated these constants to be of the order of 4×10^{-5} cgs units. For MBBA Schmidt et al.[50,51] measured $e_3 = 3.7 \times 10^{-5}$ cgs units.

The flexo-electric polarization gives rise to an additional term in the free energy and therefore causes some slight modifications in the Freedericksz transition. We will consider these modifications separately for the distortion caused by magnetic and electric fields, respectively.

4. DISTORTION BY A MAGNETIC FIELD

We have to add a new term,

$$-1/(8\pi) \int_0^d (\epsilon_\parallel \sin^2\varphi + \epsilon_\perp \cos^2\varphi)E^2 dz - \int_0^d P_z \cdot E dz$$

$$= -D_z U/(8\pi) - \frac{1}{2}\int_0^d P_z \cdot E dz,$$

to the total free energy per unit area as given by expression (7) in Section II. The condition

$$D_z = (\epsilon_\perp \cos^2\varphi + \epsilon_\parallel \sin^2\varphi)E + 4\pi P_z = 0$$

[48] Meyer, R. B., *Phys. Rev. Lett.* **22**, 918 (1969).
[49] Helfrich, W., *Z. Naturforsch., Teil A* **26** 833 (1971).
[50] Helfrich, W., *Phys. Lett. A* **35**, 393 (1971).
[51] Schmidt, D., Schadt, M., and Helfrich, W., *Z. Naturforsch., Teil A* **27**, 277 (1972).

FIG. 8.   Graph explaining the origin of flexo-electricity for the bending mode. In a bent nematic banana-shaped molecules prefer configuration (a) to configuration (b). (After Meyer.[48])

gives

$$-\frac{1}{2}\int_0^d P_z \cdot E dz = 2\pi \int_0^d P_z^2 (\epsilon_\perp \cos^2\varphi + \epsilon_\parallel \sin^2\varphi)^{-1} dz.$$

Inserting $P_z = -(e_1 + e_3)\sin\varphi\cos\varphi(d\varphi/dz)$ and introducing again φ as the independent variable we obtain for the total free energy per unit area

$$G = \int_0^{\varphi_m} \{[k_1 \cos^2\varphi + k_3 \sin^2\varphi + 4\pi(e_1 + e_3)^2 \sin^2\varphi \cos^2\varphi (\epsilon_\perp \cos^2\varphi$$

$$+ \epsilon_\parallel \sin^2\varphi)^{-1}](dz/d\varphi)^{-1} - \chi_a H^2 \sin^2\varphi(dz/d\varphi)\}d\varphi.$$

Here we see that the only effect of the flexo-electric polarization is to modify the elastic energy term.[52] The calculation now proceeds exactly as before and gives the final result[53]

$$H/H_{c1} = (2/\pi)\int_0^{\pi/2} \{(1 + \kappa\eta^2 \sin^2\varphi)/(1 - \eta^2 \sin^2\psi)$$

$$+ \Delta\eta^2 \sin^2\psi/(1 + \gamma\eta^2 \sin^2\psi)\}^{1/2}d\psi. \tag{30}$$

The parameter $\Delta = 4\pi(e_1 + e_3)^2/(k_1\epsilon_\perp)$ describes the influence of flexo-electricity on the distortion.

5.   DISTORTION BY AN ELECTRIC FIELD

In Section II we started from the balance of torques to find the distortion of a planar nematic slab caused by an electric field perpendicular to the slab. When we want to take flexo-electricity into account we find it more convenient to minimize the total free energy. The electric contribution to the total free energy per unit area of the slab is given by

$$-(8\pi)^{-1}\int_0^d (\epsilon_\perp \cos^2\varphi + \epsilon_\parallel \sin^2\varphi)E^2 dz - \int_0^d P_z \cdot E dz.$$

[52] Helfrich, W., *Mol. Cryst. Liq. Cryst.* **26**, 1 (1974).
[53] Deuling, H., *Solid State Commun.* **14**, 1073 (1974).

Using the relation $D_z = (\epsilon_\perp \cos^2\varphi + \epsilon_\parallel \sin^2\varphi)E + 4\pi P_z = \text{const}$ we can express the electric contribution as

$$-(8\pi)^{-1}D_z \cdot U - \frac{1}{2}\int_0^d P_z \cdot E\,dz.$$

Substituting for P_z we obtain

$$-\frac{1}{2}\int_0^d P_z \cdot E\,dz = 2\pi(e_1 + e_3)^2 \int_0^d \sin^2\varphi\,\cos^2\varphi(\epsilon_\perp \cos^2\varphi$$

$$+ \epsilon_\parallel \sin^2\varphi)^{-1}(d\varphi/dz)^2 dz + (1/4)(e_1 + e_3)/(\epsilon_\parallel - \epsilon_\perp)D_z$$

$$\times \log(\epsilon_\perp \cos^2\varphi + \epsilon_\parallel \sin^2\varphi)\Big|_0^d. \tag{31}$$

The first term gives again a correction to the elastic energy, and the second term depends only on the tilt angle φ at the boundaries. This term vanishes for rigid anchoring on both plates. The z-component of the dielectric displacement D_z is a functional of φ. By the relation

$$U = \int_0^d E\,dz$$

we find

$$D_z = \Big\{ U - 2\pi(e_1 + e_3)/(\epsilon_\parallel - \epsilon_\perp)$$

$$\times \log(\epsilon_\perp \cos^2\varphi + \epsilon_\parallel \sin^2\varphi)\Big|_0^d\Big\} \Big/ \int_0^d (\epsilon_\perp \cos^2\varphi + \epsilon_\parallel \sin^2\varphi)^{-1}dz. \tag{32}$$

For rigid anchoring the log term vanishes. φ will take on its maximum value φ_m at $z = d/2$. Introducing φ as the independent variable we get

$$G = \int_0^{\varphi_m} \{k_1 \cos^2\varphi + k_3 \sin^2\varphi + 4\pi(e_1 + e_3)^2 \sin^2\varphi\,\cos^2\varphi(\epsilon_\perp \cos^2\varphi$$

$$+ \epsilon_\parallel \sin^2\varphi)^{-1}\}(dz/d\varphi)^{-1}d\varphi$$

$$- (4\pi)^{-1}(U/2)^2 \Big/ \int_0^{\varphi_m} (\epsilon_\perp \cos^2\varphi + \epsilon_\parallel \sin^2\varphi)^{-1}(dz/d\varphi)d\varphi. \tag{33}$$

Minimizing this expression with respect to $z(\varphi)$ we obtain an Euler–Lagrange equation which can be integrated by standard methods. After some algebra we find[54]

$$U/U_{c1} = (2/\pi) \int_0^{\pi/2} \{(1 + \gamma\eta^2)/(1 + \gamma\eta^2 \sin^2\psi)\}^{1/2}$$

$$\times \{(1 + \kappa\eta^2 \sin^2\psi)/(1 - \eta^2 \sin^2\psi)$$

$$+ \Delta\eta^2 \sin^2\psi/(1 + \gamma\eta^2 \sin^2\psi)\}^{1/2}d\psi. \tag{34}$$

Again this expression differs from the corresponding equation for the magnetic case (Eq. 30) only by the factor $\{(1 + \gamma\eta^2)/(1 + \gamma\eta^2 \sin^2\psi)\}^{1/2}$ in the integral.

6. FLEXO-ELECTRIC INSTABILITY

Probably the most interesting consequence of flexo-electricity is the flexo-electric instability predicted by Helfrich.[55] A homeotropically aligned nematic slab with rigid anchoring at the bottom plate and with loose anchoring at the top plate should undergo a Freedericksz transition caused by flexo-electricity. The transition is polarity dependent, i.e., for reversed bias there is no instability. The angle of tilt φ will have its maximum value φ_m at $z = d$. We have to take into account the log terms in the free energy which were zero for rigid anchoring. For the total free energy per unit area we obtain

$$G = \frac{1}{2} \int_0^d \{k_3 \cos^2\varphi + k_1 \sin^2\varphi + 4\pi(e_1 + e_3)^2 \sin^2\varphi \cos^2\varphi(\epsilon_\parallel \cos^2\varphi$$

$$+ \epsilon_\perp \sin^2\varphi)^{-1}\}(d\varphi/dz)^2dz \div (8\pi)^{-1}D_z \cdot (U - \delta U). \tag{35}$$

The function D_z is given by the expression

$$D_z = (U - \delta U) \bigg/ \int_0^d (\epsilon_\parallel \cos^2\varphi + \epsilon_\perp \sin^2\varphi)^{-1}dz. \tag{36}$$

Comparing these expressions with those of the previous section we see that the only effect of the log terms is to replace U by an effective potential $U - \delta U$, where δU is given by

$$\delta U = 2\pi(e_1 + e_3)/(\epsilon_\perp - \epsilon_\parallel) \log\{1 + [(\epsilon_\perp - \epsilon_\parallel)/\epsilon_\parallel] \sin^2\varphi_m\}. \tag{37}$$

In the previous section we had the condition $d\varphi/dz = 0$ for $\varphi = \varphi_m$. This condition no longer holds for loose anchoring, but has to be replaced by the condition $dG/d\varphi_m = 0$. For small distortions we have

[54] Deuling, H., *Mol. Cryst. Liq. Cryst.* **26**, 281 (1974).
[55] Helfrich, W., *Appl. Phys. Lett.* **24**, 451 (1974).

$d\varphi/dz \approx \varphi_m/d$. Expanding G in terms of φ_m we get

$$G = -(8\pi)^{-1}\epsilon_\parallel U^2/d + (2d)^{-1}[k_3 + U(e_1 + e_3)]\varphi_m^2 + \cdots \quad (38)$$

For $U < U_c = -k_3/(e_1 + e_3)$ the φ_m^2 term has a negative sign and we get an instability caused by flexo-electricity. Substituting $k_3 = 10^{-6}$ dyn and $(e_1 + e_3) = 10^{-4}$ dyn$^{1/2}$ we obtain in practical units $|U_c| = 3V$. In all the cases considered in this section, the flexo-electric coefficients e_1 and e_2 enter only in the form $e_1 + e_3$, this sum being the only bulk flexo-electric coefficient, as has been pointed out by Prost and Pershan.[56]

IV. Observation of Distortion and Determination of Material Constants

Any anisotropic quantity can be used to detect distortion of nematic layers. Electrical and thermal conductivity data have been used as well as capacitance measurements and monitoring the shift of birefringence.

7. THERMAL CONDUCTIVITY METHOD

Pieranski et al.[57] have detected distortions of nematic layers by measuring the thermal conductivity. Let us consider a nematic layer with planar alignment enclosed in a sandwich cell of thickness d. If a heat current is maintained through the cell a temperature drop ΔT across the cell will result. This temperature difference ΔT is proportional to the inverse of the thermal conductivity $k_\perp$. When the nematic layer is distorted by a magnetic field applied perpendicular to the slab the effective thermal conductivity changes towards the value $k_\parallel$. As a result the temperature difference ΔT changes by an amount δT. δT is related to the distortion by the equation

$$1 + \delta T/\Delta T = k_\perp/k_{\text{eff}} = (k_\perp/d)\int_0^d (k_\perp\cos^2\varphi + k_\parallel\sin^2\varphi)^{-1}dz. \quad (39)$$

Using the results of Section II,1,b we can transform this integral over z into an integral over φ. The final result is

$$(1 + \delta T/\Delta T)H/H_{c1} = (2/\pi)\int_0^{\pi/2} (1 + \zeta\eta^2\sin^2\psi)^{-1}$$
$$\times \{(1 + \kappa\eta^2\sin^2\psi)/(1 - \eta^2\sin^2\psi)\}^{1/2}d\psi;$$
$$H/H_{c1} = (2/\pi)\int_0^{\pi/2} \{(1 + \kappa\eta^2\sin^2\psi)/(1 - \eta^2\sin^2\psi)\}^{1/2}d\psi. \quad (40)$$

[56] Prost, J. E., and Pershan, P. S., J. Appl. Phys. 47, 2298 (1976).
[57] Pieranski, P., Brochard, F., and Guyon, E., J. Phys. (Paris) 33, 681 (1972).

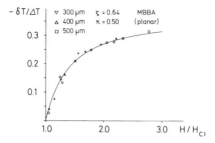

FIG. 9. Temperature difference decrease - δT as a function of magnetic field H for three different samples of MBBA. The data are taken from Ref. 57. The solid line is a theoretical curve calculated for $\kappa = 0.50$ and $\zeta = 0.64$. From Pieranski et al.[57] (Copyright by Commission des Publications Françaises de Physique. Used with permission.)

The parameter η is related to the maximum tilt angle by $\eta = \sin \varphi_m$ and $\zeta = (k_\parallel - k_\perp)/k_\perp$. Expanding in terms of η we find

$$-\delta T/\Delta T = [2\zeta/(1 + \kappa)](H/H_{c1} - 1) + \cdots. \qquad (41)$$

Figure 9 shows data by Pieranski et al.[57] obtained with three different samples of MBBA. The measured values of thermal conductivity were $k_\parallel = (5 \pm 0.25) \times 10^{-4}$ cal/cm·sec deg and $k_\perp = (3.05 \pm 0.25) \times 10^{-4}$ cal/cm·sec deg, resulting in $\zeta = 0.64$. The solid curve in Fig. 9 is a theoretical curve calculated by the present author for $\zeta = 0.64$ and $\kappa = 0.5$. Expressions (40) and (41) are valid for a sample with homeotropic alignment if we replace H_{c1} by the critical field of the bending mode H_{c3} and if we replace k_1, $k_\perp$ by k_3, $k_\parallel$ and vice-versa. We can do a high-field expansion. For $H \gg H_{c1}$ φ increases very rapidly from 0 to nearly $\pi/2$. We can approximate $\varphi(z)$ by the tangent at $z = 0$. At $z = \xi = (d/2)H_{c1}/H$ we reach $\varphi = \pi/2$. For $\xi < z < d/2$ we put $d\varphi/dz = 0$ and $\varphi = \pi/2$. In this approximation we get

$$1 + \delta T/\Delta T = (k_\perp/k_\parallel)(1 - H_{c1}/H)$$

$$+ k_\perp(H_{c1}/H)(2/\pi) \int_0^{\pi/2} (k_\perp \cos^2\varphi + k_\parallel \sin^2\varphi)^{-1}d\varphi.$$

The first term is the contribution from the interval $\xi < z < d/2$, where $\varphi = \pi/2$, and the second term is the contribution from the region $0 < z < \xi$, where $d\varphi/dz = $ const. The integral can be done by elementary methods. The final result is

$$-\delta T/\Delta T = \zeta/(1 + \zeta) - \{(1 + \zeta)^{-1/2} - (1 + \zeta)^{-1}\}(H_{c1}/H). \qquad (42)$$

Figure 10 shows the data of Fig. 9 in the form $-\delta T/\Delta T$ versus (H_{c1}/H).

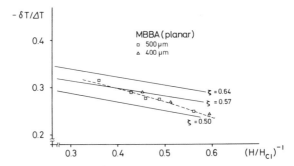

FIG. 10.   Data of Fig. 9 plotted in the form - $\delta T/\Delta T$ versus $(H/H_{c1})^{-1}$. The dotted curve is the theoretical curve of Fig. 9. The solid lines represent the high-field expansion

The solid lines correspond to (42) with $\zeta = 0.5$, $\zeta = 0.57$, and $\zeta = 0.64$. The dotted curve is the theoretical curve of Fig. 9. Although the data are already well represented by a straight line, the high-field limit has not yet been reached.

8.   ELECTRICAL CONDUCTIVITY METHOD

The electrical conductivity method has been used by Schneider *et al.*[21,58] The method is analogous to the thermal conductivity method. A difference in potential ΔV and a change in potential difference δV are measured instead of temperature differences. All the expressions hold if the thermal conductivities are replaced by electrical conductivities. Schneider *et al.* kept the field at an angle θ with the direction of alignment and varied θ from 45 to 89°. Although one does not have a critical field in this case the authors were able to determine H_{c1} and κ by fitting several curves measured for different angles θ. Keeping the field at an oblique angle to the direction of alignment has the advantage that disclinations are avoided in the distortion.

9.   CAPACITANCE METHOD

In an insulating nematic layer the z-component of the dielectric displacement D_z is constant. For the potential U across a slab with planar orientation we find

$$U = D_z \int_0^d (\epsilon_\perp \cos^2\varphi + \epsilon_\parallel \sin^2\varphi)^{-1} dz. \tag{43}$$

[58] Greulich, M. Heppke, G., and Schneider, F., *Z Naturforsch., Teil A* **30**, 515 (1975).

For the capacitance change ΔC caused by a magnetic field we find from (43) by a straightforward calculation

$$H/H_{c1} = (1 + \Delta C/C_0)(2/\pi) \int_0^{\pi/2} (1 + \gamma\eta^2 \sin^2\psi)^{-1}$$

$$\times \{(1 + \kappa\eta^2 \sin^2\psi)/(1 - \eta^2 \sin^2\psi)\}^{1/2} d\psi. \tag{44}$$

H/H_{c1} is again given by expression (40). For small distortions we can expand in terms of η and obtain

$$\Delta C/C_0 = [2\gamma/(1 + \kappa)](H/H_{c1} - 1). \tag{45}$$

For high fields we can perform an expansion in terms of H_{c1}/H as explained in Section IV,7:

$$\Delta C/C_0 = \gamma - \{(1 + \gamma)^{3/2} - (1 + \gamma)\}H_{c1}/H. \tag{46}$$

The calculation is slightly different for the capacitance change ΔC caused by an electric field. We have $(1 + \Delta C/C_0)U/U_{c1} = D_z/D_z^0$. $D_z^0 = (\epsilon_\perp/d)U_{c1}$ is the z-component of the dielectric displacement at threshold. Expressions for U/U_{c1} and D_z/D_z^0 have been given in Section II,2,a (Eq. 20). Expanding these expressions in terms of $\eta = \sin\varphi_m$ we find

$$U/U_{c1} = 1 + \tfrac{1}{4}(1 + \kappa + \gamma)\eta^2 + \cdots;$$

$$D_z/D_z^0 = 1 + \tfrac{1}{4}(1 + \kappa + 3\gamma)\eta^2 + \cdots.$$

For the capacitance change we get

$$\Delta C/C_0 = [2\gamma/(1 + \kappa + \gamma)](U/U_{c1} - 1). \tag{47}$$

For high voltage we approximate $\varphi(z)$ again by a straight line with slope $(d\varphi/dz)_{z=0} = (\pi/d)(D_z/D_z^0)$ in the interval $0 < z < \xi = (d/2)(D_z^0/D_z)$ and by $\varphi = \pi/2$ for $\xi < z < d/2$. Evaluation of Eq. (43) then leads to the expansion

$$\Delta C/C_0 = \gamma - \{(1 + \gamma)^{3/2} - (1 + \gamma)\}U_{c1}/U. \tag{48}$$

Figure 11 shows capacitance data of DIBAB by Coche and Kerllenevich. The theoretical curve is a nonlinear least squares fit to the data giving $\kappa = 1.73$ and $\gamma = 0.066$. With $\epsilon_\parallel = 4.2$ and $\epsilon_\perp = 4.0$ as measured by de Jeu[59] we would have $\gamma = 0.05$. The dotted line corresponds to the low-field expansion (47). Figure 12 shows the same data plotted in the form $\Delta C/C_0$ versus U_{c1}/U. The dotted curve is the theoretical curve calculated for $\gamma = 0.066$. The solid lines correspond to the high-field

[59] de Jeu, W. H., *Chem. Phys. Lett.* **28**, 239 (1974).

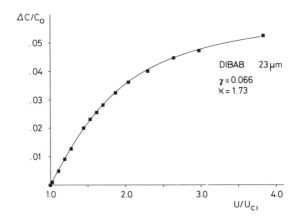

FIG. 11. Capacitance increase ΔC of a planar sample of DIBAB as a function of applied voltage U. The solid curve is a non-linear least squares fit to the data giving $\gamma = 0.066$ and $\kappa = 1.73$. (Data by courtesy of A. Coche and B. Kerllenevich.)

expansion for $\gamma = 0.066$ and $\gamma = 0.05$, respectively. Although the data have almost a linear slope, the high-field limit has not yet been reached.

10. OPTICAL METHODS

A nematic slab with planar alignment placed between crossed polarizers such that polarizer and analyzer are at 45° to the preferred direction of alignment shows a typical interference pattern consisting of a set of confocal dark and light hyperbolas. When the slab is distorted by an electric or a magnetic field the hyperbolas expand. By counting the

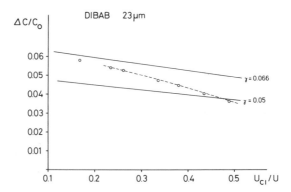

FIG. 12. Data of Fig. 11 plotted in the form $\Delta C/C_0$ versus $(U/U_{c1})^{-1}$. The dotted curve is the theoretical curve of Fig. 11. The solid lines represent the high-field expansion.

number of fringes passing a fixed point one can measure the phase difference between the ordinary ray and the extraordinary ray after passing through the sample. This method was first used by Saupe.[16] The phase difference $\Delta(H)$ between the two rays is

$$\Delta(H) = (d/\lambda)(n_{eff} - n_0),$$

where λ is the wavelength of the light and

$$n_{eff} = (1/d) \int_0^d n_e n_0 (n_e^2 \sin^2\varphi + n_0^2 \cos^2\varphi)^{-1/2} dz$$

is the average index of refraction for the extraordinary ray. A phase difference decrease $\delta(H)$ is defined as $\delta = \Delta(0) - \Delta(H)$. For δ one gets the expression

$$\delta \lambda/(n_e d) = 1 - (1/d) \int_0^d (1 + \nu \sin^2\varphi)^{-1/2} dz, \tag{49}$$

where $\nu = (n_e/n_0)^2 - 1$. We can evaluate the integral in the high-field limit and get

$$(1/d) \int_0^d (1 + \nu \sin^2\varphi)^{-1/2} dz = (1 + \nu)^{-1/2}(1 - H_{c1}/H)$$

$$+ (H_{c1}/H)(2/\pi) \int_0^{\pi/2} (1 + \nu \sin^2\varphi)^{-1/2} d\varphi.$$

The latter integral can be expressed as a complete elliptic integral of the first kind. The final result is

$$\delta \lambda/(n_e d) = 1 - (n_0/n_e)\{1 + [(2/\pi)K(k) - 1]H_{c1}/H\}, \tag{50}$$

where $k = [1 - (n_0/n_e)^2]^{1/2}$. We obtain a general expression for δ if we transform the integral over z in Eq. (49) into an integral over φ using the relation

$$d\varphi/dz = (\pi/d)(H/H_{c1})\{(\sin^2\varphi_m - \sin^2\varphi)/(1 + \kappa \sin^2\varphi)\}^{1/2}.$$

We then get[16,17]

$$[1 - \delta \lambda/(n_e d)]H/H_{c1}$$

$$= (2/\pi) \int_0^{\pi/2} \{(1 + \kappa\eta^2 \sin^2\psi)/(1 - \eta^2 \sin^2\psi)\}^{1/2}$$

$$\times (1 + \nu\eta^2 \sin^2\psi)^{-1/2} d\psi, \tag{51}$$

where H/H_{c1} is again given by Eq. (40). We obtain the low-field expansion if we expand in terms of η:

$$\delta\,\lambda/(n_e d) = [\nu/(1 + \kappa)](H/H_{c1} - 1). \tag{52}$$

Figure 13 shows data by Pieranski obtained with MBBA. The curve was calculated for $\nu = 0.29$ and $\kappa = 0.60$. For a distortion by an electric field we can obtain a similar expression using

$$d\varphi/dz = (\pi/d)(D_z/D_z^{\,0})$$

$$\times \{(\sin^2\varphi_m - \sin^2\varphi)/(1 + \kappa\sin^2\varphi)/(1 + \gamma\sin^2\varphi_m)/(1 + \gamma\sin^2\varphi)\}^{1/2}.$$

After some algebra we find[17,29]

$$[1 - \delta\lambda/(n_e d)]\int_0^{\pi/2}\{(1 + \gamma\eta^2\sin^2\psi)$$

$$\times (1 + \kappa\eta^2\sin^2\psi)/(1 - \eta^2\sin^2\psi)\}^{1/2}d\psi$$

$$= \int_0^{\pi/2}\{(1 + \gamma\eta^2\sin^2\psi)$$

$$\times (1 + \kappa\eta^2\sin^2\psi)/(1 + \nu\eta^2\sin^2\psi)/(1 - \eta^2\sin^2\psi)\}^{1/2}d\psi. \tag{53}$$

The parameter $\eta = \sin\varphi_m$ is related to U by

$$U/U_{c1} = (2/\pi)\int_0^{\pi/2}\{(1 + \gamma\eta^2)/(1 + \gamma\eta^2\sin^2\psi)\}^{1/2}$$

$$\times \{(1 + \kappa\eta^2\sin^2\psi)/(1 - \eta^2\sin^2\psi)\}^{1/2}d\psi.$$

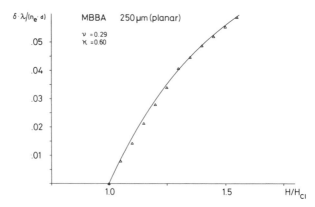

FIG. 13. Phase difference decrease of a planar sample of MBBA as a function of magnetic field H. The solid curve has been calculated for $\nu = 0.29$ and $\kappa = 0.60$. (Data by courtesy of P. Pieranski.)

In the low-voltage limit we get

$$\delta\lambda/(n_e d) = [\nu/(1 + \kappa + \gamma)](U/U_{c1} - 1).$$ (54)

A nematic slab with homeotropic alignment placed between crossed polarizers displays a typical interference pattern consisting of a set of concentric dark and bright circles with a dark cross superimposed. When a magnetic field is applied parallel to the cell the circles move outward and become distorted. Counting again the number of fringes passing a fixed point, we can measure the phase difference increase δ. All the expressions derived above remain valid if we replace k_1, n_0 by k_3, n_e and vice-versa. All the methods described so far allow one only to determine a critical field H_{c1} or H_{c3} and a parameter $\kappa = (k_3 - k_1)/k_1$. All these methods are insensitive to twist distortions. A twist distortion of a planar nematic slab will rotate the hyperbolic interference fringes by an angle ϵ. Measuring ϵ as a function of H one can determine the twist critical field H_{c2} very accurately, as was demonstrated by Cladis.[60] de Gennes[61] has derived a simple expression for the rotation angle ϵ of the conoscopic image:

$$\tan 2\epsilon = \langle\sin 2\omega\rangle/\langle\cos 2\omega\rangle.$$ (55)

The brackets stand for spatial average over z. Gabay and Guyon[62] have extended this formula to the case when splay and bend distortions are also present:

$$\tan 2\epsilon = \langle(1 + \cos 2\varphi) \sin 2\omega\rangle/\langle(1 + \cos 2\varphi) \cos 2\omega\rangle.$$ (56)

V. Dynamics of the Freedericksz Transition

In Section II we studied the static solution of the equations describing the balance of torques for every volume element of a nematic slab. Here we shall investigate the time dependence of the Freedericksz transition. In the static case we had a balance of an elastic torque and a magnetic torque per volume element. In the time-dependent problem we have in addition a viscous torque opposing any rapid rotation of the director. For a planar nematic layer in a magnetic field perpendicular to the slab the balance of torques reads

$$(k_1 \cos^2\varphi + k_3 \sin^2\varphi)\partial^2\varphi/\partial z^2 + (k_3 - k_1) \sin \varphi \cos \varphi(\partial\varphi/\partial z)^2$$
$$+ \chi_a H^2 \sin \varphi \cos \varphi - \gamma_1(\partial\varphi/\partial t) = 0.$$

[60] Cladis, P. E., *Phys. Rev. Lett.* **28**, 1629 (1972).
[61] de Gennes, P., Appendix of Cladis.[60]
[62] Gabay, J., and Guyon, E., Appendix B of Deuling *et al.*[23]

The first two terms represent the elastic torque, the third term the magnetic torque, and the last term the viscous torque. γ_1 is the rotational viscosity. If we switch on the magnetic field at $t = 0$ we may expect the system to pass through a sequence of quasistatic states. This suggests an ansatz

$$\sin \varphi(z, t) = \eta(t) \sin \psi(z), \tag{58}$$

where $\eta(t) = \sin \varphi_m(t)$ is the amplitude of the distortion. We shall see below that Eq. (58) is not an exact solution. We have to allow for a weak time dependence of $\psi(z)$, which we neglect in the following. We get for $\partial \varphi / \partial t$

$$\partial \varphi / \partial t = (\dot{\eta} / \eta) \tan \varphi. \tag{59}$$

We insert this expression into Eq. (57), multiply by $2\partial \varphi / \partial z$ and integrate once. We obtain

$$(1 + \kappa \sin^2\varphi)(\partial \varphi / \partial z)^2 + (\pi/d)^2 \{(H/H_{c1})^2 \sin^2\varphi$$
$$+ \tau_0(\dot{\eta}/\eta) \log(1 - \sin^2\varphi)\} = \text{const.}$$

Here we have defined a time constant τ_0 as $\tau_0 = (\gamma_1/\chi_a)H_{c1}^{-2}$. The constant of integration is determined by the condition that at all times $\partial \varphi / \partial z = 0$ for $z = d/2$. Switching to $\psi(z)$ as a new variable we get

$$\{(1 + \kappa\eta^2 \sin^2\psi)/(1 - \eta^2 \sin^2\psi)\}^{1/2}(d\psi/dz)$$
$$= (\pi/d)\{(H/H_{c1})^2 - \tau_0(\dot{\eta}/\eta)[\eta^2(1 - \sin^2\psi)]^{-1} \tag{60}$$
$$\times \log[(1 - \eta^2 \sin^2\psi)/(1 - \eta^2)]\}^{1/2}.$$

We expand the log term and obtain

$$\{(1 + \kappa\eta^2 \sin^2\psi)/(1 - \eta^2 \sin^2\psi)\}^{1/2}(d\psi/dz) \tag{61}$$
$$= (\pi/d\{(H/H_{c1})^2 - \tau_0(\dot{\eta}/\eta)[1 + \tfrac{1}{2}\eta^2(1 + \sin^2\psi)]\}^{1/2}.$$

Neglecting the $\sin^2\psi$ term on the right-hand side and integrating from $z = 0$ to $z = d/2$ we obtain

$$\dot{\eta} = (\eta/\tau_0) \left\{ (H/H_{c1})^2 - \left[(2/\pi) \int_0^{\pi/2} [(1 + \kappa\eta^2 \sin^2\psi)/(1 - \eta^2 \sin^2\psi)]^{1/2} d\psi \right]^2 \right\}. \tag{62}$$

We find $\dot{\eta}$ to be proportional to η. Therefore we need a small fluctuation η_0 at $t = 0$ to get the distortion started. For $t \to \infty$ we have $\dot{\eta} \to 0$ and η

$\rightarrow \eta_\infty$. We get again the static solution

$$H/H_{c1} = (2/\pi) \int_0^{\pi/2} \{(1 + \kappa\eta_\infty^2 \sin^2\psi)/(1 - \eta_\infty^2 \sin^2\psi)\}^{1/2} d\psi.$$

Expanding the integrals in terms of η and η_∞, respectively, we obtain

$$\dot{\eta} = \eta(\eta_\infty^2 - \eta^2)(1 + \kappa)/(2\tau_0).$$

This equation is easily solved. The solution is

$$\eta^2(t) = \eta_\infty^2\{1 + [(\eta_\infty^2 - \eta_0^2)/\eta_0^2]\exp(-2t/\tau_H)\}^{-1}. \tag{63}$$

The time constant τ_H is given by

$$\tau_H^{-1} = (1 + \kappa)\eta_\infty^2/(2\tau_0) = [(H/H_{c1})^2 - 1]/\tau_0.$$

Expression (63) was first derived by Pieranski *et al.*[63] $\eta(t)$ can be measured by the thermal conductivity method or by the optical method. For H not too far above H_{c1} we found the temperature difference decrease δT to be proportional to η^2. Plotting $\log[\delta T_\infty/\delta T(t) - 1]$ versus time we expect a straight-line variation with slope $2/\tau_H$. This was verified experimentally for MBBA by Pieranski *et al.*[63]

Next we consider an abrupt lowering of the field from an upper value $H_0 > H_{c1}$ at $t = 0$ to a lower value $H > H_{c1}$. We find

$$H_0/H_{c1} = 1 + (1 + \kappa)\eta_0^2/4 + \cdots;$$

$$H/H_{c1} = 1 + (1 + \kappa)\eta_\infty^2/4 + \cdots.$$

The solution of Eq. (62) takes the form

$$\eta^2(t) = \eta_\infty^2\{1 - [(\eta_0^2 - \eta_\infty^2)/\eta_0^2] \exp(-2t/\tau_H)\}^{-1}, \tag{63}$$

where the time constant τ_H is again defined by the lower-field H:

$$\tau_H^{-1} = \{(H/H_{c1})^2 - 1\}/\tau_0. \tag{64}$$

If we take the lower field to be below H_{c1}, the calculation is slightly different. Equation (62) is then of the form

$$\dot{\eta} = -[(1 + \kappa)/(2\tau_0)]\eta(a^2 + \eta^2),$$

where $a^2 = 2[1 - (H/H_{c1})^2]/(1 + \kappa)$. The solution is

$$\eta^2(t) = a^2\{(a^2/\eta_0^2 + 1) \exp(2t/\tau_H) - 1\}^{-1}, \tag{65}$$

with

$$\tau_H^{-1} = [1 - (H/H_{c1})^2]/\tau_0. \tag{66}$$

[63] Pieranski, P., Brochard, F., and Guyon, E., *J. Phys. (Paris)* **34**, 35 (1973)

Comparing Eqs. (66) and (64) we see that τ_H diverges as H approaches the critical field from below H_{c1} or from the region above H_{c1}. When we turn off the field H_0 completely, i.e., in the limit $H \to 0$, the solution (65) is no longer accurate. To calculate the turn-off mode we put $H = 0$ in Eq. (61) and integrate over z:

$$(-\tau_0 \dot{\eta}/\eta)^{1/2} = (2/\pi) \int_0^{\pi/2} \{(1 + \kappa\eta^2 \sin^2\psi)/(1 - \eta^2 \sin^2\psi)\}^{1/2}$$

$$\times \{1 + \tfrac{1}{2}\eta^2(1 + \sin^2\psi)\}^{-1/2}d\psi.$$

Expanding the integral in terms of η we obtain the equation

$$\dot{\eta} = -(\eta/\tau_0)[1 + \tfrac{1}{4}(\kappa - \tfrac{1}{2})\eta^2].$$

The solution is

$$\eta^2(t) = \eta_0^2\{[1 + \tfrac{1}{2}(\kappa - \tfrac{1}{2})\eta_0^2] \exp(2t/\tau_0)$$

$$- \tfrac{1}{2}(\kappa - \tfrac{1}{2})\eta_0^2\}^{-1}. \tag{67}$$

For $t \gg \tau_0$ we get

$$\eta(t) = \eta_0\{1 + \tfrac{1}{2}(\kappa - \tfrac{1}{2})\}^{-1/2} \exp(-t/\tau_0). \tag{68}$$

Our derivation holds also for the time dependence of a twist distortion if we put $\kappa = 0$ and replace the splay critical field H_{c1} by the twist critical field H_{c2}. Our derivation is strictly valid only for the twist distortion because we have neglected backflow. The gradient of the angular velocity of **n** induces a backflow motion of the liquid. This backflow motion in turn causes a frictional torque on the director. The balance of torques is now given by the equation

$$(1 + \kappa \sin^2\varphi)\partial^2\varphi/\partial z^2 + \kappa \sin \varphi \cos \varphi(\partial\varphi/\partial z)^2 \tag{69}$$

$$+ (\pi/d)^2\{(H/H_{c1})^2 \sin \varphi \cos \varphi - \tau_0\partial\varphi/\partial t - \lambda\tau_0\partial v_y/\partial z\} = 0.$$

The time constant τ_0 is again defined by $\tau_0 = (\gamma_1/\chi_a)H_{c1}^{-2}$. The new term in the equation due to backflow is $-(\pi/d)^2\lambda\tau_0\partial v_y/\partial z$. The parameter λ is given by $\lambda = (\gamma_1 + \gamma_2)/(2\gamma_1)$. $v_y(z, t)$ is the component of the velocity along the y-axis which is the preferred direction of alignment in the undistorted state. The velocity $v_y(z, t)$ is coupled to the gradient of the angular velocity of **n** by the Leslie–Ericksen equation of the hydrodynamic motion.[64–66] Neglecting inertial effects this equation reads

$$\partial(\eta_1\partial v_y/\partial z + \alpha_3\partial\varphi/\partial t)/\partial z = 0; \tag{70}$$

$$\eta_1 = \tfrac{1}{4}(\alpha_4 + \alpha_6 + \alpha_3).$$

[64] Leslie, F. M., *Q. J. Mech. Appl. Math.* **19**, 357 (1966).
[65] Ericksen, J. L., *Arch. Ration. Mech. Anal.* **4**, 231 (1960); **9**, 371 (1962).
[66] Ericksen, J. L, *Mol. Cryst. Liq. Cryst.* **7**, 153 (1969).

For small distortions we can linearize the equations (69) and (70). Brochard et al.[22,63] have shown that the solution of the linearized equations is of the form

$$\varphi(z, t) = \varphi_0\{\cos(qz) - \cos(qd/2)\} \exp(st);$$
$$v_y(z, t) = v_0\{\sin(qz) - (2z/d) \sin(qd/2)\} \exp(st). \tag{71}$$

The bounding plates are assumed to be at $z = \pm d/2$. q is the wavevector of the distortion and s^{-1} is the time constant. Inserting the ansatz (71) into Eq. (70) we get the condition

$$v_0 = -[s\alpha_3/(q\eta_1)]\varphi_0. \tag{72}$$

Inserting the ansatz (71) into Eq. (69) and using the relation (72) we get an equation containing terms proportional to $\cos(qz)$ and terms independent of z. To satisfy this equation for all values of z requires that therefore the two following equations hold:

$$\tau_0 s(1 - \lambda\alpha_3/\eta_1) = (H/H_{c1})^2 - (d/\pi)^2 q^2;$$
$$(H/H_{c1})^2 = \tau_0 s\{1 - 2[\lambda\alpha_3/(qd\eta_1)]\tan(qd/2)\}. \tag{73}$$

These equations give the wavevector q and the time constant s^{-1} as functions of H/H_{c1}. If the ratio $\lambda\alpha_3/\eta_1$ is small we have $q \to \pi/d$ and we recover the old results $s = \tau_0^{-1}[(H/H_{c1})^2 - 1]$. van Doorn[67] has solved Eqs. (69) and (70) numerically for the turn-off mode. For values of H several times the value of H_{c1} he found the tilt angle φ in the middle of the layer to increase initially to values larger than $\pi/2$ before finally decaying to zero. A dynamical calculation for the twist cell is even more complicated due to the twist of the director and to backflow motion in the x-direction as well as in the y-direction. Berreman[68] has solved this problem neglecting backflow. Recently van Doorn[69] and Berreman[70] independently calculated the time dependence of the transmission of light by a twist cell including backflow. The results of these calculations seem to give fair agreement with data by Baur and Meier[46,72] and by Gerritsma et al.[71]

[67] van Doorn, C. Z., J. Phys. 36-C1, 261 (1975).
[68] Berreman, D. W., Appl. Phys. Lett. 25, 12 (1974).
[69] van Doorn, C. Z. J. Appl. Phys. 46, 3738 (1975).
[70] Berreman, D. W., J. Appl. Phys. 46, 3746, (1975).
[71] Gerritsma, C. J., van Doorn, C. Z., and van Zanten, P., Phys. Lett. A 48, 263 (1974).
[72] Baur, G., Windscheidt, F., and Berreman, D. W. Appl. Phys. 8, 101 (1975).

The Dielectric Permittivity of Liquid Crystals

W. H. DE JEU

Philips Research Laboratories, Eindhoven, The Netherlands

I. Introduction

In the various liquid-crystal phases there exists long-range orientational order. The anisotropic elongated molecules are, on the average, aligned with their long axes parallel to each other.[1] Macroscopically a unique axis (the preferred direction) is defined in this way. In the nematic phase the molecules translate freely as in the isotropic liquid; the centers of mass are distributed at random. In the smectic phases there is additional positional ordering in layers. In the smectic A phase the layers are perpendicular to the preferred direction. Within the layers the distribution of the centers of mass is again random.

Because of the uniaxial symmetry in nematic and smectic A liquid crystals, the dielectric permittivity differs in value along the preferred

---

[1] See, for example; P. G. de Gennes, "The Physics of Liquid Crystals," Clarendon Press, Oxford, 1974.

axis ($\epsilon_{\parallel}$) and perpendicular to this axis ($\epsilon_{\perp}$). The dielectric anisotropy is defined as $\Delta\epsilon = \epsilon_{\parallel} - \epsilon_{\perp}$. In order to measure $\epsilon_{\parallel}$ and $\epsilon_{\perp}$ the preferred direction has to be uniform over the whole sample. Usually a plane capacitor is used with a nematic layer of about 100 μm as dielectric. With suitable surface treatments the preferred axis can be made to lie either parallel or perpendicular to the electrodes. A sample with a uniform preferred axis can also be obtained by applying a magnetic field. This has the advantage that $\epsilon_{\parallel}$ and $\epsilon_{\perp}$ can be measured on the same sample, which increases the accuracy to which $\Delta\epsilon$ is obtained. An electric field is less suitable to produce a sample with a uniform preferred axis because, due to the anisotropy of the conductivity, space charge is generated which may cause hydrodynamic flow.[2] The preferred axis in smectic liquid crystals is rather difficult to influence. This is the main reason why dielectric data on smectics are scarce.

In practice the alignment of the molecules is not perfect. The angle between the preferred direction and the long molecular axis of a molecule will be denoted by θ. The molecular orientation can be described by a distribution function $w(\theta)$, where $w(\theta)d\Omega = w(\theta)2\pi \sin \theta \, d\theta$ is the probability that the long molecular axis will form an angle between θ and $\theta + d\theta$ with the preferred direction. Since the liquid-crystal phase itself is apolar, $w(\theta) = w(\pi - \theta)$. The average degree of ordering can be described by an order parameter S which is usually defined as [1,3]

$$S = \langle 1 - \tfrac{3}{2} \sin^2\theta \rangle = 1 - \tfrac{3}{2} \int w(\theta) \sin^2\theta d\Omega. \tag{1}$$

For an isotropic liquid $S = 0$, while $S = 1$ in the case of perfectly aligned molecules.

In the molecular–statistical theory of Maier and Saupe[3] the nematic–isotropic (NI) transition is attributed to the anisotropic part of the dispersion forces (London–van der Waals forces). In a mean-field approximation this leads to a potential

$$W = -(AS/V^2)(1 - \tfrac{3}{2} \sin^2\theta), \tag{2}$$

where V is the molar volume and A is a factor related to the anisotropy of the molecular polarizability. According to this theory S is for all liquid crystals a universal function of the reduced temperature $TV^2/(T_{NI}V_{NI}^2)$. The value of T_{NI} is determined by $A/(kT_{NI} V^2) = $ constant. The distribution function $w(\theta)$ is related to W by $w(\theta) = (1/Z) \exp(-W/kT)$, where Z is a normalization constant. For the purpose of the present discussion it is usually not necessary to know W explicitly. In Section V, however,

[2] See, for example; W. Helfrich, *Mol. Cryst. Liq. Cryst.* **21**, 187 (1973).
[3] W. Maier and A. Saupe, *Z. Naturforsch., Teil A* **14**, 882 (1959); **15**, 287 (1960).

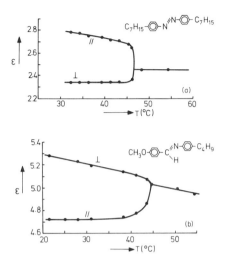

FIG. 1. Permittivities in the nematic and isotropic phases of a typical nonpolar compound[4] (a) (Copyright by Verlag der Zeitschrift für Naturforschung. Used with permission) and a typical polar compound[5] (b).

specific assumptions about the form of the nematic potential are necessary, in which case Eq. (2) will be used.

It is the purpose of dielectric theory to relate the macroscopic permittivities to the molecular properties (polarizability and dipole moment). We shall use a coordinate system (ξ, η, ζ) fixed to a specific molecule. In the nematic and the smectic A phase the molecules are assumed to rotate freely around the long molecular axis (taken as the ζ-axis). The molecular polarizability tensor α can then be described by its principal elements $\alpha_l = \alpha_{\zeta\zeta}$ and $\alpha_t = \frac{1}{2}(\alpha_{\xi\xi} + \alpha_{\eta\eta})$. For the molecules under consideration $\alpha_l - \alpha_t > 0$. For nonpolar molecules this also leads to $\Delta\epsilon > 0$. In the case of polar compounds there is an additional contribution to the permittivity from the permanent dipole moment μ. Depending on the angle β between μ and the long molecular axis, the dipole contribution can give an increase or a decrease of $\Delta\epsilon$, leading eventually to negative values of $\Delta\epsilon$. The sign and magnitude of $\Delta\epsilon$ are of the utmost importance for the applicability of the compound in devices using one of the various electrooptic effects.[2] Figure 1 gives some typical results of dielectric measurements. For nonpolar molecules $\Delta\epsilon$ is of the order of 0.4. In the case of N-(p-methoxybenzylidene)-p-n-butylaniline (MBBA) a negative anisotropy is found. In practice aniso-

[4] W. H. de Jeu and T. W. Lathouwers, *Z. Naturforsch., Teil A* **29**, 905 (1974).
[5] D. Diguet, F. Rondelez, and G. Durand, *C. R. Hebd. Seances Acad. Sci., Ser. B* **271**, 954 (1970).

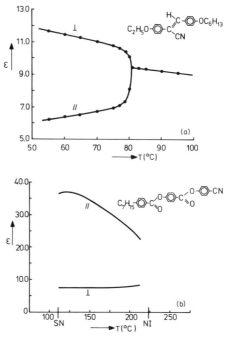

FIG. 2. Permittivities of compounds showing a large negative and a large positive dielectric anisotropy (a, Ref. 6; b, Ref. 7). [(a) Copyright by the North-Holland Publishing Company. Used with permission.]

tropies between, say, $+1$ and -1 are often observed. Compounds with a large anisotropy can be synthesized by substitution of a strongly polar group (for example a cyanide group) in specific positions. Figure 2 gives examples of the largest positive ($\Delta\epsilon \approx +30$) and the largest negative ($\Delta\epsilon \approx -5$) anisotropy observed so far.

When the existing dielectric theory is applied to liquid crystals the degree of ordering as described by S must first be taken into account. Furthermore, the question of the internal field experienced by a molecule has to be considered. For isotropic liquids, due to the contributions of the surrounding molecules, the internal field is not simply equal to the macroscopic field.[8] In the case of liquid crystals this situation is even more complicated because of the various anisotropies and the incomplete orientational order. This matter is discussed for nematics in

[6] W. H. de Jeu and J. van der Veen, *Phys. Lett. A* **44**, 277 (1973).

[7] R. T. Klingbiel, private communication; see also J. P. van Meter, R. T. Klingbiel, and D. J. Genova, *Solid State Commun.* **16**, 315 (1975).

[8] See, for example; C. J. F. Böttcher, "Theory of Electric Polarization," 2nd ed., Vol. I. Elsevier, Amsterdam, 1973.

Section II. In Section III the results are applied to various compounds of different molecular structure.

In the liquid-crystal phases short-range dipole–dipole interactions, which can often be disregarded in isotropic liquids, may become important. This is especially true when the distribution of the centers of mass of the molecules is not random (smectic phases). An extension to liquid crystals of the Kirkwood–Fröhlich theory for the permittivity of isotropic liquids[8] provides a framework in which these interactions can be treated. This matter is discussed in Section IV for nematics and smectics A. So far we have implicitly restricted ourselves to the static permittivity. In Section V the frequency dependence of the permittivity in alternating fields will be discussed.

II. The Static Permittivity of Nematic Liquid Crystals

The static electric polarization induced by an external field $\mathbf{E}$ in an isotropic liquid is given by

$$\mathbf{P} = \frac{\epsilon - 1}{4\pi} \mathbf{E}. \tag{3}$$

$\mathbf{P}$ can be divided into a part $\mathbf{P}_\alpha$ due to the molecular polarizability (induced polarization) and a part $\mathbf{P}_\mu$ due to the dipole moment (orientation polarization). In this section we shall calculate $\mathbf{P}_\alpha$ and $\mathbf{P}_\mu$ for nematic liquid crystals. In Section II,1 nonpolar molecules will first be considered ($\mathbf{P}_\mu = 0$). In Section II,2 the orientation polarization will be included for polar molecules.

1. NEMATICS WITH NONPOLAR MOLECULES

In an anisotropic medium the scalar permittivity ϵ in Eq. (3) must be replaced by a second-rank tensor $\boldsymbol{\epsilon}$. We shall use a macroscopic coordinate system (x, y, z) and take the preferred direction of the nematic along the z-axis. Then the principal elements of $\boldsymbol{\epsilon}$ are $\epsilon_\parallel = \epsilon_{zz}$ and $\epsilon_\perp = \frac{1}{2}(\epsilon_{xx} + \epsilon_{yy})$. Equation (3) can be applied to each of the directions λ which, in the case of nonpolar molecules, gives

$$P_\lambda = (\epsilon_\lambda - 1)E_\lambda/4\pi = N\langle(\boldsymbol{\alpha}\cdot\mathbf{E}^i)_\lambda\rangle, \qquad \lambda = \parallel, \perp; \tag{4}$$

N is the number of molecules per unit volume, $\boldsymbol{\alpha}$ the molecular polarizability tensor, and $\mathbf{E}^i$ the internal field that differs from the macroscopic field $\mathbf{E}$. The brackets denote a statistical average over the orientations of all molecules, hence to be performed with the nematic potential W. In principle one would expect an intricate relation between the field at a molecule and the macroscopic field, the internal field depending both on the properties and the orientation of the molecule and on the macroscopic dielectric anisotropy.[9] However, a considerable simplification of the internal field problem occurs if we rely on some experimental information.

Experimental results for both the static permittivity and the permittivity at optical frequencies $\epsilon_\lambda = n_\lambda^2$ show two important facts.[10] First, if we denote the average permittivity by $\bar{\epsilon} = (\epsilon_\parallel + 2\epsilon_\perp)/3$, we find that $\bar{\epsilon}/\rho$ (ρ being the density) is continuous at the NI phase transition. Second, the dielectric anisotropy $\Delta\epsilon$ is proportional to the product ρS; or where $N = \rho N_A /M$, N_A being Avogadro's number and M the mass number, $\Delta\epsilon \sim NS$. These results impose some constraints on the relation between internal field and macroscopic field. Since the internal field should be a linear function of the macroscopic field, this relation can be represented by

$$\mathbf{E}^i = \mathbf{K} \cdot \mathbf{E}$$

which will appear to be convenient for the consequent discussion. $\mathbf{K}$ is a normal second-rank tensor which can be used to write Eq. (4) as

$$\epsilon_\lambda = 1 + 4\pi N \langle (\boldsymbol{\alpha} \cdot \mathbf{K})_\lambda \rangle. \tag{5}$$

Consequently we find for the dielectric anisotropy

$$\Delta\epsilon = 4\pi N \{ \langle (\boldsymbol{\alpha} \cdot \mathbf{K})_\parallel \rangle - \langle (\boldsymbol{\alpha} \cdot \mathbf{K})_\perp \rangle \}.$$

It follows that $\mathbf{K}$ should behave in such a way that the statistical average on the right-hand side of this equation is proportional to the order parameter S.[10]

In order to appreciate the above proportionality we note that S itself can be expressed in the elements of the polarizability tensor alone. If an electric field $\mathbf{E}^i$ is applied along the z-axis, an electric moment $\mathbf{p}$ is induced in a molecule with components

$$p_l = \alpha_l E^i \cos \theta,$$

$$p_t = \alpha_t E^i \sin \theta.$$

This gives a component along the z-axis

$$p_\parallel = (\alpha_l \cos^2\theta + \alpha_t \sin^2\theta)E^i$$

$$\equiv \alpha_\parallel E.$$

[9] A. N. Kuznetsov, V. A. Livshits, and S. G. Cheskis, Sov. Phys.-Crystallogr. (Engl. Trans.) 20, 142 (1975).

[10] W. H. de Jeu and P. Bordewijk, J. Chem. Phys. 68, 109 (1978).

Hence we find for the average value

$$\langle \alpha_{\parallel} \rangle = \alpha_l \langle \cos^2\theta \rangle + \alpha_t \langle \sin^2\theta \rangle$$

$$= \tfrac{1}{3}\{\alpha_l(2S + 1) + \alpha_t(2 - 2S)\}, \tag{6a}$$

where Eq. (1) for S has been used. Similarly we find

$$\langle \alpha_{\perp} \rangle = \tfrac{1}{3}\{\alpha_l(1 - S) + \alpha_t(2 + S)\}. \tag{6b}$$

Combination of Eqs. (6a) and (6b) gives

$$\langle \alpha_{\parallel} \rangle - \langle \alpha_{\perp} \rangle = S(\alpha_l - \alpha_t).$$

No molecular model has been given that leads to an expression for K in agreement with the observed proportionality $\Delta\epsilon \sim NS$. A sufficient condition for it is that the tensor K for a molecule with given orientation is independent of the dielectric anisotropy of the macroscopic sample, and thus independent of the orientation of the molecule with respect to the preferred direction. It then follows that the principal axes of α and K should coincide. By a transformation similar to that leading to Eq. (6) we then find for Eq. (5):

$$\epsilon_{\parallel} = 1 + \tfrac{4}{3}\pi N\{\alpha_l K_l(2S + 1) + \alpha_t K_t(2 - 2S)\}, \tag{7a}$$

$$\epsilon_{\perp} = 1 + \tfrac{4}{3}\pi N\{\alpha_l K_l(1 - S) + \alpha_t K_t(2 + S)\}, \tag{7b}$$

where K_l and K_t are the principal values of K. We conclude that it is possible to take the effect of the internal field into account by working with an effective or "dressed" polarizability tensor $\alpha^* = \alpha \cdot K$.

To obtain in Eq. (7) reasonable values for K_l and K_t, we consider the case of ideal ordering, $S = 1$. For that case the values of $\epsilon_{\parallel}$ and $\epsilon_{\perp}$ will be approximately equal to those for the solid, if in the solid all molecules have the same orientation. Vuks[11] suggested that the internal field in a solid with less than cubic symmetry is independent of the orientation, and is given by

$$E^i = \tfrac{1}{3}(\bar{\epsilon} + 2)E.$$

This equation makes the tensor K isotropic. Substitution in Eq. (7) gives:[12]

$$(\epsilon_{\parallel} - 1)/(\bar{\epsilon} + 2) = (4\pi N/9)\{\alpha_l(2S + 1) + \alpha_t(2 - 2S)\}; \tag{8a}$$

$$(\epsilon_{\perp} - 1)/(\bar{\epsilon} + 2) = (4\pi N/9)\{\alpha_l(1 - S) + \alpha_t(2 + S)\}. \tag{8b}$$

[11] M. F. Vuks, *Opt. Spectrosc.* **20**, 361 (1966).
[12] S. Chandrasekhar and N. V. Madhusudana, *J. Phys.* **30**, C4-24 (1969).

Equation (8) can be used to calculate the average polarizability $\bar{\alpha} = (\alpha_l + 2\alpha_t)/3$, giving

$$\bar{\alpha} = (3/4\pi N)(\bar{\epsilon} - 1)/(\bar{\epsilon} + 2).$$

This is the Clausius–Mossotti equation, which is usually derived for the isotropic phase or systems with cubic symmetry where $\bar{\epsilon} = \epsilon$. In that case the internal field is called the Lorentz field. Hence the use of Vuks' approximation leads to a formula compatible with the Claussius–Mossotti equation for the isotropic phase.

There are a number of objections to Eq. (8), however. In the first place this equation lacks a molecular basis even for the case of solids, which is much simpler than the case of liquid crystals. Furthermore, in a number of cases Vuks' treatment yields rather implausible results. For p-azoxyanisole (PAA) and p-azoxyphenetole (PAP) the refractive indices in the solid state have been reported.[13] All three principal refractive indices of the crystal show normal dispersion, which is much larger for the principal axis close to the long axis of the molecules than for the other two axes. Consequently in these systems the polarizability components perpendicular to the long axis of the molecule show anomalous dispersion when calculated with Eq. (8) for $S = 1$.

On the level of the Clausius–Mossotti equation a molecule is represented by a point polarizability in a spherical cavity. A better representation of a real molecule would be an anisotropic homogeneously polarizable spheroid, filling up the cavity in an anisotropic polarized continuum. For a uniaxial solid crystal this gives[14]

$$\alpha_\lambda = \frac{1}{4\pi N} \frac{\epsilon_\lambda - 1}{1 + \Omega_\lambda(\epsilon_\lambda - 1)}, \qquad \lambda = l, t, \tag{9}$$

where the Ω_λ are shape factors that depend only on the axes ratio c/b of the spheroid. They are equal to the depolarizing factors occurring in the solution of the electrostatic problem of a dielectric spheroid in an external field (Böttcher,[8] p. 141):

$$\Omega_l = 1 - v^2 + \tfrac{1}{2}v(v^2 - 1) \ln\{(v + 1)/(v - 1)\}, \tag{10a}$$
$$\Omega_t = \tfrac{1}{2}(1 - \Omega_l), \tag{10b}$$

while $v^2 = c^2/(c^2 - b^2)$. In the case where the spheroid reduces to a sphere we have $\Omega_l = \Omega_t = 1/3$. The internal field tensor corresponding

---

[13] P. Chatelain, *C. R. Hebd. Seances Acad. Sci.* **203**, 266 and 1169 (1936); see also: P. Chatelain and M. Germain, *ibid.* **259**, 127 (1964).

[14] C. J. F. Böttcher and P. Bordewijk, "Theory of Electric Polarization," 2nd ed., Vol. II. Elsevier, Amsterdam, 1978.

to this model has principal elements

$$K_\lambda = 1 + \Omega_\lambda(\epsilon_\lambda - 1) = (1 - 4\pi N\alpha_\lambda \Omega_\lambda)^{-1}, \qquad \lambda = l, t.$$

Apart from a slight variation due to the density dependence, this equation fulfils the requirement that the internal field should not depend on the macroscopic dielectric anisotropy. With this expression for K Eq. (7) yields[10]

$$\epsilon_\parallel = 1 + \tfrac{4}{3}\pi N \left\{ \frac{\alpha_l(2S + 1)}{1 - 4\pi N\alpha_l \Omega_l} + \frac{\alpha_t(2 - 2S)}{1 - 4\pi N\alpha_t \Omega_t} \right\}, \tag{11a}$$

$$\epsilon_\perp = 1 + \tfrac{4}{3}\pi N \left\{ \frac{\alpha_l(1 - S)}{1 - 4\pi N\alpha_l \Omega_l} + \frac{\alpha_t(2 + S)}{1 - 4\pi N\alpha_t \Omega_t} \right\}. \tag{11b}$$

These expressions make it possible to infer α_l and α_t from the experimental values of $\epsilon_\parallel$ and $\epsilon_\perp$, if the density and the order parameter are known. The values obtained depend on the value of the axes ratio c/b used.

Lorentz's original calculation of the internal field in a cubic crystal has been generalized by Neugebauer.[15] He considers an arbitrary lattice in which the molecules are represented by anisotropic point polarizabilities with parallel principal axes. The result is

$$\left(A_\lambda + \frac{1}{\alpha_\lambda} \right)^{-1} = \frac{3}{4\pi N} \frac{\epsilon_\lambda - 1}{\epsilon_\lambda + 2}. \tag{12}$$

The A_λ are functions of the crystal structure, while $A_x + A_y + A_z = 0$. For uniaxial crystals this gives $A_\parallel + 2A_\perp = 0$. Writing $3A_\parallel/4\pi N = 2a$ we can rearrange Eq. (12) to the form of Eq. (9) with $\Omega_\parallel = (1 - 2a)/3$ and $\Omega_\perp = (1 + a)/3$. The difference between the two equations arises from the fact that in deriving Eq. (9) a molecule is represented by a homogeneously polarizable spheroid and in deriving Eq. (12) by an anisotropic point polarizability. In general these two situations give a different internal field (Böttcher,[8] Sec. 20).

The representation of the polarizability of a molecule by an anisotropic point polarizability is a serious limitation, since in reality the polarizability of a molecule will be distributed over its whole volume. This makes no difference for spherical or nearly spherical particles, but becomes increasingly unsatisfactory if an anisotropic shape is involved. Neugebauer's formula has been applied to nematics,[16] in which case,

[15] H. E. J. Neugebauer, *Can. J. Phys.* **32**, 1 (1954).

[16] A. Saupe and W. Maier, *Z. Naturforsch., Teil A* **16**, 816 (1961); H. S. Subramhanyam and D. Krishnamurti, *Mol. Cryst. Liq. Cryst.* **22**, 239 (1973); H. S. Subramhanyan, C. S. Prabha, and D. Krishnamurti, *ibid.* **28**, 201 (1975).

TABLE I.   ELECTRONIC POLARIZABILITIES (10^{-24} cm^3) OF
p-AZOXYANISOLE, CALCULATED FROM THE REFRACTIVE INDICES
IN THE SOLID STATE AND EXTRAPOLATED TO ZERO FREQUENCY

Formula	Equation in text	α_l	α_t	$\bar{\alpha}$	$\alpha_l - \alpha_t$
Vuks	(8)	46	21.8	30	25
Spheroid	(9)	52	21.5	32	30
Neugebauer	(12)	44	22.9	30	21

moreover, the definition of the lattice is not clear. This problem has been circumvented in the literature by treating a as a parameter. The value of a is then determined by equating $\bar{\alpha}$ in the solid state to the value obtained in the isotropic phase using the Clausius–Mossotti equation. Summarizing, we must conclude that the application of Neugebauer's formula to nematics is rather unsatisfactory.

We have calculated α_l, α_t, and $\bar{\alpha}$ of PAA from the refractive indices in the crystalline state[13] using Eq. (8) with $S = 1$, Eq. (9), and Eq. (12). Because of the dispersion of the refractive index the data thus obtained depend on the frequency. The static values of the polarizabilities can be obtained by extrapolation, using the dispersion formula $\alpha = \mathrm{const}/(\omega_0{}^2 - \omega^2)$. This procedure is valid if the frequency ω_0 where the nearest optical absorption takes place is sufficiently high. The results obtained using the different formulas do not differ much (see Table I). This is not surprising: for PAA in the denominator of Eq. (9) Ω_l is small when ϵ_l is large. Consequently the difference from Eq. (8) is numerically not very important. With Eq. (9) Ω_l and Ω_t have been calculated taking $c/b = 2.35$.[17] It will be clear that from a theoretical point of view the formulation of Eq. (9) has to be preferred to the others. Independent experimental information to test the various results is unfortunately not available.

2.   EXTENSION OF THE ONSAGER THEORY TO NEMATICS

In the case of nematics with polar molecules the orientation polarization $\mathbf{P}_\mu$ must be added to Eq. (4). This leads to

$$\frac{\epsilon_\lambda - 1}{4\pi} E_\lambda = N\{\langle (\alpha \cdot \mathbf{E}^i)_\lambda \rangle + \langle \bar{\mu}_\lambda \rangle\}, \qquad (13)$$

where $\langle \bar{\mu}_\lambda \rangle$ is the average value of the dipole component in the

[17] Calculated from the crystallographic data for PAA as indicated in de Jeu and Borde-wijk.[10]

λ-direction in the presence of the electric field. Hence two different averages are involved. The bar refers to the potential energy of the dipole moment in the electric field and the brackets to the nematic potential. Before proceeding to the calculation of $\langle \bar{\mu}_\lambda \rangle$ we shall first summarize the basic theory for the isotropic phase. For a more detailed discussion we refer to Böttcher.[8]

In Onsager's theory of the dielectric properties of isotropic polar liquids a molecule is represented by a polarizable point dipole in a spherical cavity of molecular dimensions and surrounded by a continuum with the macroscopic properties of the dielectric. The radius a of the cavity is given by

$$\tfrac{4}{3}\pi N a^3 = 1. \tag{14}$$

When evaluating Eq. (13) for the isotropic phase we use the result[8]

$$\bar{\mu} = (\mu^2/3kT)\mathbf{E}^d, \tag{15}$$

where $\mathbf{E}^d$ is the directing field. $\mathbf{E}^d$ differs from the internal field because of the presence of the reaction field. The dipole moment polarizes its surroundings, leading in turn to a reaction field at the position of the dipole. As the reaction field is always parallel to the dipole moment it cannot direct the dipole, but it does contribute to $\mathbf{E}^i$. Denoting the average reaction field by $\bar{\mathbf{R}}$, we have

$$\mathbf{E}^i = \mathbf{E}^d + \bar{\mathbf{R}}.$$

The reaction field of a dipole moment $\boldsymbol{\mu}$ with polarizability α is given by[8]

$$\mathbf{R} = f(\boldsymbol{\mu} + \alpha\mathbf{R}) = fF\boldsymbol{\mu},$$

where f is the reaction field factor

$$f = \frac{1}{a^3}\frac{2\epsilon - 2}{2\epsilon + 1} \tag{16}$$

and

$$F = (1 - f\alpha)^{-1}. \tag{17}$$

Hence

$$\bar{\mathbf{R}} = fF\bar{\mu} = fF(\mu^2/3kT)\,\mathbf{E}^d,$$

and

$$\mathbf{E}^i = \left(1 + \frac{f}{1 - f\alpha}\frac{\mu^2}{3kT}\right)\mathbf{E}^d.$$

$\mathbf{E}^d$ can be calculated as the sum of the field in a spherical cavity

$$\mathbf{E}^c = \frac{3\epsilon}{2\epsilon + 1}\mathbf{E} \equiv h\mathbf{E}, \tag{18}$$

and the reaction field of the induced dipole moment. This leads to

$$\mathbf{E}^d = \mathbf{E}^c + f\alpha\mathbf{E}^d = Fh\mathbf{E}.$$

Using Eq. (15) and the results for $\mathbf{E}^i$ and $\mathbf{E}^d$ in the basic equation, Eq. (13), we arrive for isotropic liquids at Onsager's equation

$$(\epsilon - 1)/4\pi = NhF\{\alpha + F\mu^2/(3kT)\}. \tag{19}$$

Maier and Meier[18] have extended the Onsager theory to nematic liquid crystals. The principal elements of the molecular polarizability tensor are α_l and α_t, while the dipole moment μ makes an angle β with the ζ-axis. So $\mu_l = \mu \cos \beta$ and $\mu_t = \mu \sin \beta$. In their calculation Maier and Meier follow Onsager's theory as closely as possible. This involves the following approximations:

1. The molecule under consideration is contained in a spherical cavity of radius a. Although this seems rather unrealistic for liquid crystals, the theory is in many respects the same as when an spheroidal cavity is used.

2. In the reaction field factor f and the cavity factor h the anisotropy of the permittivity is neglected. Equations (16) and (18) are used with $\bar{\epsilon}$ substituted for ϵ. This approximation is justified as long as $\Delta\epsilon \ll \bar{\epsilon}$, which is often the case.

3. In the calculation of the factor F the anisotropy of the polarizability is neglected. Equation (17) is used with $\bar{\alpha}$ substituted for α. This is in fact not correct as $\alpha_l - \alpha_t$ can be of the same order of magnitude as α_t. Nevertheless, the final results will probably not be seriously affected as F is itself a correction factor.

We shall now give the derivation of Maier and Meier's formula for $\epsilon_\parallel$. The electric field is applied along the preferred direction (z-axis). Similarly as in the case of nonpolar molecules the average induced moment in this direction is given by Eq. (6a). The calculation of the orientation polarization proceeds as follows. The potential energy of the dipole moment in the electric field is given by $U = -\boldsymbol{\mu} \cdot \mathbf{E}^d$; the total energy is $U + W$. We make the usual linear approximation, i.e.,

$$\exp\{-(U + W)/kT\} \approx (1 - U/kT)\exp(-W/kT).$$

[18] W. Maier and G. Meier, Z. Naturforsch., Teil A **16**, 262 (1961).

The dipole component along the z-axis is given by

$$\mu_\| = \mu_l \cos\theta + \mu_t \sin\phi \sin\theta,$$

where ϕ is the angle between the μ, ζ plane and the normal to the z, ζ plane. The average value of $\mu_\|$ in the presence of the electric field then is calculated to be

$$\langle \bar{\mu}_\| \rangle = \langle \mu_\| (1 + \mu_\| E^d)/kT \rangle = \langle \mu_\|^2 \rangle E^d/kT.$$

Using the expression for $\mu_\|$ and noting that $\langle \sin^2\phi \rangle = 1/2$ because of the rotational symmetry, we get

$$\langle \mu_\|^2 \rangle = \mu_l^2 \langle \cos^2\theta \rangle + \tfrac{1}{2}\mu_t^2 \langle \sin^2\theta \rangle$$

$$= \tfrac{1}{3}\{\mu_l^2(2S + 1) + \mu_t^2(1 - S)\} \qquad (20a)$$

$$= \tfrac{1}{3}\mu^2\{1 - (1 - 3\cos^2\beta)S\}.$$

After a similar calculation we find

$$\langle \mu_\perp^2 \rangle = \tfrac{1}{3}\{\mu_l^2(1 - S) + \tfrac{1}{2}\mu_t^2(2 + S)\}$$

$$= \tfrac{1}{3}\mu^2\{1 + \tfrac{1}{2}(1 - 3\cos^2\beta)S\}. \qquad (20b)$$

For $S = 1$ only the component μ_l contributes to $\langle \mu_\|^2 \rangle$ and only μ_t contributes to $\langle \mu_\perp^2 \rangle$.

Inserting the expressions for $\mathbf{E}^i$ and $\mathbf{E}^d$ analogously to the isotropic case, Eq. (13) can be written as[18]

$$(\epsilon_\| - 1)/4\pi = NhF(\langle \alpha_\| \rangle + F\langle \mu_\|^2 \rangle/kT), \qquad (21a)$$

$$(\epsilon_\perp - 1)/4\pi = NhF(\langle \alpha_\perp \rangle + F\langle \mu_\perp^2 \rangle/kT), \qquad (21b)$$

where $\langle \alpha_\lambda \rangle$ is given by Eq. (6) and $\langle \mu_\lambda^2 \rangle$ by Eq. (20). Combination of Eqs. (21a) and (21b) leads to an expression for the dielectric anisotropy:

$$\Delta\epsilon/4\pi = NhF\{(\alpha_l - \alpha_t) - F(\mu^2/2kT)(1 - 3\cos^2\beta)\}S. \qquad (22)$$

When we calculate $\bar{\epsilon}$ by combining Eqs. (21a) and (21b) Onsager's equation, Eq. (19), is reproduced with ϵ replaced by $\bar{\epsilon}$. For $\mu = 0$, Eqs. (21a) and (21b) reduce to the extension of Vuks' formula, Eq. (8). This is easily seen by substituting Eqs. (17) and (18) for F and h (with ϵ and α replaced by $\bar{\epsilon}$ and $\bar{\alpha}$) and using Eq. (14).

The above equations account satisfactorily for the essential features of the permittivity of nematic liquid crystals. This is best illustrated using Eq. (22) for $\Delta\epsilon$. If $3\cos^2\beta = 1$ ($\beta \approx 55°$) the dipole moment contributes equally to $\epsilon_\|$ and $\epsilon_\perp$. Then $\Delta\epsilon$ is determined by the (positive) anisotropy of the polarizability. The dipole contribution to $\Delta\epsilon$ is positive for $\beta < 55°$

and negative for $\beta > 55°$. In the latter case, whether $\Delta\epsilon$ itself becomes negative depends on the relative magnitudes of the two contributions. The contribution of the induced polarization to $\Delta\epsilon$ varies with temperature parallel to S; the temperature dependence of the orientation polarization goes with S/T. The temperature dependence of the other factors is weak. The experimentally observed temperature dependence of the permittivities agrees nicely with these predictions, as will be discussed in some detail in the next section.

Derzhanski and Petrov[19] have extended the theory by taking the anisotropic shape of the molecules into account and considering an spheroidal cavity. Furthermore they do not treat the continuum outside the spheroid as isotropic. This leads to functional relations for ϵ_λ that are essentially the same as Eqs. (21a) and (21b), while in addition some correction factors come in that can be numerically important. For example, the value of β where the dipole contribution to $\Delta\epsilon$ is zero now depends on the ratio of the axes of the molecular spheroid. However, contrary to the case of nonpolar molecules, where $\bar{\epsilon}/\rho$ is continuous at the NI transition and $\Delta\epsilon \sim NS$, no analogous experimental information exists for polar molecules. This makes the various approximations that were used to account for the imperfect orientational order rather arbitrary. In fact their theory does not predict $\bar{\epsilon}/\rho$ to be continuous at the NI transition in the limit $\mu = 0$.

III. Permittivity and Molecular Structure

In this section the results of Section II will be applied to various compounds that give a nematic phase. We shall start with a quantitative treatment of p-azoxyanisole (PAA). In Section III,4 the influence of various terminal substituents on the permittivity is considered. In Section III,5 the influence on the permittivity of various bridging groups is investigated for compounds with fixed terminal substituents.

3. p-Azoxyanisole

In Fig. 3 the static permittivities of PAA are given as a function of the temperature.[20] In order to interpret these data using Eqs. (21) and (22) we need additional information on either the molecular polarizabilities or the dipole moment. The electronic polarizabilities of PAA, extrapolated to static values, have been given in Table I. To be consistent with the approximations involved in Eq. (21) we shall use the results from Vuks' formula, Eq. (8), and add $\bar{\alpha}/10$ as an estimate of the ionic polarizability.

[19] A. I. Derzhanski and A. G. Petrov, *C. R. Acad. Bulg. Sci.* **24**, 569 (1971); A. G. Petrov, *ibid.* p. 573.

[20] W. Maier and G. Meier, *Z. Naturforsch., Teil A* **16**, 470 (1961).

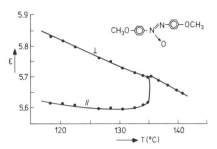

FIG. 3. Permittivities of PAA.[20] (Copyright by Verlag der Zeitschrift für Naturforschung. Used with permission.)

Using ϵ in the isotropic phase and $\bar{\alpha}$ we find $\mu = 2.22$ D from Eq. (19). A direct measurement of μ on dilute solutions of PAA in benzene gives[20] 2.30 D. Taking the first value of μ we can use Eq. (22) to obtain $\beta = 65°$. In fact the value of β is found to be not very sensitive to variations of $\alpha_l - \alpha_t$. If we use the result for $\alpha_l - \alpha_t$ from the spheroidal model, Eq. (9), we find $\beta = 67°$.

A check on the above calculation can be obtained using the Kerr constant, which also depends on $\alpha_l - \alpha_t$, μ, and β. The molar Kerr constant of PAA (measured on dilute solutions in benzene) is 57.8×10^{-12} esu as given by Tsvetkov.[21] Using his formula but our values $\alpha_l - \alpha_t = 25 \times 10^{-24}$ cm^3 and $\mu = 2.30$ D (value from benzene solutions) we find $\beta = 62°$, in good agreement with the value obtained from the dielectric measurements.

We can also calculate the total dipole moment starting from tabulated group moments. PAA exists in the trans-conformation. This leads to some ambiguity as to the position of the molecular ζ-axis. Following Maier and Meier[20] we shall simply identify this axis with the line through the two outer para-positions of the benzene rings. The ζ-axis then makes an angle $\delta \approx 10°$ with the para-axis through the para-positions of a single aromatic ring. The central dipole moment μ_M (due to the azoxy group) forms an angle γ with the ζ-axis. The components of the methoxy dipoles along the para-axis compensate each other, leaving components μ_E perpendicular to this axis. Assuming free rotation of the methoxy groups, we then find for the total dipole moment μ:

$$\mu^2 \cos^2\beta = \mu_M{}^2 \cos^2\gamma + \mu_E{}^2 \sin^2\delta,$$

$$\mu^2 \sin^2\beta = \mu_M{}^2 \sin^2\gamma + 2\mu_E{}^2(1 - \tfrac{1}{2} \sin^2\delta),$$

which can easily be solved for μ and β.

The dipole moment of azoxybenzene is 1.70 D and is usually assumed

[21] V. N. Tsvetkov and V. Marinin, *Zh. Eksp. Teor. Fiz.* **18**, 641 (1948).

to lie along the NO bond[22] ($\gamma \approx 60° + \delta = 70°$). The group moment of a methoxy group is 1.28 D at an angle of 72° to the para-axis,[23] which gives $\mu_E = 1.25$ D. Substituting these values in the above equations we find $\mu = 2.45$ D and $\beta = 75°$. The value of μ is only slightly larger than the experimental one; the value of β is much too large. However, β depends strongly on γ, which is not known exactly. Using the above group moments, $\beta = 62°$ as derived from the Kerr constant, and $\mu = 2.30$ D, we can calculate $\gamma = 52°$. The azoxy dipole is probably not directed along the NO bond.

We conclude that the theory of Maier and Meier can account at least qualitatively for the experimental results on PAA. A more quantitative test is hindered by the lack of independent information on the molecular polarizabilities and the angle between the dipole moment and the long molecular axis.

Finally, in disagreement with the theory, we note that if in Fig. 3 curves for $\bar{\epsilon}$ in the nematic phase and ϵ in the isotropic phase were extrapolated to T_{NI}, they would not coincide. This effect, though particularly evident in PAA, was also present in the examples of Figs. 1 and 2. In all cases of polar compounds $\bar{\epsilon}$ is lower than ϵ at T_{NI}. Only for the nonpolar p,p'-di-n-heptylazobenzene (Fig. 1a) do the two curves coincide at T_{NI}. We shall meet further examples of this effect in the next paragraph and discuss possible explanations at the end of Section III,5.

4. VARIATION OF THE TERMINAL GROUPS IN p,p'-DI-SUBSTITUTED AZOBENZENES

In Fig. 1a we gave the static permittivities of the nonpolar p,p'-di-n-heptylazobenzene. The influence of the terminal groups can be studied systematically by comparison with the permittivities of compounds of the type

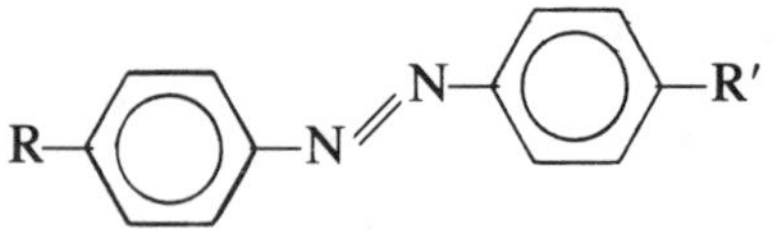

In Fig. 4 results for the permittivities are given for various end substituents R and R'.[24a]

[22] R. S. Armstrong and R. J. W. LeFèvre, *Aust. J. Chem.* **19,** 29 (1966).

[23] V. I. Minkin, D. A. Osipov, and Yu. A. Zhdanov, "Dipole Moments in Organic Chemistry," p. 91. Plenum, New York, 1970.

[24a] The liquid-crystal behavior of some of these compounds has not been reported before. The transition temperatures are:

C_7H_{15}—Ph—NN—Ph—O(CO)C_5H_{11}: K 61.4–61.7 N 75.7–75.9 I (53 S),
C_7H_{15}—Ph—NN—Ph—OC_6H_{13}: K 49.9–50.1 N 82.8–83.1 I,
C_6H_3O—Ph—NN—Ph—CN: K 99.6–99.7 N 116.6–116.8 I.

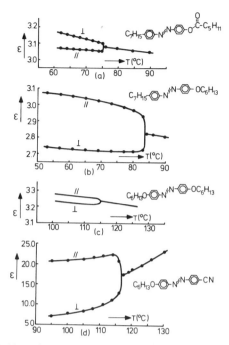

FIG. 4. Permittivities of some terminally substituted azobenzenes (c, Ref. 24). [(c) Copyright by Verlag Chemie GMBH. Used with permission.]

We shall first compare Fig. 1a with Figs. 4a and 4b. Replacing the heptyl group by an acyloxy or alkoxy group of similar size leads to a small increase of $\epsilon_\parallel$. In the case of alkoxy substitution the increase of $\epsilon_\perp$ is also small. For acyloxy substitution this increase is much larger, leading to a negative value of $\Delta\epsilon$. At a reduced temperature of $0.97T_{NI}$ $\Delta\epsilon$ has the value 0.40, -0.08, and 0.29 when R is heptyl and R' is heptyl, hexanoyloxy, and hexyloxy, respectively. The group moments of $-O(CO)CH_3$ and $-OCH_3$ are 1.69 D at 66° with the para-axis of the adjacent aromatic ring and 1.28 D at 72°, respectively.[23] We shall assume that these values are also typical in the case of longer alkyl chains than $-CH_3$. For both substituents the longitudinal component of the dipole moment counteracts the heptyl dipole (about 0.4 D along the para-axis) at the other para-position. However, exact cancellation, as in the case of di-alkyl substitution, cannot be expected. Furthermore, the polarizability along the para-axis will be somewhat larger than in the case of di-alkyl substitution. These two effects explain the increase of $\epsilon_\parallel$. The increase of $\epsilon_\perp$ is due to the transverse component of the alkoxy or acyloxy dipole moment, which is appreciably larger in the latter case.

[24] W. Maier and G. Meier, Z. Elektrochem. 65, 556 (1961).

When two alkoxy groups are introduced (Fig. 4c) the transverse component of the total dipole moment is approximately a factor of $\sqrt{2}$ larger than in the case of Fig. 4b. (The dipole moments add up quadratically, while free rotation of the alkoxy groups is assumed.) However, as $\epsilon_{\parallel}$ has also increased somewhat due to the greater polarizability along the para-axis, $\Delta\epsilon$ is still (small) positive.

Finally we consider the case of cyanide substitution (Fig. 4d). The cyanide group has a large dipole moment of 4.05 D along the para-axis.[23] The contribution to the permittivity of this dipole moment predominates over all other contributions. As the angle between the para-axis and the molecular ζ-axis is small, this leads to a large value of $\epsilon_{\parallel}$ and consequently to a large positive $\Delta\epsilon$. Recently, negative values of $\Delta\epsilon$ have been reached by incorporation of a cyanide group as part of the terminal chain rather than being directly attached to the aromatic ring.[7] Finally, we note in Fig. 4d that $\bar{\epsilon}$ decreases with decreasing temperature, contrary to the usual trend. This effect is typical of compounds that associate with their dipole moments antiparallel, the association becoming stronger at lower temperatures.

The temperature dependence of $\Delta\epsilon$ as given in the examples of Figs. 1–4 can be understood using Eq. (22):

1. For nonpolar compounds (Fig. 1a) $\Delta\epsilon$ varies parallel to S. Just below T_{NI}, where S varies strongly, $\Delta\epsilon$ increases with decreasing temperature. At lower temperatures this effect tends to saturate.

2. When $\Delta\epsilon$ is negative or strongly positive the S/T dependence of the dipole contribution to $\Delta\epsilon$ is predominant over the whole temperature range. Consequently $|\Delta\epsilon|$ increases with decreasing temperature (Figs. 3, 4a, and 4d).

3. When $\Delta\epsilon$ is positive and close to zero, the anisotropy of the induced polarization approximately equals that of the orientation polarization. Just below T_{NI} the temperature dependence of S is most important and $\Delta\epsilon$ increases with decreasing temperature. At lower temperatures the counteracting S/T dependence of the orientation polarization predominates. Then we find an almost constant $\Delta\epsilon$ (Fig. 4c), while in some cases $\Delta\epsilon$ even decreases with decreasing temperature (see next section, Fig. 5g).

5. Variation of the Bridging Group for Fixed Terminal Substituents

In this section we consider the effect of variation of the bridging group X of compounds having the structure

$$CH_3O-\langle\bigcirc\rangle-X-\langle\bigcirc\rangle-C_5H_{11}$$

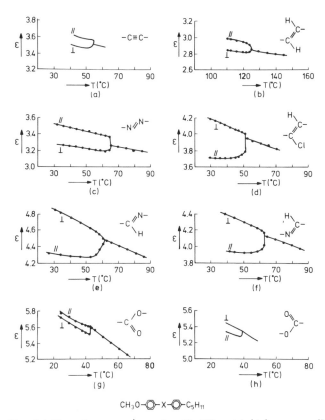

FIG. 5. Permittivities of some compounds with different bridging groups.[25] (Copyright by Verlag der Zeitschrift für Naturforschung. Used with permission.)

Di-alkyl substituted compounds would have been more suitable for studying the effect of the bridging group. However, only a few compounds of that type have been reported as displaying nematic behavior. In Fig. 5 the results for the static permittivities are given for various symmetric and asymmetric bridging groups.

The bridges —C≡C—, CH=CH—, and —N=N— are nonpolar. However, this holds only approximately in the case of asymmetric substitution. For stilbene the charge distribution is fairly uniform. In the other cases there is some accumulation of electron density between the aromatic rings, which probably makes the charge distribution more sensitive to asymmetric substitution. Further differences arise because of the *linear* form of the tolanes as compared with the *trans* conformation of the other molecules. Nevertheless, the dielectric behavior is not

[25] W. H. de Jeu and T. W. Lathouwers, *Z. Naturforsch., Teil A* **30**, 79 (1975).

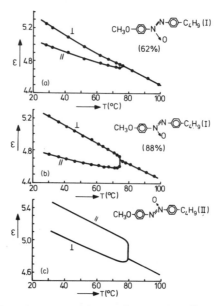

Fig. 6. Permittivities of N4 (a) and its two isomers (b,c).[27] (Copyright by the North-Holland Publishing Company. Used with permission.)

very different. In all cases $\Delta\epsilon$ is small and positive (see Figs. 5a, 5b, and 5c).

In the case of the stilbene derivative chlorine substitution in the bridge introduces an extra dipole moment that contributes more to $\epsilon_\perp$ than to $\epsilon_\parallel$, leading to a negative $\Delta\epsilon$ (Fig. 5d). The effect is much less pronounced, however, than in the case of cyanide substitution dealt with earlier (Fig. 2a). The anisotropy of the permittivity is comparable with that of the analog compound having a —CH=N— or —N=CH— bridge (Figs. 5e and 5f). In these cases the bridge dipole also contributes mainly to $\epsilon_\perp$. The substituted phenylbenzoates finally form a fairly complicated system (Figs. 5g and 5h). Apart from the two end groups, two more dipole moments are associated with the C_{ar}—O and C=O bonds in the bridging group. Although the dielectric anisotropies of the two isomers have different signs, the difference is in fact small. A more detailed comparison of various types of Schiff bases and phenylbenzoates can be found in Klingbiel et al.[26]

In the case of asymmetrically substituted azoxybenzenes a mixture of the two isomers is usually obtained that cannot easily be separated. In Fig. 6a the permittivities of such a nematic mixture are given. The

[26] R. T. Klingbiel, D. J. Genova, T. R. Criswell, and J. P. van Meter, J. Am. Chem. Soc. 96, 7651 (1974)

results of Fig. 6b are obtained when the amount of isomer I in the mixture has been increased. Assuming that the permittivities of the mixture depend linearly on those of the components, this allows a calculation of the permittivities of isomer II (Fig. 6c). The difference between the dielectric anisotropies of the two isomers can be understood as follows.[27] The dipole moment associated with the freely rotating methoxy group is approximately perpendicular to the ζ-axis. In both cases this adds up with the perpendicular component of the azoxy dipole to give about equal values of $\epsilon_\perp$. For isomer I the longitudinal component of the azoxy dipole and the butyl dipole moment counteract each other. For isomer II they add up, which explains the difference in $\epsilon_\parallel$ and thus in $\Delta\epsilon$.

Finally we note that in the examples of Fig. 5, $\bar{\epsilon}$ in the nematic phase, when extrapolated to T_{NI}, is again lower than ϵ in the isotropic phase. The difference will be denoted by $\delta\epsilon_{NI}$. The general trend in $\delta\epsilon_{NI}$ can tentatively be summarized as follows:

1. In the nematic phase of polar molecules $\bar{\epsilon}$ is lower at T_{NI} than ϵ in the isotropic phase, and independent of the sign and magnitude of $\Delta\epsilon$.

2. In the case of nonpolar molecules this effect is absent, while in the case of polar molecules $\delta\epsilon_{NI}$ increases with increasing values of $\epsilon_\parallel$.

These points can be illustrated using Fig. 2a (large negative dielectric anisotropy) and Fig. 5g (small positive anisotropy). In both cases $\delta\epsilon_{NI}$ is of the order of 0.08. In most of the other cases $\delta\epsilon_{NI}$ is somewhat smaller. These facts probably indicate an effect that decreases the dipole contribution to $\epsilon_\parallel$ and thus $\bar{\epsilon}$.

Various explanations have been proposed to account for $\delta\epsilon_{NI}$. Maier and Meier[18] attribute this difference to the neglect of the anisotropies $\Delta\epsilon$ and $\alpha_l - \alpha_t$ in the factors h and F. However, this would also affect the results for nonpolar molecules, while the effect would be expected to depend strongly on $\Delta\epsilon$. Derzhanski and Petrov[28] give an explanation based on the existence in nematic liquid crystals of flexoelectricity, the orientation dependent analog of piezoelectricity. Shape polarity of the molecules is then important. However, $\delta\epsilon_{NI}$ is not absent or very small in the case of symmetric polar molecules.[4] Finally, an explanation has been offered[29] in terms of short-range antiparallel dipole correlation of the dipole components $\mu_\parallel$, leading to a decrease of $\epsilon_\parallel$ in the nematic phase. In that case a parallel dipole correlation would be expected for the components $\mu_\perp$, leading to an increase of $\epsilon_\perp$. The net effect on $\bar{\epsilon}$ and

[27] W. H. de Jeu and T. W. Lathouwers, *Chem. Phys. Lett.* **28**, 239 (1974).

[28] A. I. Derzhanski and A. G. Petrov, *Phys. Lett. A* **34**, 427 (1971).

[29] N. V. Madhusudana and S. Chandrasekhar, *Proc. Bangalore Liq. Cryst. Conf.*, p. 57 (1975).

consequently the sign of $\delta\epsilon_{NI}$ would then depend strongly on $\Delta\epsilon$, again in disagreement with the experimental results. These correlation effects will be discussed in more detail in the next section.

The example of Fig. 4d shows that molecular association can have an important influence on the permittivities. It could be that $\delta\epsilon_{NI}$ is related to small differences in molecular association between the nematic and the isotropic phase. In that case one would indeed expect the association between molecules with μ_l antiparallel to be somewhat favored in the nematic phase, leading to a smaller value of $\epsilon_{\|}$.

IV. The Influence of Smectic Order on the Static Permittivity

Figure 7 gives the static permittivities of p,p'-di-n-heptylazoxyben-zene (HEPTAB) in the isotropic, the nematic, and the smectic A phase.[30] Just below the NI transition $\Delta\epsilon$ increases as usual with decreasing temperature. However, at lower temperatures this trend is reversed, leading finally to a change in the sign of $\Delta\epsilon$. This anomalous behavior of $\Delta\epsilon$ has been attributed to the occurrence of the smectic A phase.[30,31] In this phase the interaction of a dipole moment with the dipoles of surrounding molecules works out differently from that in the nematic phase due to the nonisotropic distribution of the centers of mass. For central dipoles the distance between the dipoles of molecules in different smectic layers is much larger than the distance between neighboring dipoles in the same layer. This leads to an increased antiparallel correlation between the dipole components along the preferred axis. Consequently, the effective dipole moment in this direction is reduced, leading to a decrease of $\epsilon_{\|}$. Similarly, an increase of $\epsilon_{\perp}$ can be explained.

In the Onsager theory the dielectric properties of polar liquids are investigated by considering a molecule in the average field of the other molecules. The environment of a molecule is treated as a continuous medium. In this way specific short-range interactions as described above are not accounted for. In the case of isotropic polar liquids the Kirkwood–Fröhlich theory[8,32] provides in principle a framework in which such interactions can be explicitly evaluated. In the next section we shall first summarize its extension to liquid crystals. In Section IV,7 we shall discuss the above-mentioned problem of dipole correlation more extensively.

[30] W. H. de Jeu, T. W. Lathouwers, and P. Bordewijk, *Phys. Rev. Lett.* **32**, 40 (1974).
[31] W. H. de Jeu, W. J. A. Goossens, and P. Bordewijk, *J. Chem. Phys.* **61**, 1985 (1974).
[32] P. Bordewijk, *Physica* **69**, 422 (1973).

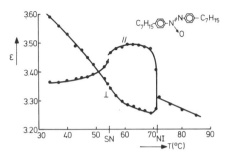

FIG. 7.   Permittivities of HEPTAB.[30] (Copyright by the American Physical Society. Used with permission.)

6. EXTENSION OF THE KIRKWOOD-FRÖHLICH THEORY TO LIQUID CRYSTALS

In the Kirkwood-Fröhlich theory a finite macroscopic sphere with volume v is considered in an infinite isotropic dielectric with permittivity ϵ. The interactions between the n molecules in the sphere can be treated explicitly. The dielectric surrounding the sphere is considered as a continuum. We shall not give a complete derivation of the result for the permittivity. Some feeling for the final equation can be obtained by realizing that when the macroscopic sphere is reduced to molecular dimensions the Onsager equation, Eq. (19), should result. It is therefore instructive first to solve Eq. (19) for μ^2. Substituting explicitly Eqs. (17) and (18) for F and h, and relating $\bar{\alpha}$ and ϵ^∞ via the Clausius-Mossotti equation we arrive at

$$(\epsilon - \epsilon^\infty) \frac{2\epsilon + \epsilon^\infty}{\epsilon(\epsilon^\infty + 2)^2} = \frac{4\pi N}{9kT} \mu^2. \qquad (23)$$

The final Kirkwood-Fröhlich equation, on the other hand, is[8,32]

$$(\epsilon - \epsilon^\infty) \frac{2\epsilon + \epsilon^\infty}{\epsilon(\epsilon^\infty + 2)^2} = \frac{4\pi}{9kTv} \langle \sum_i \sum_j \boldsymbol{\mu}^i \cdot \boldsymbol{\mu}^j \rangle, \qquad (24)$$

where the summations extend over all n molecules in the macroscopic sphere. The brackets indicate a statistical average over the sphere. In the absence of interaction between the dipole moments the term between brackets is equal to $n\mu^2$; using $N = n/v$, Eq. (23) is then reproduced. For a discussion of the various approximations involved we refer to Bordewijk.[32]

The Kirkwood-Fröhlich theory has been extended to liquid crystals by Bordewijk.[33] The derivation involves the calculation of the field in a

[33] P. Bordewijk, *Physica* **75**, 146 (1974).

spherical cavity in an anisotropic dielectric with permittivities ϵ_λ, filled with an anisotropic dielectric with permittivities ϵ_λ^∞, ($\lambda = \parallel, \perp$). In order to calculate this field the sphere must be transformed into a spheroid with axes ratio $\sqrt{\epsilon_\parallel/\epsilon_\perp}$. Hence shape factors, Ω_λ^ϵ, come in that depend on this ratio only. For $\epsilon_\parallel = \epsilon_\perp$ we have $\Omega_\lambda^\epsilon = 1/3$. The cavity field now is given by[33]

$$E_\lambda{}^c = \frac{\epsilon_\lambda E_\lambda}{\epsilon_\lambda + (\epsilon_\lambda{}^\infty - \epsilon_\lambda)\Omega_\lambda{}^\epsilon},$$

where E_λ is the component of the external field along the λ-axis. The induced part of the moment of the macroscopic sphere has originally been calculated under the assumption that either $S = 1$, which allows the use of Eq. (9) for α_λ, or that α is isotropic. These approximations can be circumvented if Vuks' formula for α_λ, Eq. (8), is used. Bordewijk's results then reduce to

$$(\epsilon_\lambda - \epsilon_\lambda{}^\infty)\frac{\epsilon_\lambda + (\epsilon_\lambda{}^\infty - \epsilon_\lambda)\Omega_\lambda{}^\epsilon}{\epsilon_\lambda(\bar{\epsilon}^\infty + 2)^2} = \frac{4\pi}{9kTv}\left\langle \sum_i \sum_j \mu_\lambda{}^i\mu_\lambda{}^j \right\rangle. \tag{25}$$

It is easily seen that in the isotropic phase Eq. (25) reduces to Eq. (24). However, in the absence of short-range correlation Eq. (25) does not reduce exactly to Maier and Meier's equations for ϵ_λ. The factorization in terms in which the anisotropy of ϵ and α is neglected or retained is slightly different. This difference is not important because the use of Eq. (25) is only relevant when short-range correlation has an appreciable influence on the orientational polarization. The theory has recently been extended[34] to incorporate the internal field calculated with a spheroidal model, Eq. (11).

7. DIPOLE CORRELATION IN NEMATICS AND SMECTICS

A similar decrease of $\Delta\epsilon$ with decreasing temperature as observed for HEPTAB (Fig. 7) is also found for p,p'-di-n-hexyl- and -octylazoxybenzene around the smectic–nematic (SN) transition.[31] On the other hand, the anisotropy of the refractive index of these compounds shows no anomaly in this region.[35] Consequently, the decrease of $\Delta\epsilon$ must be attributed to the dipole contribution to the permittivity. As this decrease of $\Delta\epsilon$ is absent in the lower members of the series where no smectic phase is observed, it must be related to the occurrence of the smectic A phase. Other examples of similar variations of $\Delta\epsilon$ have been reported by

[34] P. Bordewijk and W. H. de Jeu, *J. Chem. Phys.* **68**, 116 (1978).
[35] W. H. de Jeu, *Solid State Commun.* **13**, 1521 (1973).

Klingbiel et al.[36] However, in these cases no information about the type of smectic phase and the precise orientation of the preferred axis is available.

Equation (25) can now be used to evaluate the effect of dipole–dipole interaction in the nematic and the smectic A phase. We shall summarize the calculation for the dipole component along the preferred axis, given above Eq. (20). When calculating the average at the right-hand side of Eq. (25) the potential energy is given by

$$U_{\text{tot}} = \sum_i W_i(\theta) + \tfrac{1}{2} \sum_i \sum_{\substack{j \\ i \neq j}} V_{ij}.$$

$W_i(\theta)$ is the nematic potential and

$$V_{ij} = \boldsymbol{\mu}^i T^{ij} \boldsymbol{\mu}^j.$$

The dipole field tensor T^{ij} is given by

$$T^{ij} = \frac{1}{r_{ij}^3} - 3 \frac{\mathbf{r}_{ij}\mathbf{r}_{ij}}{r_{ij}^5}, \tag{26}$$

where r_{ij} is the distance between the dipoles of the molecules i and j. We now make the approximation

$$\exp(-U_{\text{tot}}/kT) \approx \{\exp(-\sum_i W_i/kT)\} \left(1 - \frac{1}{2kT} \sum_i \sum_{\substack{j \\ i \neq j}} V_{ij}\right).$$

The right-hand side of Eq. (25), taken for the parallel component, can then be written as

$$\frac{4\pi}{9kTv} \left\langle \sum_i \sum_j \mu_{\parallel}^i \mu_{\parallel}^j \left(1 - \frac{1}{2kT} \sum_k \sum_{\substack{l \\ k \neq l}} \boldsymbol{\mu}^k T^{kl} \boldsymbol{\mu}^l\right) \right\rangle, \tag{27}$$

where the averaging procedure now has to be performed with $\exp(-\sum_i W_i/kT)$ only. By analogy with Section II,2 the first term can be evaluated to give

$$\frac{4\pi}{9kTv} \sum_i \langle (\mu_{\parallel}^i)^2 \rangle = \frac{4\pi N}{9kT} \langle \mu_{\parallel}^2 \rangle,$$

where $\langle \mu_{\parallel}^2 \rangle$ is given by Eq. (20a).

According to Eq. (26) the elements of the tensor T^{kl} depend on the relative positions of the molecules and not on their orientations. Assuming that the positions and orientations are independent of each other, the averaging over positions and orientations can be done separately. Since the distribution of the positions of the molecules is rotationally symme-

[36] R. T. Klingbiel, D. J. Genova, and H. K. Bücher, *Mol. Cryst. Liq. Cryst.* **27**, 1 (1974).

tric with respect to the preferred direction, the nondiagonal elements of T^{kl} can then be disregarded. Working out the terms that do contribute,[31] for the second part of expression (27) we find

$$-\frac{4\pi N}{(3kT)^2} \langle \mu_\parallel{}^2\rangle^2 \langle \sum_j T_\parallel^{ij}\rangle.$$

The last quantity between brackets is the value of $\Sigma\, T_\parallel^{ij}$ averaged over all positions of the molecules j in the macroscopic sphere with respect to an arbitrarily chosen reference molecule i. The calculation of this average over the positions gives a rather different result for the nematic and the smectic A phase.

The molecules are represented by spheroids with long axis $2c$ and short axis $2b$. We shall assume that in the nematic phase a uniform density N surrounds the reference molecule i. Using standard results of electrostatic theory we find[31]

$$\langle \sum_{j\neq i} T_\parallel^{ij}\rangle = 4\pi N(1/3 - \Omega_\parallel),$$

where $\Omega_\parallel$ is a shape factor that depends on the axis ratio c/b of the spheroid only; for $c > b$ we have $0 < \Omega_\parallel < 1/3$ [see Eq. (10)]. In the smectic A phase the molecules are arranged in layers of thickness $2c$ with an isotropic distribution with two-dimensional density $2cN$ within the layers. Taking a discrete distribution perpendicular to the layers with the dipoles situated at $z = \pm p(2c)$, $p = 0, 1, 2, \ldots$, we can calculate[31]

$$\langle \sum_{j\neq i} T_\parallel^{ij}\rangle = \tfrac{8}{3}\pi N \left(\frac{c}{b} - 1\right).$$

The final result for Eq. (25) can be summarized by[31]

$$(\epsilon_\lambda - \epsilon_\lambda{}^\infty)\frac{\epsilon_\lambda + (\epsilon_\lambda{}^\infty - \epsilon_\lambda)\Omega_\lambda{}^\epsilon}{\epsilon_\lambda(\bar\epsilon^\infty + 2)^2} = \frac{4\pi N}{9kT}\langle\mu_\lambda{}^2\rangle\left(1 - \frac{4\pi N}{kT}\langle\mu_\lambda{}^2\rangle T_\lambda\right), \quad (28)$$

where T_λ is given by

$$\begin{array}{lll}
\textit{nematic phase:} & T_\parallel = 1/3 - \Omega_\parallel, & 0 < \Omega_\parallel < 1/3; \\[4pt]
& T_\perp = -(\Omega_\perp - 1/3), & 1/3 < \Omega_\perp < 1/2; \\[4pt]
\textit{smectic A phase:} & T_\parallel = \tfrac{2}{3}(c/b - 1), & c/b > 1; \\[4pt]
& T_\perp = -\tfrac{1}{3}(c/b - 1), & c/b > 1.
\end{array} \qquad (29)$$

Note that for $c = b$ the second term on the right-hand side of Eq. (28) vanishes.

From Eq. (29) we see that $T_\parallel$ is always positive. Hence the dipole–dipole interaction does tend to reduce the dipole contribution to $\epsilon_\parallel$, as is indeed observed experimentally. It is easily seen that this tendency is much stronger in the smectic phase than in the nematic phase. For HEPTAB $c/b \approx 5$ and $\Omega_\parallel$ is already very small. In the nematic phase, therefore, we have $T_\parallel \approx 1/3$ (its limiting value), and in the smectic phase $T_\parallel = 8/3$. In Eq. (27) we retained only first-order terms in V_{ij}. By doing so we calculated the tendency of $n - 1$ polar molecules to align themselves, independently of each other, parallel or antiparallel to one chosen molecule. Thus the dipole–dipole interactions between neighboring molecules that counteract such a situation to some extent are ignored. For that reason the numerical implications of Eq. (28) have to be considered with caution. (Note that $4\pi N \langle \mu_\lambda^2 \rangle / kT$ is already of the order of $\epsilon_\lambda - \epsilon_\lambda^\infty$.) The important feature is, however, that the effect of the dipole correlation is an order of magnitude larger in the smectic phase than in the nematic phase.

From Eq. (29) we see that the effect of dipole correlation perpendicular to the z-axis leads to an increase of $\epsilon_\perp$ (parallel dipole correlation). One can deduce easily that $\bar{\epsilon}$ is not affected by the dipole correlation if $\mu_\parallel^2 = \mu_\perp^2$. For HEPTAB (Fig. 7) this is approximately true. It would be interesting to investigate the effect on $\bar{\epsilon}$ for a compound in which $\mu_\parallel$ and $\mu_\perp$ differ widely.

The temperature dependence of the dipole correlation depends on the nature of the NS_A transition. In the case of HEPTAB this transition is almost second-order (the heat of transition is only 160 J/mole).[35] Consequently a marked pretransitional effect is already found in the nematic phase.

The increase of $\epsilon_\perp$ originates from the parallel correlation of the transverse components of the dipole moments. One might wonder whether a similar long-range correlation is possible in smectics. In order to get such an effect the dipole correlation should be accompanied by a freezing-out of the rotation around the molecular ζ-axis. Then the phase would be optically biaxial and therefore would probably be named smectic C. In contrast to the normally accepted smectic C phase, the molecules are not necessarily tilted with respect to the layer normal. McMillan has investigated this matter theoretically.[37] In the case of a central dipole perpendicular to the long molecular axis a second-order phase transition is predicted from smectic A to a phase called C_1. In the smectic layers this phase is a two-dimensional ferroelectric. The three-dimensional structure may be ferroelectric or antiferroelectric, depend-

[37] W. L. McMillan, *Phys. Rev. A* **8**, 1921 (1973).

ing on the total interplanar interaction. The smectic A–smectic C_1 transition temperature has to be considered as a Curie point. At this point $\epsilon_\perp$ can be expected to have a singular behavior, viz.[37]

$$\epsilon_\perp = \epsilon_\perp{}^\infty + N'\langle\mu^2\rangle/k(T - T^*),$$

where T^* is the temperature of the phase transition. The temperature dependence of $\epsilon_\perp$ in the smectic A phase of HEPTAB (Fig. 7) could be interpreted as pretransitional behavior related to such a phase transition. However, it should be emphasized that this type of phase has not yet been observed. It has been argued that for symmetry reasons a ferroelectric liquid-crystal phase must be a chiral liquid-crystal phase.[37a,b] Recently, a ferroelectric chiral smectic C phase has been reported.[37b] In that case, however, the spontaneous polarization appears as a result of a slightly nonuniform molecular rotation, and is not driving the transition. Accordingly, close to T^* only a weak divergence of $\epsilon_\perp$ is observed.[37c]

We have discussed in some detail correlation due to smectic ordering. Another type of correlation has already been noted in connection with Fig. 4d, where association of the molecules leads to an antiparallel correlation of the −CN dipoles. Finally we mention that recently in p-alkyl and p-alkoxy benzoic acids interesting correlation effects have been observed due to a change in the monomer–dimer equilibrium at the NI transition.[37d] Of course this has a considerable influence on the observed permittivities, which can in principle be discussed quantitatively within the framework of the Kirkwood–Fröhlich theory.

V. Dynamic Behavior of the Permittivity

So far we have been concerned with the permittivity in static fields. After the field is removed the orientation polarization decays exponentially with a characteristic time τ, the relaxation time. The process of (re)orientation of the permanent dipole moments connected with changes in the field requires a definite time interval. In alternating fields this leads to a time lag between the average orientation of the dipole moments and the field. This effect becomes noticeable for frequencies of

[37a] A. G. Khachaturyan, *Phys. Lett. A.* **51**, 103 (1975).
[37b] R. B. Meyer, L. Liébert, L. Strzelecki, and P. Keller, *J. Phys. Lett.* **36**, L-bg (1975).
[37c] G. Durand, private communication.
[37d] V. Breternitz and H. Kresse, *Phys. Lett. A.* **54**, 148 (1975); H. Kresse, K. H. Lücke, and H. J. Deutcher, *Z. Chem.* **16**, 55 (1976).

the order of τ^{-1}. At much higher frequencies the orientation polarization cannot follow the variations of the field any longer.[38] The residual permittivity ϵ^∞ is due to the induced polarization only. For many organic liquids the dispersion region where the relaxation of the permittivity takes place is found in the GHz range.

In nematic liquid crystals the relaxation time associated with $\epsilon_\perp$ is usually of the same order of magnitude as the relaxation time in the isotropic phase. However, in their studies of the p,p'-di-n-alkoxyazoxy-benzenes, Maier and Meier[39] observed a relaxation for $\epsilon_\parallel$ at much lower frequencies (MHz region). More recently the relaxation of $\epsilon_\parallel$ was found in other compounds at frequencies as low as 10 kHz.[36,40] In the next section we shall show that relatively low relaxation frequencies (large relaxation times) can be expected for $\epsilon_\parallel$ from Debye's classical theory of dielectric relaxation[41] if the nematic potential is included. In Section V,9 we shall discuss the experimental results in more detail and compare them with the theory.

8. Extension of Debye's Theory of Dielectric Relaxation to Nematics

For isotropic liquids the complex generalized permittivity $\epsilon^* = \epsilon' - j\epsilon''$ is given in Debye's theory by[38]

$$\epsilon^* = \epsilon^\infty + (\epsilon - \epsilon^\infty)/(1 + j\omega\tau), \tag{30}$$

where ω is the circular frequency of the alternating field and ϵ still the static value of the permittivity. When $\omega\tau$ is eliminated from the real and imaginary parts of Eq. (30) the equation of a circle is obtained. Consequently a plot of ϵ'' against ϵ' should give a semicircle if the relaxation can be described by Eq. (30).

Equation (30) was originally derived by Debye assuming that the rotation of a molecule due to an applied field is counteracted by rotational diffusion.[38,41] The potential energy of a molecule is $U = -\mu E \cos \theta$, where θ is the angle between the dipole moment and the direction of the field. The directing torque Γ is given by

$$\Gamma(\theta) = -\partial U/\partial\theta = -\mu E \sin \theta. \tag{31}$$

Due to the action of the torque the dipoles rotate at an average velocity $\langle\dot\theta\rangle = \Gamma/\gamma$, where γ is a frictional constant. The rotation of the dipoles is

[38] See, for example; V. V. Daniel, "Dielectric Relaxation." Academic Press, New York, 1967.

[39] W. Maier and G. Meier, Z. Naturforsch., Teil A 16, 1200 (1961).

[40] W. H. de Jeu and T. W. Lathouwers, Mol. Cryst. Liq. Cryst. 26, 225 (1974).

[41] P. Debye, "Polar Molecules," 1929, Chapter V (reprinted in Dover, New York).

counteracted by diffusion. Let $f(\theta,t)d\Omega$ be the fraction of molecules with their dipole moment in the direction $d\Omega = 2\pi \sin \theta \, d\theta$. Taking the diffusive process into account then Debye's equation for $f(\theta,t)$ is obtained:[41]

$$\dot{f}(\theta, t) = \frac{D}{\sin \theta} \frac{\partial}{\partial \theta} \left[\sin \theta \left\{ \frac{\partial f(\theta, t)}{\partial \theta} - \frac{\Gamma(\theta)}{kT} f(\theta, t) \right\} \right],$$

where $D = kT/\gamma$ is the rotational diffusion constant. When this equation is solved for an alternating field the orientation polarization can be calculated, and one arrives at Eq. (30). The macroscopic relaxation time τ is related to an intrinsic relaxation time $\tau' = 1/2D$ in a way that depends on the particular model of the internal field.

Debye calculated the frictional resistance to rotation experienced by a molecule by treating the molecule as a sphere rotating in a continuous viscous medium. Using the viscosity of the liquid in bulk, according to Stokes' law we get $\gamma = 8\pi\eta a^3$, where a is the radius of the sphere. This leads to

$$\tau = 4\pi\eta a^3/kT. \tag{32}$$

This relation often gives approximately the order of magnitude of the relaxation time, usually overestimating it by a factor of between 5 and 10. In view of the crudeness of the assumption of molecules rotating in a continuous medium this is hardly surprising. Equation (32) predicts that τ will show the same temperature dependence as η/T, which is often true.

In order to extend Debye's theory to the case of nematic liquid crystals we consider a compound with a dipole moment along the long molecular axis. The electric field is applied along the preferred direction. In Eq. (31) the nematic potential W must then be added to U. Using Eq. (2), this leads to a total torque

$$\Gamma(\theta) = -\mu E \sin \theta - (3AS/V^2) \sin \theta \cos \theta. \tag{33}$$

The distribution function $f(\theta, t)$ can be written as $f(\theta, t) = w(\theta)F(\theta, t)$, where $F(\theta, t)$ describes the deviation from the nematic distribution function $w(\theta)$ due to the presence of the electric field. With these changes Debye's equation leads to a fairly complicated eigenvalue problem for $F(\theta, t)$, the solution of which is given in Martin et al.[42] From the result for $F(\theta, t)$ the orientation polarization and thus τ can be calculated as a function of q/kT, where $q = 3AS/2V^2$ is the strength of the nematic potential. The resulting relaxation time in the nematic phase

[42] A. J. Martin, G. Meier, and A. Saupe, *Symp. Faraday Soc.* **5**, 119 (1971).

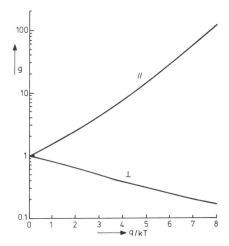

FIG. 8.   Retardation factors as a function of q/kT, where q is the nematic potential parameter (after Ref. 42). (Copyright by The Chemical Society. Used with permission.)

can be written as

$$\tau_{\parallel} = g_{\parallel}\tau_{\parallel}(0) \tag{34}$$

where $\tau_{\parallel}(0)$ is the relaxation time calculated with $S = 0$, and $g_{\parallel}$ is called the retardation factor. In Fig. 8 $g_{\parallel}$ is given as a function of q/kT. Using further results of Maier and Saupe's molecular–statistical theory[3] we have at the NI transition $S = 0.429$ and $A/(kT_{NI} V^2) = 4.541$. This leads to $q/kT = 2.92$ and, from Fig. 8, to $g_{\parallel} = 4$. In Table II we give theoretical results for various values of S as a function of the reduced temperature $TV^2/(T_{NI} V_{NI}^2)$. We see that $g_{\parallel}$ increases with decreasing temperature, as is to be expected with increasing S.

In an earlier discussion of dielectric relaxation in nematics the explicit solution of the differential equation for the distribution function was

TABLE II.   RETARDATION FACTORS FOR THE RELAXATION OF $\epsilon_{\parallel}$ AND $\epsilon_{\perp}$ IN THE NEMATIC PHASE CALCULATED FROM MAIER AND SAUPE'S POTENTIAL

$TV^2/(T_{NI} V_{NI}^2)$	S	q/kT	$g_{\parallel}$	$g_{\perp}$
1.000	0.429	2.92	4.0	0.52
0.985	0.479	3.31	5.0	0.48
0.970	0.516	3.62	6.0	0.43
0.941	0.569	4.12	8.2	0.39
0.912	0.610	4.55	11	0.37
0.884	0.643	4.95	15	0.35

circumvented by assuming $F(\theta, t) \approx F(\theta, 0) = 1 + a(E, t) \cos \theta$.[43] The result for $g_\parallel$ is

$$g_\parallel = (kT/q)\{\exp(q/kT) - 1\}. \tag{35}$$

Usually $\exp(q/kT) \gg 1$, and the last term can be neglected. The values of $g_\parallel$ calculated from Eq. (35) are approximately a factor of 2 higher than those from Fig. 8.

In the above extension of Debye's theory to nematic liquid crystals we have restricted ourselves to the dipole contribution to $\epsilon_\parallel$ from the component μ_l. Due to the imperfect orientational order there is also a contribution to $\epsilon_\parallel$ from the component μ_t. However, a reorientation of μ_t can be accomplished, to first order in $\mathbf{E}$, by a rotation around the molecular ζ-axis uninfluenced by the nematic potential. Consequently, in the nematic phase in general two relaxation regions will be observed for $\epsilon_\parallel$, of which only the one associated with μ_l is shifted to lower frequencies.

When we consider $\epsilon_\perp$, the main dipole contribution is usually due to the component μ_t; its relaxation region will again not be influenced by the nematic potential. However, now we have to consider also the contribution to $\epsilon_\perp$ from μ_l. In this case a reorientation of μ_l is accomplished by rotation through an angle π over the latitude $\theta =$ constant. The smaller the angle θ, the easier this is. Hence the nematic potential now leads to a decrease in relaxation time, or to an increase of the relaxation frequency ($g_\perp < 1$, see Fig. 8). Again using Maier and Saupe's molecular–statistical theory we calculate $g_\perp = 0.52$ at T_{NI}. The results for other temperatures are given in Table II. In general it will be difficult to observe this effect because of the overlap with other relaxations.

9. Experimental Results and Discussion

Experimentally, values of $g_\parallel$ and $g_\perp$ can be obtained in a straightforward way from the frequency dependence of $\epsilon_\parallel$ and $\epsilon_\perp$ in the nematic phase and of ϵ in the isotropic phase when a compound is used in which the total dipole moment is parallel to the long molecular axis ($\mu = \mu_l$, $\mu_t = 0$). In that case, simply, $\tau_\parallel(0) = \tau_{is}$, the relaxation time in the isotropic phase. The permittivities have been reported for a mixture of cholesteryl chloride and cholesteryl myristate,[44] of which the first compound approximately fulfils this requirement. The authors give a value $g_\parallel = 2$ at the NI transition, which agrees with the theory in order

[43] G. Meier and A. Saupe, *Mol. Cryst.* **1**, 515 (1966).
[44] H. Baessler, R. B. Beard, and M. M. Labes, *J. Chem. Phys.* **52**, 2292 (1970).

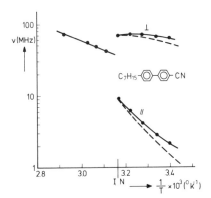

FIG. 9 Relaxation frequency $\nu = 1/2\pi\tau$ in the nematic and isotropic phases of p-n-heptyl-p'-cyanobiphenyl. Filled dots: experimental points;[45] broken line: calculated from the extrapolated isotropic relaxation frequency and the theoretical values of $g_{\parallel}$ and $g_{\perp}$ from Table II (assuming $\rho = 1.000$ g·cm^{-3} at T_{NI} and $d\rho/dt = 0.8 \times 10^{-3}$ g·cm^{-3}·K^{-1}). Copyright by The Chemical Society. Used by permission.

of magnitude. Better candidates for measurements of this type are p-alkyl-p'-cyano or -halogen substituted biphenyls or tolanes. As these compounds are linear there is no ambiguity about the direction of the long molecular axis. In Fig. 9 we give relaxation data of the permittivities of p-n-heptyl-p'-cyanobiphenyl.[45] From the jump in the relaxation frequency at T_{NI} we calculate $g_{\parallel} = 4.2$ at this temperature, in excellent agreement with the theoretical value from Table II. For the relaxation of $\epsilon_{\perp}$ we find, similarly, $g_{\perp} = 0.55$ at T_{NI}, indeed a value smaller than 1. The temperature dependence of the relaxation frequencies in the nematic phase is also well reproduced. On the other hand, the relaxation for $\epsilon_{\perp}$ is strongly distributed,[45a] which is not predicted by the theory.

One might be somewhat surprised that the extension of Debye's rotational diffusion equation to nematics works so well. Probably this equation is appropriate for compounds forming liquid crystals, because the molecular shape permits neither free rotation nor reorientation by instantaneous jumps. Theories incorporating these other models of rotational motion are given in Nordio *et al.* and Luckhurst and Zannoni.[45b] Furthermore, as the short-range order in the isotropic phase and that in the nematic phase do not differ much, all conceivable types of

45 M. Davies, R. Moutran, A. H. Price, M. S. Beevers, and G. Williams, *J. Chem. Soc., Faraday Trans. 2* **72**, 1447 (1976).

45a D. Lippens, J. P. Parneix, and A. Chapoton, *J. Phys.* **38**, 1465 (1977).

45b P. L. Nordio, G. Rigatti, and U. Segré, *Mol. Phys.* **25**, 129 (1973); G. R. Luckhurst and C. Zannoni, *Proc. Roy. Soc. Ser. A* **343**, 389 (1975).

corrections will in a first approximation be the same in both phases. The different relaxations in the two phases must then again be attributed to the nematic potential. Therefore one can expect that in spite of the limitations of the Debye model the calculation of the retardation factors is fairly reliable.

Data on dielectric relaxation in the nematic and the isotropic phase are also available for PAA.[46] In this case the situation is much more complicated due to the fact that various group moments contribute to the total dipole moment. Several of the relaxation regions that occur should be described by a distribution of relaxation times. Then in Eq. (30) $j\omega\tau$ has to be replaced by $(j\omega\tau)^{1-h}$, where h measures the width of the distribution around τ.[38] For two temperatures the data for PAA can be summarized as follows:[46]

$$125°C \qquad \tau_\| = 4.3 \times 10^{-9} \text{ s}, \qquad h_\| = 0$$

$$(= 0.957 T_{NI}) \qquad \tau_\|' = 2.5 \times 10^{-11} \text{ s}, \qquad h_\|' = 0.15$$

$$\tau_\perp = 3.3 \times 10^{-11} \text{ s}, \qquad h_\perp = 0.18$$

$$135.5°C \qquad \tau_{is} = 3.2 \times 10^{-11} \text{ s}, \qquad h_{is} = 0.23.$$

For $\epsilon_\|$ there are two relaxation regions. The relaxation indicated by $\tau_\|$ occurs at relatively low frequencies and is associated with the contribution to $\epsilon_\|$ from μ_l. As the longitudinal components of the end dipoles compensate, μ_l is due to the longitudinal component of the dipole moment of the azoxy group only. We can interpret $\tau_\|$ as a single relaxation time ($h_\| = 0$). We ascribe the dispersion region of $\epsilon_\|$ at higher frequencies (denoted by $\tau_\|'$) to the contribution to $\epsilon_\|$ from μ_t. Both the azoxy dipole moment and the methoxy dipoles contribute to μ_t. Because of the internal freedom of rotation of the methoxy groups these contributions may have a somewhat different relaxation time, leading to $h_\|' > 0$. Finally we note that only one dispersion region has been observed for $\epsilon_\perp$. The associated value of $h_\perp$ is relatively large, indicating several contributions to this relaxation.

The problem in determining $g_\|$ for PAA is to obtain a value for $\tau_\|(0)$. The common procedure has been to use τ_{is} as an approximation for $\tau_\|(0)$, which gives $g_\| = 135$ at 125°C. However, in fact τ_{is} is an average of $\tau_\|(0)$, $\tau_\|'(0)$, and $\tau_\perp(0)$. The latter two values are related to molecular reorientations that can be expected to be faster than that associated with $\tau_\|(0)$, as is also indicated by the relatively large value of h_{is}. Consequently it is not surprising that the above procedure gives a value of $g_\|$ that is an

[46] A. Axmann, Z. Naturforsch. Teil A 21, 615 (1966).

order of magnitude higher than predicted by the theory. From nuclear magnetic resonance measurements of the proton spin-lattice or longitudinal relaxation time T_1 molecular correlation times τ_c can be obtained that should be comparable with the dielectric relaxation times.[47] Using methyl deuterated PAA the contribution from the central part of the molecule can be separated from that due to the methoxy groups. After correction for the intermolecular contribution to T_1 one obtains $\tau_c = 1.25 \times 10^{-10}$ s at 136°C.[48] In a simple diffusion model $\tau_\parallel(0)$ should be equal to $3\tau_c$,[47] which leads to $g_\parallel = 11.5$ at 125°C.[49] This result indicates that for PAA also values of $g_\parallel$ of the order of magnitude predicted by the theory are obtained if the correct $\tau_\parallel(0)$ is used.

Further experimental results on the dynamic behavior of the permittivities of liquid crystals are relatively scarce. Apart from PAA and its higher homologs the frequency dependence of the permittivities in both the nematic and the isotropic phase has also been measured for MBBA.[50,51] Furthermore, Eq. (32) has been used in combination with Eq. (35) for $g_\parallel$ in order to determine approximately the value of q, the height of the potential barrier.[40,52] As the temperature dependence of the viscosity can be described by $\eta \sim \exp(q_{\text{visc}}/kT)$ this combination leads to

$$\tau_\parallel \sim (4\pi a^3/q) \exp\{(q + q_{\text{visc}})/kT\}. \tag{36}$$

The value of q can be determined by measuring the temperature dependence of $\tau_\parallel$ and η. In the temperature range studied (not close to T_{NI}), the linear relation between $\ln \tau_\parallel$ and $1/T$ predicted by Eq. (36) was well obeyed, leading to a value of q independent of the temperature. For different compounds little variation was found: $q \approx 30$–50 kJ/mole.[40] MBBA seemed to be an exception, but the various authors[50,51] give a rather different temperature dependence for $\tau_\parallel$. The values quoted for q are fairly large, as was to be expected from the use of the approximate equation (35). The low relaxation frequencies in the kHz region found in some cases [36,40] are due to a combination of a high viscosity [reducing $\tau_\parallel(0)$] and a low reduced temperature T/T_{NI} (increasing S and thus $g_\parallel$). The case in which $\Delta\epsilon$ is initially positive is especially interesting. The

[47] See, for example; J. G. Powles, *Chem. Soc., Spec. Publ.* **20**, 127 (1966).

[48] E. T. Samulski, C. R. Dybowski, and C. G. Wade, *Mol. Cryst. Liq. Cryst.* **22**, 309 (1973).

[49] P. L. Nordio, G. Rigatti, and U. Segré, *Chem. Phys. Lett.* **19**, 295 (1973).

[50] F. Rondelez, D. Diguet, and G. Durand, *Mol. Cryst. Liq. Cryst.* **15**, 183 (1971); for corrected values, see F. Rondelez and A. Mircéa-Roussel, *ibid* **28**, 173 (1974).

[51] V. K. Agarwal and A. H. Price, *J. Chem. Soc., Faraday Trans. 2* **70**, 188 (1974).

[52] M. Schadt, *J. Chem. Phys.* **56**, 1494 (1972).

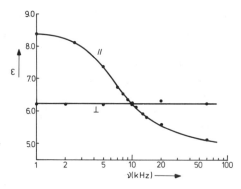

FIG. 10.   Change of sign of $\Delta\epsilon$ as a function of the frequency in the nematic phase of a mixture of p, p'-di-substituted phenylbenzoates (room temperature[53]).

relaxation of $\epsilon_{\parallel}$ can then reverse the sign of $\Delta\epsilon$ in the experimentally easily accessible kHz region (see Fig. 10). This offers the possibility of studying the behavior of the nematic in external electric fields as a function of the dielectric anisotropy.[53-56] More recently some more measurements of the relaxation of the permittivity of liquid crystals have become available.[57-63]

We conclude that the dispersion region of $\epsilon_{\parallel}$ at low frequencies in the nematic phase is due to (1) the nematic potential that leads to a retardation factor $g_{\parallel}$, (2) the fact that the relaxation considered is associated with the slowest molecular reorientation and is separated from other possible relaxations. Each factor may shift the relaxation frequency approximately an order of magnitude downward compared with the relaxation in the isotropic phase.

[53] W. H. de Jeu, C. J. Gerritsma, P. van Zanten, and W. J. A. Goossens, Phys. Lett. A **39** 355 (1972).

[54] W. H. de Jeu and T. W. Lathouwers, Mol. Cryst. Liq. Cryst. **26**, 235 (1974).

[55] G. Baur, A. Stieb, and G. Meier, in "Liquid Crystals and Ordered Fluids." (J. F. Johnson and R. S. Porter, eds), Vol. II, p. 645. Plenum, New York, 1974.

[56] H. K. Bücher, R. T. Klingbiel, and J. P. van Meter, Appl. Phys. Lett. **25**, 186 (1974).

[57] Y. Galerne, C. R. Hebd. Seances Acad. Sci., Ser. B **278**, 347 (1974).

[58] H. Kresse, D. Demus, and C. Krinzer, Z. Phys. Chem. (Leipzig) **256**, 7 (1975); H. Kresse, P. Schmidt, and D. Demus, Phys. Status Solidi A **32**, 315 (1975).

[59] L. Bata and G. Molnár, Chem. Phys. Lett. **33**, 535 (1975).

[60] A. Mircéa-Roussel and F. Rondelez, J. Chem. Phys. **63**, 2311 (1975).

[61] M. Schadt and C. von Planta, J. Chem. Phys. **63**, 4379 (1975).

[62] J. P. Parneix, A. Chapoton, and E. Constant, J. Phys. **36**, 1143 (1975).

[63] G. R. Luckhurst and R. N. Yeates, Chem. Phys. Lett. **38**, 551 (1976).

ACKNOWLEDGMENT

The author owes much to cooperation from his colleagues of the Philips Liquid Crystal group. He also wishes to thank Dr. P. Bordewijk (University of Leiden) for valuable discussions on dielectric theory, and Dr. W. van Haeringen and Dr. W. J. A. Goossens for a critical reading of the manuscript.

SOLID STATE PHYSICS, SUPPLEMENT 14

Instabilities in Nematic Liquid Crystals

E. Dubois-Violette, G. Durand, E. Guyon, P. Manneville,* and
P. Pieranski

Laboratoire de Physique des Solides, Université Paris-Sud, Orsay, France

I. Introduction

Nematic liquid crystals can flow easily (viscosities are in the range of 10^{-1} to 1 poise) and their hydrodynamic study can be done as an extension of that of ordinary isotropic fluids. However, the flow properties are complex: The friction coefficients are anisotropic; the orientation, defined by a unit vector along the optical axis, the director $\mathbf{n(r)}$, is coupled with the flow velocity field $\mathbf{v(r)}$. The equations of

---

* Centre d'Etudes Nucléaires de Saclay, Gif sur Yvette, France.

147

hydrodynamics of the coupled fields $v(r)$ and $n(r)$ have been written in detail by Ericksen[1] and Leslie,[2] and simplified by Parodi[3] (ELP). Another approach by the Harvard group[4] has been shown to be equivalent to the first one[5] which will be used in the following. In principle, the hydrodynamic behavior can be studied for various laminar flows. However, it is important to work with channels in which: (a) the alignment of the director is well defined—this can be done through the influence of strong fields (FA: field alignment) or by inducing the bulk alignment by suitable surface treatment (SA: surface alignment). (b) The geometry is simple enough. For example, channels of circular cross section may lead to complicated director fields and are not convenient. (c) The channel thickness is sufficiently small ($d \sim .01$ to 1 mm) to have a transparent sample and to be able to monitor the alignment optically. Many of the flow experiments in nematics up to now do not satisfy these requirements and the flow properties are controlled by the presence of defects. In the following we will present and discuss the framework of the ELP description, some hydrodynamic experiments on nematics in which the conditions a, b, and c are satisfied: laminar flows and convective flows near a threshold such that a linear hydrodynamic approach can be used. We will only indicate the more complex problems raised by nonlinear contributions (an extension of the $(v \cdot \nabla)v$ term in the Navier–Stokes equation for ordinary fluids) and by the approach of turbulence. Much remains to be done before a satisfactory picture can be developed.

In particular, statistical properties of the defects which are likely to be produced in nonlinear flows should be considered along with those of the turbulent flow.

II. Linear Hydrodynamics of Isotropic and Nematic Liquids

The hydrodynamic behavior of isotropic fluids[6] is characterized by the spatially dependent velocity field $v(x_i)$ ($i = 1, 2, 3$ labels the three spatial coordinates). The second-rank stress tensor σ_{ij} gives the jth component of the force per unit area exerted across a plane surface element normal

[1] Ericksen, J. L., *Arch. Ration. Mech. Anal.* **4**, 231 (1960).
[2] Leslie, F. M., *Q. J. Mech. Appl. Math.* **19**, 357 (1966).
[3] Parodi, O., *J. Phys. (Paris)* **31**, 581 (1970).
[4] Forster, D., Lubenski, T., Martin, P. S., Swift, J., and Pershan, P. S., *Phys. Rev. Lett.* **26**, 1016 (1971).
[5] de Gennes, P. G., "The Physics of Liquid Crystals." Oxford Univ. Press (Clarendon), London and New York, 1974.
[6] Landau, L. D., and Lifshitz, E. M., "Fluid Mechanics." Pergamon, Oxford, 1959.

to the i direction and is a linear function of the deformation rate:

$$2A_{ij} = (\partial v_i/\partial x_j + \partial v_j/\partial x_i); \tag{II.1}$$

$$\sigma_{ij} = \eta A_{ij}. \tag{II.2}$$

As σ_{ij} is a symmetric tensor, no torques develop in the bulk of such a fluid.

In a nematic characterized by the unit vector $\mathbf{n}(x_i)$, σ_{ij} is also a function of $\mathbf{n}$ and of its derivatives.[6a]

It is convenient to use the vector $\mathbf{N}$ which represents the relative angular velocity of the director with respect to the fluid:

$$\mathbf{N} = \mathring{\mathbf{n}} - \boldsymbol{\omega} \times \mathbf{n}, \tag{II.3}$$

where the vorticity $\boldsymbol{\omega}$ is defined by

$$2\omega_{ij} = (\partial v_i/\partial x_j - \partial v_j/\partial x_i) = \varepsilon_{ijk}\omega_k. \tag{II.4}$$

The dynamic behavior is characterized by the data of the Leslie–Ericksen stress tensor which is the most general linear expression in A, ω, and $\mathbf{n}$ consistent with the symmetry of the nematic order (invariance when $\mathbf{n} \to -\mathbf{n}$):

$$\sigma_{ij} = \alpha_1 n_i(n_k A_{kl} n_l)n_j + \alpha_2 n_i N_j + \alpha_3 n_j N_i \tag{II.5}$$

$$+ \alpha_4 A_{ij} + \alpha_5 n_i n_k A_{kj} + \alpha_6 n_j n_k A_{ki}.$$

The stress tensor is asymmetric and the viscous torque Γ_v exerted on the molecules is

$$\Gamma_{vi} = \varepsilon_{ijk}\sigma_{jk}. \tag{II.6}$$

The meaning of the different terms in σ is given in Fig. 1. The α_4 term corresponds to the usual viscosity of an isotropic fluid. The α_1 term is symmetric and expresses the stretching effect of an irrotational flow (such as a stagnation point flow). The four other asymmetric terms express the torques exerted on the molecules by a rotational (α_2, α_3) or irrotational (α_5, α_6) flow. The torque (II.6) can be written as

$$\Gamma_v = -\mathbf{n} \times (\gamma_1 \mathbf{N} + \gamma_2 \mathbf{n}A),^{7a} \tag{II.7}$$

[6a] The coupling of the variables $\mathbf{n}$ and $\mathbf{v}$ is clearly demonstrated in the Freedericksz experiment.[7] When a large enough field is applied at right angles to a well aligned nematic, the time rate of deformation is controlled by a geometry-dependent viscosity. This viscosity is modified by backflow effects associated with the rate of change of $\mathbf{n}$, $\mathring{\mathbf{n}}$. However, in the geometry where the field induces a twist in the plane of the layers no backflow term appears and the dynamic behavior depends only on $\mathbf{n}(x_i, t)$. Thus, one cannot reduce the hydrodynamic description to the data of $\mathbf{v}(x_i, t)$ alone.

[7] Deuling, H. J., this volume, Chapter 3.

[7a] Where $(\mathbf{n}A)_i = n_j A_{ji}$.

where $(\mathbf{n}A)_i = n_j A_{ji}$ and

$$\gamma_1 = \alpha_3 - \alpha_2 ; \qquad\qquad (\text{II.8})$$

$$\gamma_2 = \alpha_6 - \alpha_5 = \alpha_2 + \alpha_3 . \qquad\qquad (\text{II.9})$$

The last equality was obtained by Parodi[3] by applying the Onsager relations. It reduces the number of parameters characterizing a nematic flow to five. The meaning of the different terms can be more easily understood by considering a typical shear flow experiment in which the velocity v_y is a linear function of z:

$$\partial_z v_y = \tfrac{1}{2}\partial_z v_y + \tfrac{1}{2}\partial_z v_y = \tfrac{1}{2}(A + \boldsymbol{\omega})_{zy} .$$

One deduces immediately, from consideration of Fig. 1, the values of the three main viscosities measured with an homogeneous alignment in space (FA) (Fig. 2):

$$\eta_a = \alpha_4/2, \qquad \mathbf{n} \text{ along } x; \qquad\qquad (\text{II.10a})$$

$$\eta_b = (\alpha_4 + \alpha_3 + \alpha_6)/2, \qquad \mathbf{n} \text{ along the flow axis } (y); \qquad (\text{II.10b})$$

$$\eta_c = (\alpha_4 + \alpha_5 - \alpha_2)/2, \qquad \mathbf{n} \text{ along the velocity gradient } (z). \quad (\text{II.10c})$$

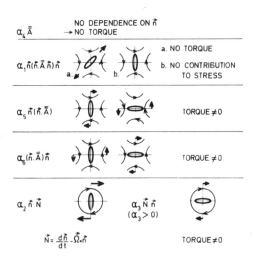

FIG. 1. Meaning of the different α terms in the viscous stress tensor σ. α_4 corresponds to the usual viscosity; α_1 corresponds to a stretching effect; α_5, α_6 correspond to torques exerted on the molecules by an irrotational flow; α_2, α_3 express the torque induced by a rotational flow.

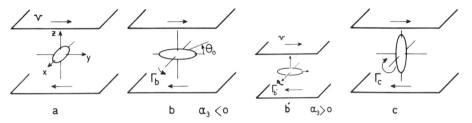

FIG. 2.   Different geometries used in shear flow experiments: (b) Sign of the viscous torque Γ_b induced on the molecules for $\alpha_3 < 0$; (b') Sign of the viscous torque $\Gamma_{b'}$ for samples where $\alpha_3 > 0$.

In geometries b and c, torques along x are exerted on the molecules:

$$\Gamma_c = +\alpha_2 s \quad\quad\quad\quad\quad\quad\quad\quad\quad\quad\quad\quad\text{(II.11)}$$

$$\left.\begin{array}{c} \\ \\ \end{array}\right\} \text{with } s = \partial_z v_y .$$

$$\Gamma_b = -\alpha_3 s \quad\quad\quad\quad\quad\quad\quad\quad\quad\quad\quad\quad\text{(II.12)}$$

Results obtained by Gähwiller[8] on one nematic, HBAB, together with our values[9] on the same compound are given in Fig. 3.

a. For $T_c - T < 10°$ the torque coefficients α_2 and α_3 are negative. If a large enough shear is applied on molecules initially oriented parallel to y or z under the effect of surface alignment (SA), the director aligns within the bulk in the plane yz at an angle θ_0 with y given by (see Fig. 2b)

$$\tan^2\theta_0 = \alpha_3/\alpha_2 .$$

Wahl and Fischer[10] have carried out a careful study of flow alignment by shearing a homeotropic nematic contained between two superimposed disks in relative rotation. The shear rate increases linearly with the distance to the center. From a study in monochromatic light they deduce the ratio of the elastic to the viscous constant as well as the saturation angle θ_0 obtained for large shears. In MBBA, α_2 and α_3 are negative and $\theta_0 \sim 10°$. This result is consistent with direct determinations of α_2 and α_3.

b. For $T_c - T > 10°$ (see Fig. 3), the small torque Γ_b is positive and $\alpha_2\alpha_3 < 0$.

The torque contributions due to α_2 and α_3 act in the same direction and no equilibrium angle is obtained (see Fig. 2b').

[8] Gähwiller, C., Mol. Cryst. Liq. Cryst. 20, 301 (1973).
[9] Pieranski, P., and Guyon, E., Phys. Rev. Lett. 32, 924 (1974).
[10] Wahl, J., and Fischer, F., Mol. Cryst. Liq. Cryst. 22, 359 (1973).

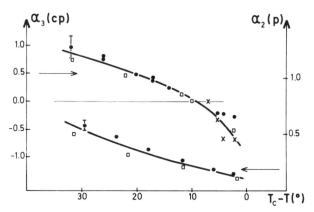

FIG. 3. Plot of the two viscosities α_2 and α_3 in nematic HBAB as a function of the temperature T ($T_c \sim 102°$ is the transition to the isotropic state). The left scale (and upper curve) corresponds to the viscosity α_3, the right one (and lower curve) to α_2.

This condition prevails in nematics near a second-order phase transition to a low-temperature smectic phase[11] (and can be understood as being related to the formation of elongated cybotactic groups perpendicular to the shear velocity in the planar configuration and experiencing a large torque of sign opposite to that of Fig. 2b). We will consider in detail the important changes in the shear flow instabilities which depend on this change of sign.

1. INTERMEDIATE GEOMETRIES

If the director, fixed in space (FA), is at an angle θ with the horizontal y in the plane of the flow, the shear viscosity is

$$\eta = \eta_b \cos^2\theta + \eta_c \sin^2\theta + \alpha_1 \sin\theta \cos\theta,$$

which reduces to the expressions (II.10a), (II.10b) in the symmetric configurations. Similarly, if the alignment is at an angle φ with the flow direction in the horizontal plane,

$$\eta = \eta_b \cos^2\varphi + \eta_a \sin^2\varphi.$$

In these configurations, the stress tensor is anisotropic but is uniform in space and no forces are present in the samples, as the body forces are given by

$$F_i = \partial\sigma_{ji}/\partial x_j. \tag{II.16}$$

A more interesting case is obtained in a Poiseuille flow between two fixed plates where the velocity varies parabolically across the sample

[11] Brochard, F., Thesis, Orsay (1974).

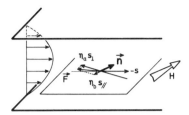

FIG. 4.   By considering the effects of the two components of the shear gradient parallel ($\parallel$) and perpendicular ($\perp$) to the director **n** one obtains the longitudinal and transverse components of the viscous force; the resulting force is not parallel to the shear **s**.

(Fig. 4):

$$v_y = v_m(d^2/4 - z^2); \qquad (\text{II}.17)$$

$$\partial v_y / \partial z = -2z v_m = s' \cdot z. \qquad (\text{II}.18)$$

If the intermediate alignment is uniform in space (FA) bulk forces, transverse to the flow, develop. If the director is in the horizontal plane, using Eqs. II.5 and II.16, we obtain transverse forces given by

$$F_x = \partial \sigma_{zx} / \partial z = |(\eta_b - \eta_a)/2|(\partial / \partial z)(\partial v_y / \partial z \sin 2\phi). \qquad (\text{II}.19)$$

Evidence of transverse pressure forces and of a resulting transverse flow in quantitative agreement with these results has been provided by Poiseuille flow experiments on MBBA.[12] (Note that, in isotropic fluids, bulk forces are parallel to the flow.)

Note 1. Bulk forces also occur in uniform shears if the nematic alignment varies in space (SA) across the sample thickness. They give rise to destabilizing mechanisms in the study of hydrodynamic instabilities involving a periodic distortion of **n** (see Section V,2,*a,i*).

Note 2. We have discussed the viscosity problem using steady-state experiments. In dynamic measurements such as those involving the attenuation of shear waves in nematics or in the dynamics of a Freedericksz transition,[7,13] viscosities must be renormalized to account for the backflow effect[11] [role of the **ṅ** term in Eq. (II.3)].

III. Bénard–Rayleigh Thermal Convection in Isotropic Fluids

This classical example contains many of the features discussed later in nematic convection, and we review it briefly. Convective effects are observed when an infinite horizontal layer of an isotropic fluid contained

[12] Pieranski, P., and Guyon, E., *Phys. Lett. A* **29**, 237 (1974).
[13] Brochard, F., Guyon, E., and Pieranski, P., *J. Phys. (Paris)* **34**, 35 (1973).

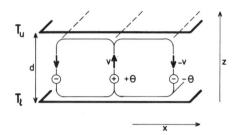

FIG. 5. Bénard–Rayleigh instability. A temperature difference $T_1 - T_u > 0$ is maintained between the lower and upper plates of the sample. The instability sets up when the upward motion of a locally warmer ($+$) fluid under the buoyancy force compensates the viscous force.

between two solid conducting boundaries ($z = \pm d/2$) and heated from below is submitted to a large enough temperature gradient $-\partial_z T = \beta = (T_1 - T_u)/d > 0$, where l and u refer to the lower and upper plates (Fig. 5). The velocity of the convective rolls, $\mathbf{v}$, goes continuously to zero as the temperature gradient β decreases to the critical value β_c and $|\mathbf{v}| \propto (\beta - \beta_c)^{1/2}$.[14] It is possible to characterize the value β_c using the linear hydrodynamic description, valid for small $\mathbf{v}$, where nonlinear effects are neglected.

In addition it is observed that the wavelength of the convective rolls is very nearly equal to $2d$. An exact two-dimensional analysis leads to $\lambda_c = 2.015d$ and $q_c = 2\pi/\lambda \sim \pi/d$. In a *one-dimensional* description of the convection, the variation of the fluctuating parameters across the layer thickness is neglected and the thickness d is included in the model by considering only spatially dependent properties along one direction x with a wavevector q_x. One takes into account the boundary conditions by expressing q_x as a function of d. At threshold β_c the production of internal energy, due to the upward motion of a locally warmer fluid under the buoyancy force, compensates exactly the viscous losses (in an inviscid fluid the threshold β_c is zero). The linear one-dimensional problem is governed by two coupled equations expressing the conservation of energy and linear momentum. θ represents a local fluctuation of temperature with respect to the average temperature T_0, $T = T_0 + \theta$, $\theta \ll T$:

$$\theta(x, t)/\theta_0(t) = v(x, t)/v_0(t) = \exp(iq_x x); \qquad \text{(III.1)}$$

$$\dot{\theta}_0 + \theta_0/T_\theta - \beta v_0 = 0; \qquad \text{(III.2)}$$

$$\dot{v}_0 + v_0/T_v - \alpha g \theta_0 = 0. \qquad \text{(III.3)}$$

[14] Berge, P., and Dubois, M., *Phys. Rev. Lett.* **32**, 1041 (1974).

The first equation (III.2) expresses the relaxation of a thermal fluctuation (Fourier equation). The thermal time constant is given by

$$T_\theta^{-1} \sim \kappa q_x^2, \tag{III.4}$$

where the thermal diffusivity κ is related to the heat conductivity k and the specific heat C by

$$\kappa = k/\rho C. \tag{III.5}$$

βv_0 gives a destabilizing contribution of a hot spot moving upwards in a slightly colder surrounding environment.

The second equation (III.3) comes from the linearized Navier–Stokes equation and expresses the time evolution of a fluctuation of the vertical velocity v_z. In a linear theory one neglects viscous dissipation, which is quadratic in fluctuations. The time constant for the diffusion of velocity is given by

$$T_v^{-1} \sim \nu q_x^2 \tag{III.6}$$

with the kinematic viscosity

$$\nu = \eta/\rho, \tag{III.7}$$

where η is the viscosity. The coupling term $\alpha g \theta_0$, where $\alpha > 0$ is the expansion coefficient, is a destabilizing buoyancy force.

We state that, just above threshold, v_0 and θ_0 are proportional to $\exp(st)$ with the real quantity $s = 0^+$ (principle of exchange of stabilities):[15,15a]

$$\theta_0(t) \quad \text{and} \quad v_0(t) \propto \exp(st). \tag{III.8}$$

The compatibility condition for the coupled equations (III.2), (III.3) can be expressed as

$$\frac{T_v T_\theta}{1/\alpha g \beta_c} = \frac{T_v T_\theta}{\tau^2} = 1 \tag{III.9}$$

or

$$\text{Ra}_c = \frac{\alpha g d^4 \beta_c}{\kappa \nu} = \pi^4. \tag{III.10}$$

The exact two-dimensional analysis, including the effect of viscosity and heat transport across z, gives a critical Rayleigh number, $\text{Ra}_c = 1704$

[15] Chandrasekhar, S., "Hydrodynamic and Hydromagnetic Stability." Oxford Univ. Press (Clarendon), London and New York, 1961.
[15a] Velarde, G. M., Lecture Notes, Les Houches. Gordon & Breach, New York, 1973.

(rigid and conducting boundaries) or 120 (stress-free insulating boundaries, which is a condition closer to our simplified model) instead of π^4.

Loosely speaking, the equation (III.9) expresses that convection takes place when the diffusion time constants are long enough compared to the time τ associated with the destabilizing motion of a local hot spot from bottom to top. This analysis also implies that low convection thresholds are associated with slow diffusive processes.

Equation (III.10) gives the classical dependence of the temperature difference $\Delta T_c (=\beta_c d)$ as d^{-3}. For a 1 cm thick liquid layer having properties comparable to the average properties of the nematics used in the following ($\kappa \sim 10^{-3}$ cgs, $\alpha \sim 10^{-3}$, $\nu \sim 1$ cgs):

$$T_v = 10^{-1} \text{ sec}; \quad T_\theta = 10^2 \text{ sec}; \quad \text{and} \quad \Delta T \sim 10°.$$

In the instabilities of nematics to be discussed next we will find many features found in the Bénard problem:

Constructive interplay of fluctuating variables (with equations similar to III.2 and III.3).

Linear onset expressed by an equality between stabilizing and destabilizing terms (time constants or energy).

Rolls with a diameter roughly equal to the thickness [a naïve explanation is the following: Narrower rolls would imply larger dissipative viscous forces, whereas flat ones would mean an inefficient transport, considering the large viscous effects due to $v_x(z)$].

Note. Near the second-order phase transition taking place at β_c, the time constant s^{-1} diverges as $(\beta - \beta_c)^{-1}$. The behavior above threshold (determination of the amplitude of the θ and v fluctuations) implies a much more complex nonlinear description and is not considered here.

IV. Thermal Convection in Nematics

2. PLANAR CASE[16,17]

We first consider the case of a *planar* film: The alignment is parallel to the horizontal limiting plates (along x) and is induced in the bulk by a suitable surface treatment (SA). When a vertical negative temperature gradient is applied ($\beta = -\partial_z T > 0$), the following features presented in the photograph in Fig. 6 are observed.

a. Convection appears above a threshold 10^2 to 10^3 smaller than that for an isotropic fluid of comparable average (labeled with overbar)

[16] Dubois-Violette, E., *C.R. Hebd. Seances Acad. Sci.* **21**, 923 (1971).
[17] Dubois-Violette, E., Guyon, E., and Pieranski, P., *Mol. Cryst. Liq. Cryst.* **26**, 193 (1974).

FIG. 6.   The lower temperature is near T_c and the temperature gradient has a ∗small horizontal component along Oy. The wavy shape of the limit between regions with (left part) and without (right part) a nematic isotropic interface allows an estimate of the horizontal modulation of the temperature in the presence of convection.

properties. In particular, if one keeps the cold plate at a fixed temperature T_u and increases the lower one T_1 to a value which exceeds the critical temperature to the isotropic phase, T_{NI}, the convection disappears (although $\Delta T = T_1 - T_u$ has increased and both $\bar{\kappa}$ and $\bar{\nu}$ have decreased).

b. The rolls are periodic with the axis being perpendicular to x. In an isotropic fluid the direction of the rolls is controlled by the boundaries of the container or by residual (geometric and thermal) fluctuations. The wavelength of the roll pattern is twice the thickness, as in the isotropic case.

c. An ambiguity by a factor of 2 in the wavelength is resolved by the following experiment: The temperature T_1 is raised to a value very close to T_{NI} and a small horizontal temperature gradient $|\beta'| \ll |\beta|$ is applied. The isotherm $T = T_c$ is the wavy line in Fig. 6. The amplitude of this profile with respect to the average straight line obtained if convection had not taken place can be related to the temperature fluctuation in a horizontal direction across the rolls.

d. The rolls are observable only in extraordinary light (polarized along x). The sharp contrast comes from the large variation of the effective index of the extraordinary light.

Points b and d indicate that the convection involves an orientation effect: $\mathbf{n}$ is lifted off the xy-plane: $\mathbf{n} = \mathbf{n}_0 + \mathbf{n}_z$ (see Fig. 7). Close to the

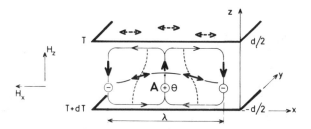

FIG. 7.   Thermal instability in a nematic sample. Due to the anisotropy of the thermal conductivity a fluctuation of the nematic orientation (↔)induces a deviation of the heat fluxes(- - -) and creates a temperature fluctuation [(+) and (−) correspond to warmer and cooler regions]. The gravity forces give rise to opposite hydrodynamic velocities in regions (+) and (−). The resulting shear induces a viscous torque (shown in Fig. 2c) which increases the initial fluctuation (destabilizing effect).

threshold ΔT_c the optical contrast disappears progressively. This indicates a linear instability with the distortion n_z going progressively to zero. We will show how n_z is coupled with the temperature fluctuation $\theta(x)$ and with the vertical convective velocity $v_z(x)[v_z(x)$ and $\theta(x)$ defined as in the isotropic case]. We look for a one-dimensional solution expressed in Fourier space and retain only the component $q_x = \pi/d$ since the three parameters have the same spatial periodicity. Let φ be the angle between the director and the x-axis. Then,

$$e^{iq_x x}(\dot{v}_0 + v_0/T_v - \alpha g\theta_0) - (\alpha_2/\rho)(\partial/\partial t)[\partial_x\varphi(x)] = 0; \qquad \text{(IV.1)}$$

$$e^{iq_x x}(\dot{\theta}_0 + \theta_0/T_\theta - \beta v_0) + \kappa_a\beta\partial_x\varphi(x) = 0; \qquad \text{(IV.2)}$$

$$e^{iq_x x}(\dot{\varphi}_0 + \varphi_0/T_{0z}) + (\alpha_2/\gamma_1)\partial_x v_z(x) = 0. \qquad \text{(IV.3)}$$

The parameter $n_z(x)$ enters through a curvature term (bend distortion):

$$\partial_x n_z = iq_x\varphi_0\,\exp(iq_x x).$$

[In this intermediate notation, subscript 0 refers to the amplitude of the Fourier component q_x as in Eq. (III.1).]

We present here a simplified solution of the coupled equations (IV.1) to (IV.3). A complete theoretical description[18] involving the precise form of the viscous constants, to be discussed next, agrees with the simplified approach. Equation (IV.1) is the analog of the flow equation (III.3). The additional term expresses the "backflow effect"[19] which is introduced in the Leslie hydrodynamic description. Due to the time dependence of the curvature, the viscosity is renormalized (see the article by Deuling in

[18] Dubois-Violette, E., *Solid State Commun.* **14**, 767 (1974).
[19] Brochard, F., *Mol. Cryst. Liq. Cryst.* **23**, 51 (1973).

this volume).[7] The heat equation (IV.2) introduces the effect of the heat diffusivity anisotropy of nematics, $\kappa_a = \kappa_\parallel - \kappa_\perp$ ($\parallel$ and $\perp$ refer to the orientation of the heat flux with respect to **n**). In all the nematics we have studied, κ_a is positive. The term with κ_a expresses a "heat focusing" effect. The upward heat flux lines (dotted lines in Fig. 7) are deflected toward regions in which the curvature is downward.

In the new equation (IV.3) the first two terms introduce the notion of an orientation time constant

$$T_{oz} = (\gamma_1 / K q_x{}^2), \tag{IV.4}$$

where K is the bend elastic constant and γ_1 is the orientational viscosity defined in Eq. (II.7). In the absence of coupling, T_{oz} describes the spontaneous relaxation of a fluctuation of orientation, n_z. The destabilizing torque $\alpha_2 \partial_x v_z$ [see Eq. (II.11)] is a shear torque induced by the convection currents.

Qualitatively, one obtains the following destabilizing sequence: In the presence of a fluctuation of curvature (point A), an inhomogeneous horizontal temperature distribution is produced due to the anisotropy of κ. The buoyancy effects cause vertical (up and down) currents to appear. The resulting shear tends to further increase the initial fluctuation of **n**.

The solutions of the equations (IV.1)–(IV.3) can be obtained easily if we neglect $\dot{v}$ in front of v/T_v in Eq. (IV.1) (the time constant for the relaxation of vorticity is the shortest one and we assume that v instantaneously follows the fluctuation of θ and φ):

$$\nu q_x{}^2 v_0 = \alpha g \theta_0 + (\alpha_2 / \rho)\psi, \tag{IV.5}$$

where $\dot{\varphi}_x = \partial_x \dot{\varphi}$ and ν is an effective kinematic viscosity for the nematic flow.

We eliminate the velocity between Eqs. (IV.2) and (IV.3). The resulting equations give the coupling between φ_x and θ. At the threshold, where $\dot{\varphi} = \dot{\theta} = 0$,

$$[-\beta T_v g \alpha + T_\theta^{-1}]\theta_0 + \kappa_a \beta \psi = 0; \tag{IV.6}$$

$$(-\alpha_2/\gamma_1)\rho(g\alpha\theta_0/\eta_2) + \psi/T_{oz} = 0. \tag{IV.7}$$

The two equations indicate clearly the coupling role (heat focusing) between the anisotropic heat conducting term κ_a and the torque given by the α_2 term.

The homogeneous equations (IV.6) and (IV.7) have a nontrivial solution for

$$(T_v T_\theta / \tau^2)[1 + T_{oz}/T_a] = 1, \tag{IV.8}$$

with the anisotropic time constant

$$T_a^{-1} = \kappa_a q_x^2 \gamma_1 / (-\alpha_2).$$ (IV.9)

In an isotropic liquid ($\kappa_a = 0$) we recover the Bénard threshold.

In a 1 mm thick nematic film the following values are obtained:

$$T_{oz} \sim 10^3 \text{ sec} \gg T_\theta \sim T_a \sim 1 \text{ sec} > T_v \sim 10^{-2} \text{ sec}.$$

The large value of the ratio T_{oz}/T_a leads to a strong decrease of the threshold as observed experimentally.

In general, we will find that the instability threshold is much lower than in isotropic liquids due to the long relaxation time of the director, T_{oz}, since slowly relaxing fluctuations also mean reduced thresholds [see Eq. (III.9)].

If we look for time-dependent solutions which grow in time as exp st far from saturation, we obtain

$$s \sim A T_{oz}^{-1},$$

with a numerical factor A going linearly to zero at threshold. The growth rate is small and its value is controlled by the slow development of the distortion.

The value of T_{oz} and, consequently, the threshold can be monitored by application of a destabilizing ($-$ in the following equation) vertical field or stabilizing ($+$) horizontal magnetic field H along $\mathbf{n}$:[10]

$$T_{oz}(H) = T_{oz}(H = 0)[1 \pm (H/H_c)^2]^{-1},$$ (IV.10)

where H_c is the Freedericksz critical field:

$$H_c = (\pi/d)(K_1/\chi_a)^{1/2}.$$

The effect of the anisotropic diamagnetic susceptibility $\chi_a > 0$ tends to align the molecules along H. If H is along z, we must restrict it to values below H_c, since spontaneous distortion in the field takes place for $H > H_c$. The effect of the field has been checked quantitatively using a wedge geometry.[17] As the field was changed, the temperature gradient also was adjusted to maintain the boundary between the thick part of the cell, where convection was observed, and the thinner part at the same position. The boundary is sharp due to the rapid change of threshold with d.

The dynamic behavior can be studied easily, as it implies long time constants. First, a strong stabilizing field is applied for a standard time, once a given temperature difference has been obtained. At time $t = 0$ the field is suppressed and the time t needed to establish a standard distortion is measured. The experiment was done by taking a slow speed movie (1 image per second) of the convection cell and studying the dynamics by projection at ordinary speed. The time constant has a divergence near threshold in $(\beta - \beta_c)^{-1}$, for $1.1 < \beta/\beta_c < 2.2$. For $d = 1$ mm and $\beta = 2\beta_c$ the time constant is of the order of 1 hr. Such data are only indicative and should not be used to characterize the critical behavior close to threshold. Much remains to be done on the description of the behavior above threshold. Some possible experiments are:

a. Study of the velocity field using optical techniques (motion of dust particles and Doppler shift of a monochromatic laser light due to the velocity field).

b. Study of the amplitude of the distortion by the diffraction of light by the convection rolls (the wavelength of the rolls is large ($\lambda_c \simeq 0.5$ to 1 mm) and, in order to avoid measurements at small angles, the diffraction can be studied on a reduced image).

c. Measurement of the Nusselt number giving the ratio of convective to conducting heat-transfer terms.

We will now briefly review some additional convective effects in planar nematics.

1. In a nematic of negative anisotropic heat conductivity ($k_a < 0$) heated from below, the anisotropic mechanism would be strongly stabilizing compared with the isotropic case. Equation (IV.8) indicates that heat convection should be obtained when heating from above. This has not been checked experimentally.

2. Convection can also be obtained using the destabilizing action of the centrifugal force.[20] A nematic is contained between two concentric cylinders (radii R and $R + dR$ with $dR \ll R$) in a solid-body rotation at an angular velocity ω. The temperature difference between the two cylinders $\Delta T = T_R - T_{R+dR} < 0$ leads to an unstable density stratification.

When the direction of alignment is along the azimuthal axis, convection is observed with rolls of vertical axis z and the threshold corresponds precisely to that in the plane geometry with the replacement of the gravitational buoyancy acceleration $\mathbf{g}$ by $\omega^2 R(\mathbf{R}/R)$.

When molecules are parallel to the axis of rotation, convection appears at a larger threshold (in the same conditions). The behavior is strongly hysteretic. In addition, the rolls are aligned parallel to z. A new coupling mechanism is involved where the heat focusing is associated with a splay plus twist distortion rather than bend. The coupling mechanism between the distortion and the convective shear flow in the plane perpendicular to $\mathbf{n}$ is more complex and becomes effective only for

[20] Guyon, E., and Carrigan, C., *J. Phys. Lett.* **36**, L145 (1975).

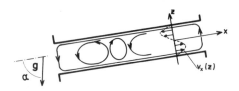

FIG. 8. Convection in an oblique geometry. A large convection current loop appears for a very low temperature threshold in a wide oblique cell. If one raises the thermal gradient β above the critical threshold β_c where the roll convection appears, one sees a superimposed pattern resulting in an alternation of wide and narrow rolls: The wide rolls have the same circulation as the large convection loop and the small ones the opposite circulation.

a *finite* value of the shear, as will be discussed in Section V. A resulting finite amplitude (first-order-like) instability is consequently obtained, rather than just a linear instability which would involve an infinitely small velocity field at threshold. The prevalence of rolls with axis parallel to the axis of rotation originates from the action of the Coriolis force, which strongly prevents the formation of horizontal rolls (the Taylor–Proudman theorem states that hydrodynamic solutions in a rotating geometry should be two-dimensional in a plane perpendicular to z).

3. Planar convection can be studied in an oblique geometry (α is the angle between the xy-plane and the vertical). In an isotropic fluid, the presence of a temperature gradient causes a large convection current loop circulating across the cell (Fig. 8). The temperature threshold for convection is practically zero in this problem, as the characteristic distance is the length of the cell L, which is large compared with the thickness d (the aspect ratio $d/L \ll 1$).

The velocity profile is

$$v_x(z) = \frac{1}{6} \frac{\alpha\beta g \cos \alpha}{\gamma} z[z^2 - d^2/4], \qquad \text{(IV.11)}$$

with a destabilizing buoyancy acceleration $g \cos \alpha$. This shear has been used with nematics to reproduce the shear flow effect in the three geometries of Fig. 2, using a small enough value of β ($<\beta_c$) to avoid small roll convection.[21a]

If the cell is nearly horizontal ($\alpha \sim \pi/2$) and β is larger than β_c, the convection rolls are modified by the superimposed large-scale convection. If the orientation is such that the roll axis is horizontal, the application of such a shear leads to the formation of alternately wide rolls (of circulation corresponding to that of the large convection cell)

[21a] Horn, D., Guyon, E., and Pieranski, P., *Rev. Phys. Appl.* **11**, 139 (1976).

and narrow ones of opposite circulation. The disappearance of the small rolls takes place typically when the velocity in the convection cell (which varies rapidly with β near threshold) compensates that of the large loop (which varies only slowly with β).[21b]

4. Currie[22] has pointed out that in a strong downward temperature gradient, nematics of positive anisotropy k_a would tend to align along ∇T, whereas, in a upward gradient, they would align at a right angle with v. The first effect is clearly understood from an extension of the study of the alignment effect in the convective shear flow. Such an extension, however, neglects the effect of turbulence in large temperature gradients.

The second prediction is consistent with old experiments by Stewart and co-workers.[22a] They found that a disordered nematic (PAA) aligns with the average "swarm" (director) axis perpendicular to the heat flow when heated from above.[22b] Such a result has stimulated us to study convection in homeotropic nematics (n perpendicular to the limiting plates).

3. Homeotropic Convection[23]

Experiments on homeotropic samples *heated from above* indicate that the alignment becomes unstable above a critical threshold of the same order of magnitude as that for a planar film heated from below. This suggests that the anisotropic mechanism also plays a role. The rolls obtained have no preferred orientation since the director is initially perpendicular to them. A one-dimensional approach is not sufficient to describe the homeotropic convection. Consider the two components of the viscous torques acting on the director. The first one, Γ_3, is due to the fluid velocity component parallel to the molecules and is proportional to α_3 (see Fig. 1). The second one, Γ_2, due to the transverse velocity component, is proportional to α_2. The role of the two terms is schematized in Fig. 9; in practice $|\alpha_2| \gg |\alpha_3|$ and $|\Gamma_2| \gg |\Gamma_3|$. In the planar configuration, a one-dimensional model takes into account the torque Γ_2 = $\alpha_2(\partial v_z / \partial x)$ which is the dominant one and leads to the focusing effect favorable to the instability [Fig. 9a(P)]. In contrast, in the homeotropic

[21b] Guyon, E., Pieranski, P., and Boix, M. *J. Phys. (Paris)* **39** (1978).

[22] Currie, P. K., *Rheol. Acta* **12**, 165 (1973).

[22a] Steward, G. W., Holland, D. O., and Reynolds, L. M., *Phys. Rev.* **58** (1940).

[22b] They also found that the heat conductivity is decreased due to the flow alignment effect (as the heat conductive effect measured by $k_\perp$ is smaller than $\bar{k}$). Such a result gives a warning concerning the use of extrema principles: The convection does not mean necessarily an increase of the heat transfer!

[23] Dubois-Violette, E., Pieranski, P., and Guyon, E., *Phys. Rev. Lett.* **30**, 736 (1973).

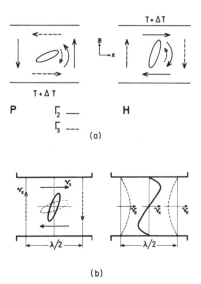

FIG. 9. (a) The viscous torque Γ_2 is associated with the α_2 viscosity and Γ_3 with α_3; $|\Gamma_2|$ $\gg |\Gamma_3|$. In a planar (P) configuration, a one-dimensional model takes into account the dominant viscous torque $\Gamma_2 = \alpha_2 \partial v_z / \partial x$. In a homeotropic configuration (H) one must introduce two-dimensional effects and a gradient term $\partial v_x / \partial z$ since $\Gamma_2 = \alpha_2 \partial v_x / \partial z$. (b) Considering the velocity profiles $v_x(z)$, $v_z(z)$ (on the right part) one sees that in the planar case (dotted line) the dominant shear effect $\partial v_z / \partial x$ keeps the same sign across the thickness, and that in the homeotropic geometry (full line) the shear $\partial v_x / \partial z$ is destabilizing only in the central region.

configuration a one-dimensional model takes into account the torque Γ_3 $= \alpha_3(\partial v_z / \partial x)$ which is not the dominant one and would be stabilizing anyway [Fig. 9a(H)]. Therefore we must introduce two-dimensional effects and terms such as $\partial v_x / \partial z$ since now $\Gamma_2 = \alpha_2(\partial v_x / \partial z)$. A complete description of the conservation laws has been given in Dubois-Violette et al.[18,23] We shall write here only the torque equation which emphasizes the importance of the two-dimensional treatment (φ is now the small angle between the molecules and the z-axis):

$$\Gamma_y = - \left(K_1 \frac{\partial^2}{\partial x^2} + K_3 \frac{\partial^2}{\partial z^2} \right) \varphi + \gamma_1 \frac{\partial \varphi}{\partial t} - \alpha_2 \frac{\partial v_x}{\partial z} - \alpha_3 \frac{\partial v_z}{\partial x} . \quad \text{(IV.12)}$$

The conservation laws lead to two equations (Γ_y, Γ_z) for angular momentum conservation; two equations (F_x, F_z) for linear momentum conservation; one equation (heat equation) for energy conservation; one equation (div v = 0) for the incompressibility condition.

To write down this system, one expands the fluctuations in a Fourier series:

$$\varphi = \sum_j \varphi_j \cos q_x x e^{iq_{jz}z} e^{st};$$

$$\mathbf{v} = \sum_j \left(\frac{iq_{jz}}{q_x} \cos q_x x, 0, \sin q_x x \right) v_j e^{iq_{jz}z} e^{st};$$

$$\delta p = \sum_j p_j \sin q_x x e^{iq_{jz}z} e^{st}; \qquad \text{(IV.13)}$$

$$\theta = \sum_j \theta_j \sin q_x x e^{iq_{jz}z} e^{st};$$

δp is the pressure fluctuation. The expression for the velocity takes into account the incompressibility.

After elimination of the pressure between the two force equations, one gets a system of four equations linear in v_x, v_z, θ, and φ. The compatibility condition for this system reads:

$$\beta = - \frac{q_x{}^4}{\rho g \alpha} \frac{\bar\kappa \bar K \bar\eta_H}{\bar K + \kappa_a(\alpha_2 r^2 - \alpha_3)}. \qquad \text{(IV.14)}$$

The average quantities are functions of conductivity, elasticity, and viscosity, and depend on the relative distortions along x and z, measured by the ratio $r = q_z/q_x$:

$$\bar\kappa = \kappa_\perp + \kappa_\| r^2;$$

$$\bar K = K_1 + K_3 r^2;$$

$$\bar\eta = \eta_2 r^4 + \mu r^2 + \eta_1;$$

the viscosity $\mu = \alpha_1 + \alpha_3 + \alpha_4 + 2\alpha_5$.

For a fixed mode (β and q_x fixed), the above expression defines a relation $q_z(q_x)$ between the z and x components of the wavevector. Expression (IV.14) is an eighth-order equation in q_z but it does not lead to any analytical form of the eight roots $q_{jz}(q_x, \beta)$, $j = 1, \ldots, 8$. The wavevector q_x is determined by consideration of the boundary conditions

$$v_z(\pm d/2) = v_x(\pm d/2) = \theta(\pm d/2) = \varphi(\pm d/2) = 0, \qquad \text{(IV.15)}$$

which are expressed as a function of the roots $q_{jz}(q_x, \beta)$ via Eqs. (IV.13). The resolution of the set of eight linear homogeneous equations (IV.15) depending on the eight variables $q_{jz}(j = 1, \ldots, 8)$ defines $q_x(\beta)$ [through the implicit relations $q_{jz}(q_x)$]. Equations (IV.14) and (IV.15) have been solved numerically. Typical solutions are given in Fig. 10.

The threshold β_c corresponds to the minimum value of β and defines the critical wave number q_c. The numerical solution leads to a higher threshold $T_u - T_1 = \beta_{ch} d$ for the homeotropic configuration than for the planar one, $T_1 - T_u = \beta_{cp} d$. For a sample thickness $d = 1$ mm, one finds $(T_u - T_1)_{ch} = 5°$ and $(T_1 - T_u)_{cp} = 2.6°$, in very good agreement with experimental results.

The velocity profile and the critical wavelength are very similar in both configurations:

$$\lambda_{ch} = 2.18d \qquad \lambda_{cp} \sim 2.09d.$$

Consideration of the spatial dependence of the velocity profile, schematized in Fig. 9b, can explain qualitatively the higher threshold in the homeotropic case. In the planar case, the dominant shear term $\partial v_z/\partial x$ keeps the same sign across the thickness, whereas in the homeotropic geometry the dominant term is given by the shear $\partial v_x/\partial z$, which is destabilizing only in the central portion of the film.

The existence of a regular array of crossed rolls observed slightly above the convection threshold[23] is indicative of nonlinear effects; in fact, it is observed that very close to threshold only one set of rolls will be observed in any region of the sample, although there is no general direction of alignment for the rolls. It is possible to obtain organized roll convection by applying a horizontal magnetic field H which lifts the orientational degeneracy and leads to rolls perpendicular to H.

Extrapolating the measured threshold to the limit $H = 0$ (see Fig. 11), one obtains a limit threshold practically equal to that for the crossed rolls formation.

Recently Lekkerkerker (*J. Phys. Lett.* **38**, 277 (1977)) has pointed out that an usual ($k_a > 0$) homeotropic nematic heated from below should

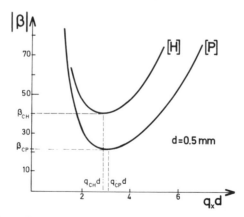

FIG. 10.   The thermal gradient β as a function of the dimensionless wavevector $q_x d$. β_{CP} and β_{CH} are the thresholds for the thermal convection problem in a planar (P) and homeotropic (H) nematic.

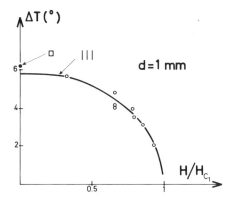

FIG. 11. The curve corresponds to a roll instability in a homeotropic sample in the presence of a horizontal destabilizing magnetic field, which in addition lifts the orientational degeneracy. The point, for $H = 0$, corresponds to a regular array of crossed rolls and is slightly above the convection threshold (limit of the preceding curve for $H = 0$).

become unstable above a threshold typical of an average isotropic fluid. The stabilizing effect of the heat focusing is neutralized by having an oscillatory convection (near threshold $s = 0^+ + i\omega$; see form III.8). Recent experiments (E. Guyon, P. Pieranski, and J. Salan, submitted to *J. Phys. (Paris)*) support completely this prediction.

V. Hydrodynamic Instabilities in Nematics

4. EXPERIMENTS[24a,25a]

We now return to the hydrodynamic problems where the flow is produced by a relative motion of the limiting plates containing the liquid crystal (shear flow) or by a flow taking place between fixed plates (Poiseuille flow).

Consider the shear flow geometries b and c of Fig. 2 with an orientation obtained by surface alignment. The distortion of the director varies progressively with the shear rate $s = \partial_z v_y$. Due to the reduced symmetry of the initial fluctuations (distorted states θ and $-\theta$ along the velocity axis are different) there is no threshold for distortion.

The case of geometry a is more interesting from the point of view of instabilities as there are two equivalent distortions which are images of each other with respect to the plane of the velocity. At low shear the orientation remains undistorted and the nematic behaves as an isotropic fluid of viscosity η_a. However, above a critical shear rate s_c this state is

[24a] Pieranski, P., and Guyon, E., *Solid State Commun.* **13**, 435 (1973).
[25a] Pieranski, P., and Guyon, E., *Phys. Rev.* A **9**, 404 (1974).

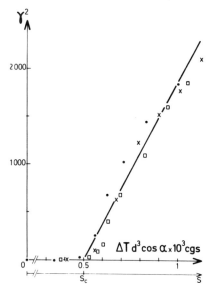

FIG. 12. The curve gives the variation of the twist distortion above the critical threshold in a hydrodynamic instability experiment (the geometry corresponds to Fig. 2a) on MBBA. γ is the rotation angle of the conoscopic image. The shear is obtained by applying a horizontal temperature gradient between vertical plates.

unstable with respect to distortions involving both twist [a small component $n_y(z)$ in the plane of the plate] and splay [with $n_z(z)$ in the vertical plane]. There are two equivalent solutions above a branching point s_c (second-order-like transition point).[24a]

Experimental results are presented in Fig. 12 for a $d = 200$ μm thick sample. (In this experiment the shear was obtained by applying a horizontal temperature gradient between vertical plates (see Section IV,2, p. 162; $\alpha = 0$), the nematic axis being horizontal and perpendicular to both the velocity and velocity gradient).[21a] The rotation angle γ of the conoscopic set of hyperbolas characteristic of a planar nematic is simply connected to the maximum twist angle φ_m by $\gamma = \varphi_m/2\pi$, where

$$\varphi(z) = n_y(z) = \varphi_m \cos(\pi z/d)$$

[which satisfies $\varphi(\pm d/2) = 0$].

The angle γ is measured as a function of the shear rate [proportional to ΔT in Eq. (IV.11)]. It goes progressively to zero[24b] at s_c as $(s - s_c)^{1/2}$ (Fig. 12; see also Fig. 14, $H = 0$).

[24b] In Pieranski and Guyon,[24a] a roughly linear variation was reported. In order to obtain a single domain distortion a small angle was kept between the flow and the normal to **n**. This also created a smeared transition which was seen as a linear behavior corresponding to the effect of a slightly oblique field in the Freedericksz problem.[7]

The critical shear rate is $s_c = v_c d = 3.6 \times 10^{-3}$ mm²/sec in MBBA. The dimensionless parameter corresponding to the Reynolds number

$$Re = vd/\nu \qquad (V.1)$$

is easily found by dimensional analysis. Re can be expressed as the ratio of a time constant for convective diffusion (d^2/ν) over that for direct transport (d/v), where d is the typical size of the system. Here the convective effects are governed by the slow time $T_0 = \gamma/Kd^2$ and a dimensionless Ericksen number can be introduced:

$$Er = \gamma vd/K = sd^2/D \qquad (V.2)$$

with a diffusivity of orientation $D = K/\gamma$.

For a given nematic, the critical shear rate s_c is given by a critical value Er_c. This simplified analysis is confirmed by the detailed description given in the next paragraph. The *homogeneous* instability mode obtained is characterized by a uniform distortion over the sample area. The presence of irregular domains related to the two possible signs of distortion can be easily distinguished from a regular convective pattern.

The mechanism leading to the instability can be understood as follows. We assume that at a fluctuation $n_z > 0$ is present (see Fig. 17a). A viscous torque Γ_z proportional to n_z is created by the flow and tends to induce a fluctuation $n_y > 0$. The flow now induces a viscous torque $\Gamma_y < 0$ (for materials with $\alpha_3 < 0$) proportional to n_y (see Fig. 17b). This tends to create a fluctuation $n_z > 0$. So the initial fluctuation $n_z > 0$ has been increased due to the interplay with the other fluctuating variable n_y.

The above argument holds for an infinite sample. For a film of finite thickness with **n** anchored at both limiting plates, the distortion is not uniform along z and requires some elastic energy. The instability occurs when the destabilizing hydrodynamic torques exceed the restoring elastic ones. There will be a threshold rate s_c below which the system stays unperturbed. The problem is analogous to the distortion of a nematic slab under the influence of a perpendicular magnetic field (Freedericksz transition).

The homogeneous mode does not exist in materials where the small torque Γ_y (Fig. 2b') is positive and has a stabilizing effect. However (see the note in paragraph 6b), if such a material is placed in a large enough shear (we keep the geometry of Fig. 2a) the distortion appears, above a critical threshold, as a linear convective instability with *rolls* aligned parallel to v. The properties of the rolls[25b] are strongly reminiscent of those found in the thermal convection problem (Section IV): (1) the rolls

[25b] Guyon, E., Janossy, I., Pieranski, P., and Jonathan, J. M., *J. Ortics* **8**, 357 (1977).

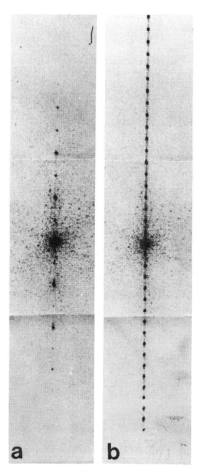

Fig. 13. The diffraction pattern of a laser beam impinged on a roll structure, produced in a shear flow instability, is characteristic of the period of the distortion $n_z(x)$ of the rolls. a and b were obtained for the two different Y and Z modes in the same MBBA film (note the change of wavelength by a factor of nearly 2).

are perpendicular to $\mathbf{n}_0$ and are observable only with extraordinary light; (2) their period is of the order of twice the thickness. The photograph (Fig. 13) shows the sharp diffraction pattern of a laser beam shone across the sample, indicative of the high regularity of the structure. The existence of high interference orders can be used to evaluate the Fourier components of the distortion above threshold. Just above threshold the pattern reduces to the fundamental peaks associated with $q_x \sim 2\pi/d$.

This instability mode is related to the existence of "hydrodynamic focusing terms '' (Section II, Note 1) which give a destabilizing vertical torque greater than the stabilizing (when $\alpha_3 > 0$) α_3 contribution.

In materials where the small torque α_3 is negative one can also obtain convective rolls instabilities: (1) by applying a large enough shear (s typically 2 or $3s_c$). However, this is a nonlinear effect above threshold and we will not discuss it here. (2) If a strong stabilizing magnetic field is applied, the range of existence of the homogeneous mode decreases and for large fields ($H \gtrsim 3$ kG), only the roll instability modes are obtained. The stabilizing elastic effect, which is a rapidly increasing function of the wavevector q of the distortion, is replaced by the stabilizing diamagnetic one independent of q. The formation of rolls, which implies a strong distortion of **n**, is favored in an external field. The variation of the distortion above threshold is given for different values of fields in Fig. 14. The effect of field will be discussed at length in Section V,5,b,ii. (3) If an alternating shear is applied, the homogeneous mode is also replaced by rolls. We will now discuss further the features of this instability and also analyze them theoretically.

A careful observation of the rolls obtained in an alternating flow shows two types of roll instabilities:

i. In the first regime, the bright and dark lines do not alternate from one half-period to the next. This proves that the quantity n_z, which characterizes the vertical distortion, does not change sign.

The formation of double images of dust points present below the lower face of the liquid crystal can also be used to study the distortion of the director. In this regime, the direction of the alternative displacement of the extraordinary image can be used to show that the horizontal component n_y changes sign at each period. We will call this regime Y.

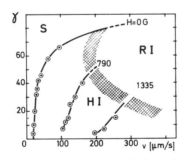

FIG. 14. The angle γ of rotation of the conoscopic image is measured as a function of the shear velocity $v = sd$. When a magnetic field is applied one first observes an increase of the threshold of the homogeneous instability (HI) mode. For large shear, a roll instability mode (RI) is obtained. For larger fields (typically above 2 kG) the RI mode would be obtained at threshold.

ii. In a second regime, Z, we see an interchange of bright and dark lines over each half-period. In this regime n_z, but not n_y, changes sign.

It is also possible to characterize the two regimes from a study of the time dependence of the intensity of one diffraction spot.

Figure 15 gives a schematic representation of the two regimes. For a constant shear amplitude, the threshold is studied as a function of the frequency T^{-1} of the shear.

The effect of a magnetic field along the x-axis and of an electric field along z has been studied systematically. $E_z = V/d$ is produced using a 10 KHz ac voltage supply up to 100 V. At this high frequency and low voltage the electric field by itself does not give rise to instabilities but only tends to align the molecules perpendicular to it (as $\epsilon_a = \epsilon_{\parallel} - \epsilon_{\perp} < 0$). Figure 16 gives a series of experimental threshold curves as a function of V and T^{-1} at different shear amplitudes for a sine wave motion. The magnetic field was kept constant ($H = 3200$ G). To the left of the curves is the domain of existence of rolls: At larger frequencies there are no instabilities (the instabilities have no time to develop in half a period). The most spectacular feature of the curve, strongly reminiscent of the electrohydrodynamic instabilities, is the existence of two branches. We keep the frequency constant ($T^{-1} = 1$ Hz on Fig. 16) and increase V. In the absence of an electric field the instability present is of the Y type. At a certain value of field E_1 the instability disappear. For an even larger value E_2 the instability reappears, but it is now of the Z type. Finally, for the value E_3, the Z instability disappears. The spatial period of the rolls remains always of the order of the sample thickness in the three branches but it is usually larger (by 20%) in branch 2.

At a frequency slightly smaller than that of the cusp ($T^{-1} = 0.8$ Hz), the instability is still of the Y type for $E = 0$. In larger fields, it is Z-type. It disappears at a critical field E_3'. This branch is the continuation of the E_3 branch. The switching between the Y and Z regimes takes place in a domain which is defined by the extrapolation of the threshold branches 1

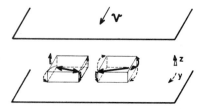

FIG. 15. Two types of roll instabilities in an alternating shear: Z regime—n_z changes sign (and not n_y) at each half-period; Y regime—n_y changes sign (and not n_z) at each half-period.

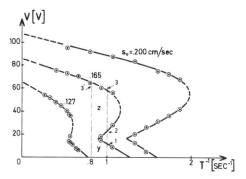

FIG. 16. Experimental threshold curves as a function of the applied voltage and frequency T^{-1} of the alternating sine-wave shear. The different curves correspond to different shear amplitudes s. The lower branches correspond to Y regimes and the upper one to Z regimes.

and 2 to the left. This intermediate domain YZ is characterized by some "exotic" structures; the rolls do not extend over long distances but break in a rather regular way. This corresponds to the interchange between the Y- and Z-type instabilities along a given roll. The diffraction pattern, instead of showing a single set of points along the axis Ox, now contains satellite spots in the direction Oy. In addition, the diffraction pattern shows peaks twice as closely spaced as in the regular case, indicating the existence of a doubled spatial period along Ox.

5. THEORETICAL APPROACH

We keep the geometry of Fig. 2a and turn to a more quantitative discussion of these linear flow instabilities. Let us denote the orientation fluctuation by $\mathbf{n} = (0, n_y, n_z)$, the velocity fluctuation $\mathbf{v} = (v_x, v_y, v_z)$, and the pressure fluctuation p. The boundary conditions imposed by the plates (strong anchoring) are:

$$n_y(z = \pm d/2) = n_z(z = \pm d/2) = 0;$$

$$v_x(z = \pm d/2) = v_z(z = \pm d/2) = v_y(z = \pm d/2) = 0;$$

$$p(z = \pm d/2) = 0.$$

Experiments show that (a) the unperturbed state is unstable against small fluctuations at $v \gtrsim v_c$; (b) the distorted system is translationally invariant along the direction of the flow. Consequently, we shall discuss the stability of the flow within the framework of a linearized theory where all quantities are taken to be independent of the y coordinate. We now write the equations governing the time and space dependence of the fluctuations retaining only first-order terms.

i. Torque equations. The total torque exerted on the director (viscous, plus elastic, plus that induced by external fields H along Ox, and E along Oz) must be equal to zero. This gives

$$\Gamma_y = 0 = - \left(K_1 \frac{\partial^2}{\partial z^2} + K_3 \frac{\partial^2}{\partial x^2} - \chi_a H^2 + \frac{\epsilon_a}{4\pi} E^2 \right) n_z + \alpha_3 s n_y$$

$$+ \alpha_3 \frac{\partial v_x}{\partial z} + \alpha_2 \frac{\partial v_z}{\partial x} + \gamma_1 \frac{\partial n_z}{\partial t} ; \qquad \text{(V.3-1)}$$

$$\Gamma_z = 0 = \left(K_2 \frac{\partial^2}{\partial z^2} + K_3 \frac{\partial^2}{\partial x^2} - \chi_a H^2 \right) n_y$$

$$- \alpha_2 s n_z - \alpha_2 \frac{\partial v_y}{\partial x} - \gamma_1 \frac{\partial n_y}{\partial t} . \qquad \text{(V.3-2)}$$

From left to right we recognize: the elastic torque; the magnetic and electric torque components [notice that H has a stabilizing effect for n_y and n_z but E acts only on n_z and either tends to align $\mathbf{n}$ along Oz if $\epsilon_a > 0$ (destabilizing effect) or to keep $\mathbf{n}$ in the xy-plane if $\epsilon_a < 0$ (stabilizing effect: case of MBBA)]; the main destabilizing contributions coming from the viscous torque; and finally additional terms due to shear induced, by a nonuniform distortion, in the flow and to the time variation of the director itself.

ii. Force equations. The force equations express linear momentum conservation:

$$\frac{\partial p}{\partial x} = \left(b \frac{\partial^2}{\partial x^2} + \eta_b \frac{\partial^2}{\partial z^2} \right) v_x + \alpha_3 \frac{\partial^2}{\partial z \partial t} n_z + \frac{\alpha_3 + \alpha_6}{2} \frac{\partial}{\partial z} s n_y ; \qquad \text{(V.3-3)}$$

$$\rho s v_z = \left(\eta_c \frac{\partial^2}{\partial x^2} + \eta_a \frac{\partial^2}{\partial z^2} \right) v_y + \frac{\alpha_5 - \alpha_2}{2} \frac{\partial}{\partial x} s n_z + \alpha_2 \frac{\partial^2}{\partial x \partial t} n_y ; \qquad \text{(V.3-4)}$$

$$\frac{\partial p}{\partial x} = \left(\eta_c \frac{\partial^2}{\partial x^2} + g \frac{\partial^2}{\partial z^2} \right) v_z + \alpha_2 \frac{\partial^2}{\partial z \partial t} n_z + \frac{\alpha_5 + \alpha_2}{2} \frac{\partial}{\partial x} s n_y , \qquad \text{(V.3-5)}$$

where

$$b = \tfrac{1}{2}(2\alpha_1 + \alpha_3 + \alpha_4 + 2\alpha_5 + \alpha_6);$$

$$g = \tfrac{1}{2}(\alpha_4 - \alpha_5 - \alpha_2).$$

In these equations, we have neglected inertial terms of the form $\rho \partial v_\alpha / \partial t$ in front of the terms involving viscosity coefficients. Indeed, for the frequencies under consideration ($T^{-1} < 10$ Hz), with $\eta \sim 0.1$ cgs, $d \sim$

10^{-2} cm, and $\rho \sim 1$ g/cm^3, we have $\rho/T \sim 1$ and $\eta/d^2 \sim 10^3$, so that $\rho \partial v_\alpha /\partial t \ll \eta \partial^2 v_\alpha /\partial x^2$. The same analysis shows that $\rho v_z s$ will be negligible unless very high shears are applied.

iii. The incompressibility condition reads

$$\partial v_x /\partial x + \partial v_z /\partial z = 0. \tag{V.3-6}$$

A detailed analysis of the properties of this set of partial differential equations (V.3) has been given elsewhere;[26] we shall give here only an approximate discussion and quote exact expressions when necessary.

a. Time Stationary Instability ($\partial/\partial t \equiv 0$)

i. Homogeneous Mode. For a solution uniform in the x-direction (all variables depending only on z) one easily shows from Eqs. (V.3-4)–(V.3-6) that $v_z = v_y = 0$ and $p = 0$. A fluctuation $n_y(z)$ induces a velocity gradient $\partial v_x /\partial z$ through Eq. (V.3-3) which reads:

$$\eta_b \frac{\partial^2 v_x}{\partial z^2} = -\frac{\alpha_3 + \alpha_6}{2} s \frac{\partial n_y}{\partial z}.$$

This velocity gradient gives rise to a component along y of the viscous torque not introduced in the qualitative description given in Section V,4:

$$\delta\Gamma_y = -\frac{\alpha_3(\alpha_3 + \alpha_6)}{2n_b} sn_y,$$

but $(\alpha_3 + \alpha_6)/2 = \eta_b - \eta_a < 0$ and this additional torque reinforces the term $\alpha_3 sn_y$. We may write Eqs. (V.3-1) and (V.3-2) in the approximate form

$$0 = \left(-K_1 \frac{\partial^2}{\partial z^2} + \chi_a H^2 - \frac{\epsilon_a}{4\pi} E^2\right) n_z + \alpha_3 \frac{\eta_a}{\eta_b} sn_y; \tag{V.4-1}$$

$$0 = \left(K_2 \frac{\partial^2}{\partial z^2} - \chi_a H^2\right) n_y - \alpha_2 sn_z = 0. \tag{V.4-2}$$

In order to find the threshold, we suppose a distortion containing only the dominant wavevector $q_z = \pi/d$:

$$n_z /A = n_y /B = \cos q_z z.$$

Equations (V.4-1), (V.4-2) above take on the form

$$\left(K_1 q_z{}^2 + \chi_a H^2 - \frac{\epsilon_a}{4\pi} E^2\right) A + \alpha_3 \frac{\eta_a}{\eta_b} sB = 0;$$

$$\alpha_2 sA + (K_2 q_z{}^2 + \chi_a H^2)B = 0.$$

[26] Manneville, P., and Dubois-Violette, E., *J. Phys. (Paris)* **37**, 285 (1976).

This system has a nontrivial solution if

$$\left(K_1 q_z{}^2 + \chi_a H^2 - \frac{\epsilon_a}{4\pi} E^2\right) (K_2 q_z{}^2 + \chi_a H^2) - \alpha_2 \alpha_3 \frac{\eta_a}{\eta_b} s^2 = 0 \qquad (V.5)$$

when $H = 0$ and $E = 0$, Eq. (V.5) reduces to

$$\pi^2 = sd^2 \sqrt{\alpha_2 \alpha_3 \eta_a / K_1 K_2 \eta_b} , \qquad (V.5')$$

where the right-hand side has the form expected for an Ericksen number [see Eq. (V.2).]

   Remark: The above result takes into account the boundary conditions on **n** only approximately and neglects those on the transverse velocity v_x. Nevertheless, the result obtained is not very different from the exact one. For example, in Eq. (V.5') above, π^2 should be replaced by $1.23\pi^2$. Agreement with experiment results is quite satisfactory for MBBA.

   ii. Roll Instability. The experiments have shown that for a large enough magnetic field H the homogeneous mode is no longer the one which appears at threshold, but that convective rolls develop (see Fig. 14). The axis of the rolls is parallel to the flow, and the wavelength of the structure is of the order of the sample thickness. This instability results from a new mechanism. Let us assume fluctuations depending only on x. A "bend" fluctuation (Fig. 17d) $n_y = n_y(x)$ is coupled to the velocity gradient $\partial v_z / \partial x$ through Eq. (V.3-5), which reads

$$\eta_c \frac{\partial^2 v_z}{\partial x^2} + \frac{\alpha_2 + \alpha_5}{2} s \frac{\partial n_y}{\partial x} = 0$$

and may be understood as expressing a balance between a "focusing" force (Section II, Note 1)

$$F_z = \frac{\alpha_2 + \alpha_5}{2} s \frac{\partial n_y}{\partial x}$$

and a viscous force $\eta_c \partial^2 v_z / \partial x^2$. The velocity gradient $\partial v_z / \partial x$ then produces an additional torque

$$\delta \Gamma_y = -\frac{\alpha_2}{\eta_c} \frac{\alpha_2 + \alpha_5}{2} s n_y ;$$

since $(\alpha_2 + \alpha_5)/2 = (\eta_c - \eta_a + \alpha_2)$ is negative in the case considered here, then $\delta \Gamma_y$ is destabilizing (Fig. 17d). Denoting the quantity $-(\alpha_2/\eta_c)(\eta_c - \eta_a + \alpha_2)$ by α_3', the total y component of the viscous torque reads

$$\Gamma_y = (\alpha_3 + \alpha_3')s n_y .$$

The case of a "bend" fluctuation $n_z = n_z(x)$ is similar (Fig. 17c). Such a

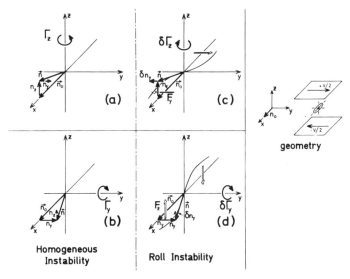

FIG. 17. Left: Instability mechanism for homogeneous distortion. (a) Suppose a fluctuation $n_z > 0$. The flow induces a viscous torque $\Gamma_z > 0$ such that a fluctuation $n_y > 0$ appears. (b) Suppose a fluctuation $n_y > 0$. The flow induces a viscous torque Γ_y such that the initial fluctuation n_z is increased if $\alpha_3 < 0$ ($\Gamma_y < 0$) (see Fig. 2). Right: Instability mechanism for roll instability. (c) A bend distortion $n_z(x)$ induces a shear flow in the y-direction such that the viscous torque $\delta\Gamma_z > 0$ is stabilizing. (d) A bend distortion $n_y(x)$ induces a shear flow in the z-direction which, in turn, induces a destabilizing viscous torque $\delta\Gamma_y$.

fluctuation is coupled to the velocity gradient $\partial v_y / \partial x$ through Eq. (V.3-4):

$$\eta_c \frac{\partial^2 v_y}{\partial x^2} + \frac{\alpha_5 - \alpha_2}{2} s \frac{\partial n_z}{\partial x} = 0,$$

which expresses the balance between the "focusing" force

$$F_y = \frac{\alpha_5 - \alpha_2}{2} s \frac{\partial n_z}{\partial x}$$

and the viscous one $\eta_c \partial^2 v_y / \partial x^2$. The velocity gradient gives rise to a torque component

$$\delta\Gamma_z = + \frac{\alpha_2(\alpha_5 - \alpha_2)}{2\eta_c} s n_z < 0,$$

as $(\alpha_5 - \alpha_2)/2 = \eta_c - \eta_a > 0$ [see Eq. (II.10).] This additional torque is stabilizing (Fig. 17c). However, the total z component of the viscous torque is positive:

$$\Gamma_z + \delta\Gamma_z = -\alpha_2 \left(1 - \frac{\eta_c - \eta_a}{\eta_c}\right) sn_z = \alpha_2 \frac{\eta_a}{\eta_c} sn_z \,,$$

which leads finally to the instability but is smaller than in the case of the homogeneous instability by a factor $\eta_a/\eta_c \sim 0.5$.

We may now write Eqs. (V.3-1) and (V.3-2) in the approximate form

$$\left(-K_3 \frac{\partial^2}{\partial x^2} + \chi_a H^2 - \frac{\epsilon_a}{4\pi} E^2\right) n_z + (\alpha_3 + \alpha_3{}')sn_y = 0$$

$$\left(+K_3 \frac{\partial^2}{\partial x^2} - \chi_a H^2\right) n_y - \alpha_2 \frac{\eta_a}{\eta_c} sn_z = 0$$

and perform an analysis analogous to the one for the homogeneous mode. Here we assume

$$n_z/A = n_y/B = \cos q_x x.$$

The threshold is given by

$$\left(K_3 q_x{}^2 + \chi_a H^2 - \frac{\epsilon_a}{4\pi} E^2\right)(K_3 q_x{}^2 + \chi_a H^2) = \alpha_2 \frac{\eta_a}{\eta_c}(\alpha_3 + \alpha_3{}')s^2. \quad \text{(V.6)}$$

This simplified analysis of Eqs. (V.3-1)–(V.3-6) gives only an idea of the detailed mechanism since it neglects the effects of the distortion along z. These effects may be partly taken into account in the elastic contribution; Eq. (V.6) is replaced by

$$\left(K_1 q_z{}^2 + K_3 q_x{}^2 + \chi_a H^2 - \frac{\epsilon_a}{4\pi} E^2\right)(K_2 q_z{}^2 + K_3 q_x{}^2 + \chi_a H^2) \quad \text{(V.6')}$$

$$= \alpha_2 \frac{\eta_a}{\eta_c}(\alpha_3 + \alpha_3{}')s^2.$$

Moreover, α_2 and $\alpha_3 + \alpha_3{}'$ are renormalized by back flow effects and should be replaced by the values function of the ratio $r = q_z/q_x$ [as in Eq. (IV.14)].

iii. Discussion. Two solutions are in competition; the first one [Eq. (V.5)] corresponds to an instability uniform in the x-direction, the second one [Eq. (V.6)] to an instability with convective rolls parallel to the flow (Oy). Taking into account the values of viscosity given earlier for MBBA and the elastic constants K_1, K_2, $K_3 = 6 \times 10^{-7}$, 3×10^{-7}, and 7×10^{-7} cgs, respectively, we compare the thresholds given by Eqs. (V.5) and (V.6) when H is varied. This comparison is made in Fig. 18, where we plot s versus H^2 ($E = 0$). We see that, for fields lower than 400

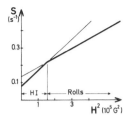

FIG. 18.   Instability threshold as a function of the applied magnetic field H. At low fields an homogeneous instability (HI) sets up; for higher fields the rolls appear (see also Fig. 15).

G, a homogeneous instability (HI) is obtained at threshold, whereas for higher fields the rolls appear. This is in qualitative agreement with experimental results.

A quantitative agreement is obtained using the exact two-dimensional solution, which, in addition, gives the field dependence of the wavelength of the rolls at threshold. At $H = 0$, $(d/2)q_x \simeq 2$. However, the corresponding threshold is higher than for an homogeneous instability. On the other hand, at $H = 1200$ G, $(d/2)q_x \simeq 3.1$ and a roll instability is expected. The crossover between the homogeneous mode and the rolls is obtained for $H \sim 1000$ G, $(d/2)q_x \simeq 2.9$. Again the agreement with experiment is excellent.

b. Time-Dependent Shears

i. Basic Equations. Experiments with alternating shear show that, except for very low frequencies, the roll instability is preferred even when no stabilizing field is applied. This can be understood by the following argument: One finds from Eqs. (V.3-1), (V.3-2) that the term $\gamma_1 \partial/\partial t$ plays a role similar to $\chi_a H^2$ and is independent of wavevectors q_x, q_z. Let us define $T_0^{-1}(H) = \chi_a H^2/\gamma_1$. (This frequency is related to the relaxation of a distortion under a magnetic field H.)[13] Frequencies of interest in the experiments are $T^{-1} \sim 1$ Hz. Then $T \sim T_0(H)$ implies H $\sim$ 3×10^3 G. Experiments on stationary flow show that rolls appear for fields greater than $\sim 10^3$ G. Thus we conclude that the homogeneous instability under an alternating shear is expected only for very low frequencies $<1/3$ Hz. When all unknown quantities except n_y and n_z are eliminated, from Eqs. (V.3-1)–(V.3-6) one obtains

$$\dot{n}_y + n_y/T_y + Asn_z = 0 \qquad \text{(V.7-1)}$$

$$\dot{n}_z + n_z/T_z + Bsn_y = 0, \qquad \text{(V.7-2)}$$

where $1/T_y$, $1/T_z$, A, B, have a complicated analytical form. When H is large enough so that elastic effects may be neglected, and for low enough shear rates, we have

$$\frac{1}{T_y} = \frac{\chi_a H^2}{\gamma_y}, \frac{1}{T_z} = \left(\chi_a H^2 - \frac{\epsilon_a}{4\pi} E^2\right)/\gamma_z,$$

where γ_y and γ_z are effective viscosities. Coefficients in these equations will be considered as phenomenological quantities.[25a]

This form of coupled equations is rather general in the discussion of the time dependence of linear instabilities. In the appendix we will discuss briefly an analogical solution of these equations which emphasizes the physical character of the different terms.

ii. Threshold Equations for Y- and Z-Type Instability. Eqs. (V.7-1), (V.7-2) are strongly reminiscent of those governing electrohydrodynamic instabilities:[27,28] the electric field E plays the same role as the shear rate s, the curvature ψ and the charge q stand for n_y and n_z, respectively.

A nematic in an alternating electric field may show two types of instabilities. In the first one, called the "conduction regime" (low frequency), q oscillates in time while ψ is nearly constant. On the other hand, ψ oscillates and q is nearly constant in the "dielectric regime" (high frequency).

This is the complete analog of the Y-type (n_z nearly constant and n_y oscillating) and of the Z-type regimes shown in Fig. 15. The solution for the case of a square-wave excitation has been developed in Pieranski and Guyon,[25] Dubois-Violette,[28] and Galerne.[29] We shall obtain here similar results for a sinusoidal time dependence of the shear rate $s = s_0$ cos ωt ($\omega = 2\pi/T$) corresponding to the experiments of Fig. 16. When the steady state is reached, n_y and n_z must be periodic functions of time. Let us first assume $T_z > T_y$ and $T_z > T$; n_z may be considered as constant. From Eq. (V.7-2) we obtain

$$n_z \simeq -BT_z s_0 \frac{1}{T} \int_0^T n_y(t) \cos \omega t dt. \tag{V.8}$$

Indeed, s and n_y are periodic functions of t; in addition to the oscillatory components, the product $s(t)n_y(t)$ will contain a constant term. Since the relaxation time T_z is long, n_z will respond only to the static part of $s(t)n_y(t)$, which is expressed by Eq. (V.8) above (a more formal

[27] A recent review article giving more references is Smith, I. W., Galerne, Y., Lagerwall, S. T., Dubois-Violette, E., and Durand, G., *J. Phys.* **36-C1**, 237 (1975).

[28] Dubois-Violette, E., *J. Phys. (Paris)* **33**, 95 (1972).

[29] Galerne, Y., Thesis, 3rd cycle, Orsay (1973).

derivation may be adapted from Dubois-Violette *et al.*[38]). Now Eq. (V.7-1) can be integrated as

$$n_y = -\exp(-t/T_y)As_0n_z \int_{-\infty}^{t} \cos \omega t' \exp(t'/T_y)dt'$$

or

$$n_y = -As_0n_z \left(\frac{1}{T_y}\cos \omega t + \omega \sin \omega t\right) \frac{T_y^2}{1 + \omega^2 T_y^2}. \qquad (V.9)$$

Self-consistency of Eqs. (V.8) and (V.9) leads to

$$ABs_0^2 T_z T_y /(2[1 + (\omega T_y)^2]) = 1. \qquad (V.10\text{-}1)$$

This is the threshold condition for a Y-type regime.

n_y and n_z enter the original equations in symmetrical fashion. If we suppose that $T_y > T_z$ and $\omega T_y > 1$, we get the threshold condition for a Z-type instability:

$$ABs_0^2 T_y T_z /(2[1 + (\omega T_z)^2]) = 1. \qquad (V.10\text{-}2)$$

iii. Discussion. From a phenomenological point of view we may consider the system as depending on three parameters:

$$x = \omega T_y, \quad y = (T_y/T_z)^{1/2}, \quad C = ABs_0^2 T_y^2/2,$$

where x is a reduced frequency and C defines the reduced shear. Using these variables, the threshold conditions read

Y-type (Fig. 19, curve A):

$$y^2 = \frac{C}{(1 + x^2)}, \quad x \gg y^2, \quad y > 1; \qquad (V.11\text{-}1)$$

Z-type (Fig. 19, curve B):

$$x^2 + y^4 - Cy^2 = 0, \quad x \gg 1, \quad y < 1. \qquad (V.11\text{-}2)$$

For comparison the curve C corresponds to the case where the amplitude of the shear is kept constant over each half-period.

Let us keep all parameters constant except the frequency. For large frequencies, no instability has enough time to develop. The instability is present only if the frequency is lower than the value given by Eq. (V.11-1) or (V.11-2) above. Then at the left of the curve A (resp B) the system is unstable against a fluctuation of the Y type (resp Z type). The two curves divide the xy-plane into four regions:

Region S: The system is stable;

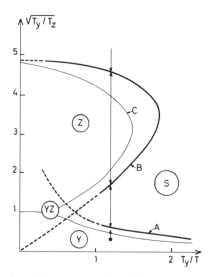

FIG. 19. Theoretical instability curves. The different curves correspond to different applied shear s. Curve B is obtained for a sine-wave signal, curve A for a square-wave signal. Region S: Stability for large reduced frequencies T_y/T; no instability has time to develop. Region Y (resp Z): The system is unstable against a Y (resp Z)-type fluctuation, but is stable against a fluctuation of the opposite type.

Region Y (resp Z): The system is unstable against a Y (resp Z)-type fluctuation but is stable against a fluctuation of the opposite type;

Region YZ: Theoretical curves [Eqs. (V.11-1) and (V.11-2)] are valid for the full line parts of curves A ($y < 1$) and B ($y > 1$). This excludes the description of the region YZ ($y \sim 1$). The theoretical curve is similar to the experimental one (Fig. 16), but this is fortuitous. The theoretical description deals with dimensionless parameters such as T_y and T_y/T_z. In fact, T_y depends on the wavevectors q_x and q_z of the distortion. A diagram plotted with nonnormalized variables (V, T^{-1}) can have a very different aspect, in particular because the wavevectors, at threshold, are different in the two regimes.

If we assume negligible elastic effects as compared to those caused by an external magnetic field, we get

$$y^2 = \frac{\gamma_y}{\gamma_z} \frac{\chi_a H^2 - \dfrac{\epsilon_a}{4\pi} E^2}{\chi_a H^2}.$$

The effective viscosities γ_y and γ_z may be evaluated from two-dimensional analysis.[26] Supposing roughly circular rolls (i.e., $q_x \sim q_z \sim \pi/d$)

one obtains

$$\gamma_y \sim 0.45\gamma_1; \quad \gamma_z \sim 0.70\gamma_1,$$

so that when $E = 0$,

$$(T_y/T_z)_0 = y_0^2 \sim 0.65,$$

and at least in a certain frequency domain one should observe a pure Y-type instability. This has indeed been found experimentally. Now if we keep the frequency constant and increase the electric field (that is to say, increase y since $\epsilon_a < 0$) we can follow the description of the experiments given at the end of Section V,4.

6. OTHER FLOWS

a. Poiseuille Flow

The problem of a dc flow between parallel plates with the velocity profile being characterized by Eq. (II.17) presents many analogies with the simple shear flow problem. However, there are two main differences:

1. The shear rate varies across the thickness. The analysis of the stress tensor involves more terms. In particular, transverse flow effects[12] discussed in Section II are essential.

2. The nucleation of instabilities takes place near the boundaries where the shear is largest. Interference between the distortions in the two surface sheaths must be considered and involves elastic as well as flow coupling.

A detailed theoretical analysis of the problem is yet to be done and we restrict ourselves to the homogeneous case, neglecting the coupling with velocity (in fact, this is a possible solution if one uses very narrow cells where the transverse flow effect is quenched). Equations (V.4-1) and (V.4-2) can be simplified in zero field. We write them in a decoupled form as

$$\partial^2 u/\partial z^2 + (s'z/D)u = 0; \tag{V.12}$$

$$\partial^2 v/\partial z^2 - (s'z/D)v = 0, \tag{V.13}$$

using the new variables $u(+)$ and $v(-) = n_y \pm (K_1\alpha_2/K_2\alpha_3)^{1/2}n_z$ and the shear rate $\partial v_x/\partial z = s'z$.

The two equations have the form of a wave equation with a potential linear in z (Fig. 20, upper curve). An approximate solution can be obtained using the WKB approximation. A numerical estimate also gives

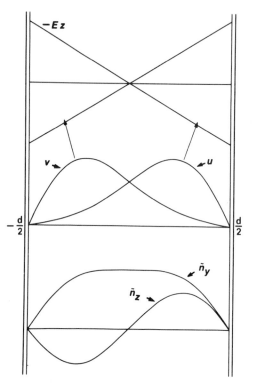

FIG. 20.   Poiseuille flow. Two possible distortions n_y and n_z are obtained from symmetrical solutions u and v of the potential E_z. The solution drawn on the lower curve is antisymmetrical in splay $n_z(z) = -n_z(-z)$ and symmetrical in twist. Another equally probable solution is symmetrical in splay and antisymmetrical in twist.

the following threshold:[30a,b]

$$Er_c = s_c' d^3/D \sim 10^2.$$

Two possible distortions are obtained from the form of the symmetrical solutions for u and v (Fig. 20, intermediate curve).

A first solution is antisymmetric in the splay variable $[(n_z(z) = -n_z(-z)]$ and symmetric in twist (Fig. 20, lower curve). Another equally probable solution is symmetric in splay and antisymmetric in twist. The different symmetries of curves $\bar{n}_y$ and $\bar{n}_z$ come from the change of sign of the shear in the two halves of the layer. Both solutions can be produced by inducing a single domain distortion by the application of a magnetic field for a short time. A field in the plane of the plate at an oblique angle

[30a] Guyon, E., and Pieranski, P., *J. Phys. (Paris)* **36-C1**, 203 (1975).
[30b] Mannerville, P., and Dubois-Violette, E., *J. Phys. (Paris)* **37**, 1115 (1976).

to the director will favor the first solution, which has a nonzero average twist, and a field oblique in the vertical plane of **n** will favor the other solution with an average finite splay. The two solutions are found equally often in narrow cells. In wide cells the solution with finite total twist is found to be more stable, probably due to the contribution, not considered here, of the coupling with the flow.[30b]

Two types of homogeneous instabilities are found: In one mode the dc flow, visualized from the motion of dust particles, is not affected by the distortion. The other mode is characterized by a regular sequence of domains along the length of the sample. The spatial period is much larger than d and depends on the width of the cell. The average twist and the transverse flow are uniform in each domain and change sign over a short distance (of the order of the thickness d) from one domain to the next. A detailed description of the domains and of the walls separating them remains to be done.

In large enough ac flows we also observe convective rolls along the flow axis.[30a] Figure 21 gives the domain of existence of the modes as a function of the pressure head applied across the sample ($\Delta p \propto$ average velocity) and of the frequency of the flow. At large enough frequencies ($f = 0.5$ Hz), a Y mode is obtained at threshold. The vertical focusing force introduced in Fig. 17d is of different sign for two points symmetri-

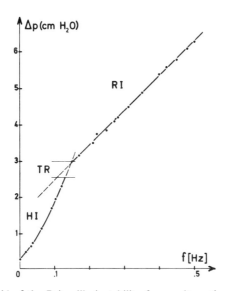

FIG. 21. Threshold of the Poiseuille instability for an alternating flow frequency as a function of the pressure difference across the cell (cell thickness $d = 200$ μm; length $= 10$ cm). For low frequencies a homogeneous mode (HI) is obtained. For larger frequency a roll instability (RI) is found with a wavelength decreasing as Δp increases.

cal with respect to the $z = 0$ plane. This suggests the existence of convective motions symmetrical with respect to this plane and decoupled solutions in the two sheaths. In fact, when the flow frequency increases, the wavelength of the convection rolls decreases (as $\Delta p^{-1/2}$), indicating that the instability modes occurring near the two plates are indeed decoupled.

At large amplitudes the instability mode is of the Z type. The vertical focusing force keeps the same sign over the two halves of the sample. This suggests that the convection currents circulate across the total thickness, which is confirmed by the finding that the wavelength of this mode is of the order of $2d$ and relatively independent of frequency.

This Poiseuille flow problem shows well the variety of convective problems in nematics.

b. Other flows

In the cylindrical Couette flow problem the shear is applied between a rapidly rotating inner cylinder 1 and a concentric outer cylinder 2. If the director **n** is parallel to the axis of rotation z, the problem should be similar to the shear flow problem, except possibly for the Coriolis effect (also found in the experiment of ref. 20), which should oppose the formation of convective rolls along the azimuthal direction. Predictions of an original convection mode have been obtained in the planar geometry with **n** in the azimuthal direction.[31] The coupled effect of the Coriolis force and vertical hydrodynamic focusing should lead to a low threshold in the domain of relative angular velocities where the classical isotropic Taylor instability is not possible:[32]

$$1 < \Omega_1/\Omega_2 < R_2{}^2/R_1{}^2.$$

(R are the radii, Ω the angular velocities.)

We note also the effect of the shear flow induced by thermal convection between vertical plates on a planar nematic with **n** horizontal [see discussion of Eq. (IV.11)]. The problem can be solved by analogy with the Poiseuille problem as a Schrödinger equation in a potential $v(z) = (3\bar{z}^2 - 1)$ (this is the form of the shear corresponding to a velocity profile $\bar{z}^3 - \bar{z}$ (where $\bar{z} = 2z/d$). Unlike the Poiseuille problem, the shear does not keep a constant sign over the thickness. In the absence of a field, distortion is dominated by the shear in the central region of the film where the effect of the elasticity due to the boundary is smaller. However, in large magnetic fields, the instability occurs near the boundaries where the shear is largest. The intermediate case where an

[31] Guyon, E., and Pieranski, P., *Adv. Chem. Phys.* **32**, 151 (1975).
[32] Guyon, E., and Pieranski, P., *Physica (Utrecht)*, **73**, 184 (1974).

instability could be nucleated in three different regions across the thickness is rather curious.

The action of an elliptically polarized shear ($s_x = s_{x_0} \cos \omega t$; $s_y = s_{y_0} \sin \omega t$) on an homeotropic nematic film ($\mathbf{n} \,// z$) leads to an interesting linear steady instability emphasizing the role of nonlinearities coming from the coupling between ac distortion and ac shear [P. Pieranski and E. Guyon, *Phys. Rev. Lett.* **39**, 1280 (1977); E. Dubois-Violette and F. Rothen, to appear in *J. Phys. (Paris)*].

c. Instabilities in Material where $\alpha_3 > 0$

The situation in which a liquid crystal having a positive small torque coefficient $\alpha_3 > 0$ is sheared with the director parallel to v (Fig. 2b') has been investigated extensively and we will only mention qualitatively some results obtained experimentally[9,33] in agreement with a theoretical description.[34] At low shears, the director will rotate in the plane of symmetry of the director but in a direction opposite to the $\alpha_3 < 0$ case. For these low shears the α_3 destabilizing torque is balanced by the elastic splay term.

However, as the rotation angle θ increases (typically as $|\alpha_2|\theta^2$ gets larger than $|\alpha_3|$), the large torque measured by α_2 tends to increase the effect of α_3. A threshold leading to a nonlinear instability in which molecules tumble around while remaining in the plane of symmetry of v is given by

$$\frac{v_{c1} d(|\alpha_2 \alpha_3|)^{1/2}}{K_1} = 4.8(\alpha_2 / \alpha_3)^{1/2}.$$

One can also estimate the threshold of a *linear* instability involving a coupled distortion out of the plane of symmetry (polar angle ϕ) and one around the equilibrium value θ. It is given by

$$(|\alpha_2 \alpha_3| / K_1 K_2)^{1/2} v_{c2} d = 9.51.$$

The form is very reminiscent of Eq. (V.5') and the mechanism is complementary. It turns out that the second threshold v_{c2} is slightly larger than v_{c1}, where a nonstationary instability is obtained. However, this mode turns out to be unstable with respect to a mode where the director gets out of the plane of symmetry (possibility corresponding to a state of finite amplitude above the threshold v_{c2}). When the shear increases further, ϕ increases to $\pi/2$, whereas θ goes to zero. We are taken back to the geometry 2a which is stable for the case $\alpha_3 > 0$ against a homogeneous mode. However, for even larger shear a roll instability can develop, which involves a focusing effect as discussed in Section V,4.

[33] Guyon, E., Pieranski, P., Pikin, S. A., *J. Phys. (Paris)* **37** C1, 3 (1976).
[34] Pikin, S. A., *Sov. Phys.—JETP (Engl. Transl.)* **38**, 1246 (1974).

VI. Electrohydrodynamic (EHD) Instabilities in Nematics

7. INTRODUCTION

The EHD instabilities of nematics have been known since the beginning of the century, i.e., almost since the very discovery of liquid crystals. However, for several reasons, they have not played the same historical role as the Bénard–Rayleigh instability reviewed in Section III. One of these reasons is the reluctance of physicists to accept new models. By analogy with the magnetic field case, one expected to be able to align nematics with electric fields, using the anisotropy of their dielectric constants. At high frequency, a static alignment was achieved, but at low frequencies (or dc), EHD instabilities were dominant. This led to strong controversies between groups which used detection techniques (like X-rays or dielectric measurements) insensitive to motion. With direct optical observations, hydrodynamic motions in an electric field were observed, forgotten, and periodically rediscovered. The difficulty, of course, derives from the fact that a steady cellular flow of the nematic molecules is associated with a *static* distortion of the nematic "director." In particular, the so-called "Williams domains" have been interpreted as being "ferroelectric" domains. Another cause of confusion is related to the geometry of the experiment. As in the Bénard problem of Sections III and IV, the instability rolls have a general tendency (at least in the low-frequency regime) to adjust their spatial period to the sample thickness independently of a specific instability mechanism.

We are not giving historical references here. A more detailed review of the works discussed in this chapter can be found in Durand.[35] Let us say that the mechanism of EHD instabilities of nematics under the action of ac electric fields is now reasonably well understood. It has in fact played an essential role in the further studies of thermal and hydrodynamic instabilities discussed in the previous chapters.

In addition to the effect of the anisotropy of the dielectric constant of the nematic, the new ingredient is the existence of electric space charges in the bulk of the material. Space charges, distortion of the director, and velocity of the fluid are parametrically coupled through the applied electric field. Above a given threshold, fluctuations of any of these quantities are amplified and EHD instabilities develop. In this paper, we sketch the Carr,[36] Helfrich,[37] and Dubois-Violette, de Gennes, and

[35] See Durand, G., *In* "Les Houches Summer School Lectures" (with a more detailed bibliography). Gordon & Breach, New York, 1973.

[36] Carr, E. F., *Mol. Cryst. Liq. Cryst.* **7**, 253 (1969).

[37] Helfrich, W., *J. Chem. Phys.* **51**, 4092 (1969); **52**, 4318 (1970).

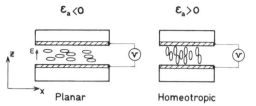

FIG. 22. The two geometries used to demonstrate pure EHD instabilities, with stabilizing dielectric torque.

Parodi[38] (DGP) models for the simple situation leading to the "dynamic scattering mode" (DSM) used for applications in displays. We discuss the behavior of such a system close to threshold in terms of "pretransitional" properties, using the phase transition terminology. (We will not speak much on the behavior above threshold which is just beginning to be systematically studied.)

8. THE HELFRICH AND DGP MODEL

Classical electroelastic instabilities occur in the Freedericksz experiment, where, due to the anisotropy of the dielectric constant, small angular fluctuations can be destabilized by an electric field. To observe *pure* EHD instabilities, we choose geometries where the torques, induced by dielectric anisotropy, *stabilize* the small angular distortions of the director. The nematic is contained between two thin glass plates. The typical geometries are the planar (rubbed) geometry ($\mathbf{n}$ along x) for "negative" materials (dielectric constant anisotropy $\epsilon_a < 0$) and the homeotropic geometry for "positive" materials ($\mathbf{n}$ along z) ($\epsilon_a > 0$) (see Fig. 22). In addition to the dielectric anisotropy, an anisotropic conductivity is assumed. In classical experiments, $\sigma_{\parallel}$ is larger than $\sigma_{\perp}$ ($\sigma_a = \sigma_{\parallel} - \sigma_{\perp}$; $\sigma_a/\sigma_{\perp}$ can be as large as 50%). A typical material like MBBA with $\epsilon_a < 0$ and $\sigma_a > 0$ is called an ($\epsilon_a -$, $\sigma_a +$) material. For some materials which undergo a smectic phase transition, σ_a can go through 0 and become negative.[39] For simplicity, we also assume $\gamma_1 = |\gamma_2| = \gamma$.

We now concentrate on the planar geometry. As in Sections III and IV,2, we use a one-dimensional model where the quantities of interest depend only on x with a variation of the form $\exp iq_x x$; when necessary we shall assume that the period of rolls is of the order of $2d(q_{xc} = \pi/d)$.

[38] Dubois-Violette, E., de Gennes, P. G., and Parodi, O., *J. Phys. (Paris)* **32**, 305 (1971).
[39] Rondelez, F., *Solid State Commun.* **11**, 1675 (1972).

The relevant quantities are:

1. The applied field $E(t)$ (sine- or square-wave function with an angular frequency ω);

2. The distortion of the director field, here a bend mode $n_z(x, t) = \varphi$ or the curvature of the bend mode $\partial\varphi/\partial x = \psi(x, t)$;

3. The space-charge density $Q(x, t)$ in the bulk of the nematic. The material is assumed to be neutral ($Q \equiv 0$) in the absence of field, but of finite conductivity. We assume ideal electrodes which do not contribute to Q by charge injection.

4. The velocity field $v_z(x, t)$, associated with the rolls of hydrodynamic flow in the nematic. Equations of motion for these quantities are obtained from the laws of conservation for the angular momentum, linear momentum, and charge density.

a. Angular Momentum Conservation: Curvature Equation

Inertial effects are negligible for molecular rotations (the relaxation frequencies $\sim 10^6$–10^{10} are much larger than ω). We write that the sum of all torques (along y) acting on the director is zero. We add the viscous torque $\gamma[\dot{\varphi} - \partial v_z/\partial x]$, the elastic torque $K_3\partial^2\varphi/\partial x^2 \equiv K_3 q_x^2\varphi$ and the dielectric torque $-(\epsilon_a/4\pi E^2)(\varphi + E_x/E)$. Here the tilt angle $(\varphi + E_x/E)$ is measured (see Fig. 23) from the direction of the local field E_r coming from the superposition of the applied field E and of the transverse field E_x induced by space charges Q. E_x is deduced from the Maxwell equation: div $\mathbf{D} = 4\pi Q$. We call

$$T_0^{-1} = \frac{K_3 q_x^2 - \dfrac{\epsilon_a}{4\pi}\dfrac{\epsilon_\perp}{\epsilon_\parallel} E^2}{\gamma}$$

the bend relaxation rate, in the presence of the field, but in absence of back flow. We finally obtain

$$\psi + \psi/T_0 = (\epsilon_a/\epsilon_\parallel)(QE/\gamma) - q_x^2 v_z.$$

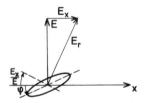

FIG. 23. The relative distortion of the director compared to the local field E_r.

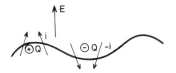

FIG. 24.   In a periodic distortion of curvature amplitude ϕ_x, an applied field E accumulates space charges $\pm Q$.

b. Charge Conservation Equation

The conservation of charges can be written as

$$\partial Q/\partial t + \text{div } \mathbf{j} = 0,$$

where $\mathbf{j} = \sigma\mathbf{E}$ is the current density.

We have also the Maxwell equation div $\mathbf{D} = \text{div } \varepsilon\mathbf{E} = 4\pi Q$.

In isotropic materials, σ and ε are the ordinary numbers σ and ϵ; one derives easily from these two equations the charge relaxation equation $\dot{Q}$ + $Q/\tau = 0$, where τ is defined by: $\tau^{-1} = 4\pi\sigma/\epsilon$.

In nematics, one finds the same simple field-independent relaxation equation for Q only for a very special case in which σ and ε are proportional. In general σ and ε have the same principal axis (the director), but their eigenvalues are not proportional. In the case of interest, the equation for charge conservation is

$$\dot{Q} + Q/\tau + \sigma_H E\psi = 0,$$

where $\sigma_H = \sigma_\|(\epsilon_\perp/\epsilon_\| - \sigma_\perp/\sigma_\|)$.

This equation shows that an applied field can give rise to space charges in a distorted nematic. More specifically, the space charges Q appear in the region of large curvature ψ (see Fig. 24). The positive anisotropy of conductivity and the negative dielectric anisotropy both contribute to the charge source term. σ_a is usually the dominant term in σ_H.

c. Momentum Conservation

We write the Navier–Stokes equation for the v_z component in the form

$$\rho\dot{v}_z = (\partial/\partial x)\Gamma_v + \nu(\partial^2 v_z/\partial x^2) + QE,$$

where we recognize the viscous coupling between the director rotation and the flow, the usual viscous force, and the drag force from space charges. We rewrite this equation as

$$\dot{v}_z + v_z/T_v = (QE/\rho) - (\gamma\dot{\psi}/\rho).$$

where $1/T_v = (\nu + \gamma)q_x{}^2/\rho$ is the damping rate of a roll of wavevector q_x, in the nematic.

d. Simplified Equations

We have three homogeneous equations for the quantities v_z, Q, and ψ. We are looking for a threshold value of applied field E_c which allows nonzero solution for v_z, Q, and ψ. As E is time-dependent, we have to compare the damping rate of each variable with ω. An important simplification can be made: ωT_v is usually very small, with typical sample thicknesses used (10 to 100 μm): The diffusivity of vorticity η/ρ is always much larger than that of the orientation K_3/η in nematics.

One can drop the $\dot{v}_z$ term and eliminate v_z as in the thermal convection problem with Eq. (IV.1). Although flow is always implied in the EHD instabilities (one verifies easily that $v_z \propto (QE - \gamma\dot\psi)$ is always different from zero for the solutions of the problem), the interesting variables are Q and ψ. They obey the two equations

$$\dot{Q} + Q/\tau + \sigma_H E\psi = 0; \qquad \text{(VI.1-1)}$$

$$\dot{\psi} + \psi/T_E + (1/\eta_H)QE = 0, \qquad \text{(VI.1-2)}$$

the source of curvature due to space charges is represented by a coupling constant η_H having the dimension of a viscosity. T_E is the orientation time constant (bend damping) in presence of the field and back flow.

These equations show that the space charge Q and the bend curvature ψ are parametrically coupled by the field E in the same formal way as were the velocity v and the temperature fluctuation θ in the isotropic Bénard problem [Eqs. (III.2) and (III.3)]. It is now simple to understand how parametric oscillations can occur: Assume a bend fluctuation ψ; then the anisotropy of σ and ϵ causes the appearance of space charges [Eq. (VI.1-1)] (Fig. 24). These space charges create a transverse field which tilts the local field E_r. The director is forced into a more tilted direction by the dielectric torque. In addition, the induced flow also increases the tilt through the viscous torque. Both effects accumulate to amplify the initial fluctuation and are included in η_H (Fig. 25). A naïve dc analysis of Eqs. (VI.1-1), (VI.1-2) gives the instability condition

$$(1/\tau)(1/T_E) \le (\sigma_H/\eta_H)E^2 \qquad (\dot\psi = \dot{Q} = 0),$$

which means, in the language of electronic engineering, that the "open loop" gain $(\sigma_H/\eta_H)E^2$ must be larger than the open loop losses. This condition corresponds to that given in Eq. (III.9) for the Bénard case.

Remembering now that T_E is field-dependent: $1/T_E = \lambda(E^2 + E_o{}^2)$ with $\lambda = -(\epsilon_a/4\pi)(\epsilon_\perp/\epsilon_\parallel)(1/\eta_B)$ (η_b = bend viscosity) and $E_o{}^2 = K_3 q_x{}^2/\lambda\eta_B$

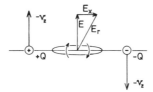

FIG. 25.   In a periodic distribution of charges $\pm Q$, an applied field induces an angular distortion: by dielectric alignment on the tilted local field E_r; or by the hydrodynamic shear induced by the flow of the charges along E.

gives the effect of the elastic torque in field units. One obtains

$$E_{th}^2 = (K_3 q_x{}^2/\lambda \eta_B)1/(\xi^2 - 1) \qquad \text{with } \xi^2 = \sigma_H \tau/\eta_H \lambda;$$

ξ^2 is a dimensionless parameter introduced by Helfrich and should be larger than 1 for instability to occur. It characterizes the ability of the system to sustain parametric instability. If q_x were allowed to vanish, E_{th} would go to zero. In practice, one uses a cutoff wave vector π/d which results in a *voltage* threshold ($V_c = E_c d$), of the order of 7 V for a typical negative nematic like MBBA. The cutoff value π/d has been previously justified in the analogous problem of thermal instabilities.

8.   TIME-DEPENDENT ANALYSIS

We must now understand how an ac excitation can also produce an EHD instability. The simplest ac signal is a square wave $\pm E$, of period T. Assume that at time $t = 0$ we are slightly above threshold ($V > V_c$). Some space charge Q has been accumulated in the regions of bend curvature ψ. Then we reverse the applied field instantaneously from E to $-E$. The effect of $-E$ on Q is now *stabilizing*. Q and φ_x begin to decrease. Two situations can occur:

a. $\tau \ll T$ (more generally, $\omega \tau \ll 1$); the space charges relax rapidly to zero. Usually the bend damping time is much larger than τ and ψ decreases only weakly. For most of the remaining period ($\sim T - \tau$), space charges are rebuilt, but with an opposite sign, to follow the inversion of the field. The field effect $\sim EQ$ is destabilizing again and increases the curvature ψ which has kept the same sign. Apart from transient impulses at the switching time (0, T, $2T$, . . .), steady bend distortion and flow develop in the form of "Williams domains." The axis of the rolls is perpendicular to the director (Fig. 26a) as in the other linear convective phenomena in nematics. This is the so-called "conduction regime."

This time-dependent analysis is very directly reproduced in the analog calculation given in the Appendix where the two time constants T_y and

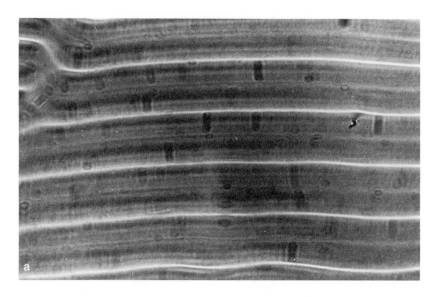

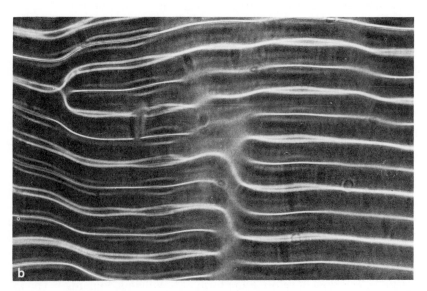

FIG. 26. Williams striations. (a) Close to threshold; (b) above threshold (courtesy of Y. Galerne).

$T_z(<T_y)$ reproduce the role of the time constants for the relaxation of orientation and charge, respectively.

b. $\tau \gg T(\omega\tau \gg 1)$. The space charge cannot oscillate at the frequency of the field. If oscillations occur, Q can only grow steadily. At each reversal of the field, the forces QE change sign. They can excite oscillations of a bend mode, provided that the bend relaxation time T_E compares with T. The bend relaxation rate $T_E^{-1} = \lambda(E^2 + E_o^2)$ is field dependent and the value of E_0 introduced above is defined by the condition $T \sim T_E$ ($\omega T_E \sim 1$). The square of the threshold field is a linear function of ω. An original property of this system is that it can adjust its spatial period: A bend distortion of wavevector q_x larger than π/d shortens T_E and can decrease the threshold, although the elastic restoring torque may be larger. The optimum value for q_x^2 is found to be a linear function of ω. At higher frequencies, diffusion currents quench the oscillations and limit the q_x increase.[40] An exact one-dimensional solution for the square wave problem is given in Galerne[41] and Durand,[35] in terms of the Helfrich parameter ξ^2:

$$\lambda E_c^2/\omega = (2.403 \ldots)/\xi^2;$$

$$K_3 q_x^2/\eta_B\omega = (1.022 \ldots)- (2.403 \ldots)/\xi^2.$$

The bend oscillation is accompanied by a shear flow oscillating in time, in the form of parallel oscillating sheets normal to the plates. The one-dimensional analysis is quite appropriate for this geometry, because $q_x d > 1$. At threshold, the optical aspect of the bend oscillations is a parallel array of thin closely spaced lines, perpendicular to the director (Fig. 27). Slightly above threshold this situation becomes unstable. The molecules tilt alternatively in domains of size d, where larger cylindrical hydrodynamic rolls appear. The distortion is a superposition of bend and splay and presents the aspect of "chevrons" (Fig. 28). This is the so-called "dielectric regime," previously called the "fast turn-off mode."

One can summarize the threshold dependence on frequency on a $V(\omega)$ diagram (Fig. 29) for a sample of given thickness. This diagram shows the low-frequency conduction regime (V_c = cte) and the high frequency dielectric regime ($V_c \propto \omega^{1/2}$).[45,46] They are separated in ω scale by the

[40] Galerne, Y., Durand, G., and Veyssié, M., *Phys. Rev. A* **6**, 484 (1972).
[41] Galerne, Y., Thesis, 3rd cycle, Orsay.
[42] Smith, I. W., Galerne, Y., Lagerwall, S. T., Dubois-Violette, E., and Durand, G., *J. Phys.* **36-C1**, 237 (1975).
[43] Orsay Liquid Crystal Group, *Phys. Rev. Lett.* **25**, 1642 (1970).
[44] Orsay Liquid Crystal Group, *Mol. Cryst. Liq. Cryst.* **12**, 251 (1971).
[45] Orsay Liquid Crystal Group, *Phys. Lett. A* **39**, 181 (1972).
[46] de Jeu, W. H., and Van der Veen, J., *Phys. Lett. A* **44**, 277 (1973).

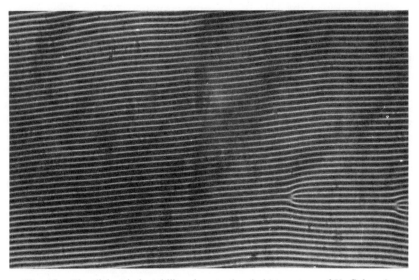

FIG. 27. The dielectric instability close to threshold (courtesy of Y. Galerne).

charge relaxation frequency τ^{-1}. For high fields and low frequency both instabilities can occur. To understand what happens in this region, where ψ and Q have comparable relaxation times, one has to consider the following special case.

c. $T_E \sim \tau$. The system chooses one or the other of the two modes of oscillation by "remembering" after half a cycle $T/2$ its previous history through the variable Q or ψ of larger lifetime (the bend curvature ψ for the conduction regime or the space charge Q for the dielectric regime).

When the curvature and space charge have equal relaxation times ($T_E = \tau$), one can show that the system is perfectly stable (see ref. 42 and the analog simulation in the Appendix). This result is of the same origin as the cusp observed in the hydrodynamic instability problem (Fig. 19). It appears to be a particular case of a more general result in instabilities described from coupled equations which are not necessarily linear.[47] Because of the ξ^2 dependence of the field threshold, the S-type threshold curve[45] is more easily observed with low ξ^2 materials, of large negative dielectric anisotropy.[46] It may happen that the mechanical contribution to the bend rate $\lambda E_0^2 \sim (K_3/\eta_B)(\pi/d)^2$ is large enough to fulfil the condition $T_E = \tau$. This occurs for thin enough materials of high

[47] See Lefèvre, R. J. W., in "Instabilities and Fluctuations" (T. Riste, ed.). Plenum, New York, 1975.

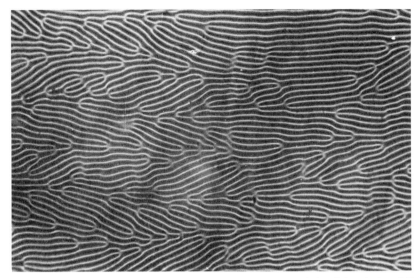

FIG. 28. The appearance of chevrons, above the threshold of the dielectric instability (courtesy of Y. Galerne).

resistivity. In this case, the conduction regime no longer exists; the dielectric regime is observed down to zero frequency; the value of its wavevector $q_x \sim \pi/d$ explains the possible confusion with "Williams domains." In all the usual cases where the conduction regime exists, the "dynamic scattering" mode used in displays corresponds to a field excitation well inside the instability domain in the $V(\omega)$ diagram (Fig. 29).

10. GENERALIZATION TO DIFFERENT MATERIALS AND GEOMETRIES

We have considered the planar case and a material of negative dielectric anisotropy ($\epsilon_a{}^-$) and of positive conductivity anisotropy ($\sigma_a{}^+$),

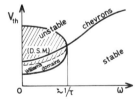

FIG. 29. The domains of instability, in a $V(\omega)$ diagram, for a sample of given thickness. The two regimes (Williams domains and chevrons) are separated by the dielectric relaxation frequency τ^{-1}.

which resulted in a strong source term for space charges. Materials of the $(\epsilon_a{}^-, \sigma_a{}^-)$ type have been discovered among nematics which show pretransitional effects close to a smectic phase. Fluctuations of smectic order (the "cybotactic groups") decrease the parallel conductivity and change the sign of σ_a.[39] By decreasing the temperature in the nematic phase, one first finds a normal unstable material $(\epsilon_a{}^-, \sigma_a{}^+)$; at a lower temperature, σ_H vanishes and the instability is no longer obtained. For typical nematics, this happens for a slightly negative value of σ_a (a few percent of σ). As the temperature is lowered further, one observes another instability, which is difficult to understand, because electric and hydrodynamic torques are now stabilizing. An explanation could be related to the fact that these materials undergo a smectic C phase transition.

The other situation of interest is in the (ϵ^+) materials in the homeotropic configuration. Consider first $(\epsilon_a{}^+, \sigma_a{}^-)$ materials which lead to the strongest anisotropic effect. The instability is now expected to appear as a splay mode (Fig. 30a); an approximate description of the effect is obtained simply in the one-dimensional model by writing K_1 (the splay elastic constant) instead of K_3 (the bend constant) and exchanging $\parallel$ and $\perp$ in the indices of ϵ and σ. The new σ_H is positive;[47a] space charges are induced as shown in Fig. 30a; the dielectric torque caused by the transverse field is destabilizing; the hydrodynamic torque originates now not only from the velocity gradient $\partial v_z / \partial x$ described in a one-dimensional model, but also from the component $\partial v_x / \partial z$. We do not repeat here the discussion on the relative values of these two torques, which has been given already in the discussion of Fig. 9a. One can predict a splay instability with, perhaps, a spatial period larger than usual (the sample thickness) to minimize the stabilizing part (due to $\partial v_z / \partial x$) of the hydrodynamic torque.

The other interesting materials are the $(\epsilon_a{}^+, \sigma_a{}^+)$ ones. When the relative anisotropy of σ is smaller than that of ϵ, nothing is changed from the previous situation. The interesting (and very common) case is the one of materials of small positive dielectric anisotropy and much larger positive conductivity anisotropy. σ_H is now negative; space charges accumulate as shown in Fig. 30b. The dielectric torque due to the transverse field is here stabilizing. The vertical part of the velocity gradient $\partial v_x / \partial z$ gives a strong stabilizing torque proportional to α_2. The only destabilizing torque comes from the velocity gradient $\partial v_z / \partial x$, parallel to the plates, but is weak (proportional to α_3). Following an argument already introduced in the discussion of Fig. 9, we expect instabilities to occur in the form of flattened vortices, with wavevectors

[47a] σ_H is now defined as $\sigma_H = \sigma_\perp (\epsilon_\parallel / \epsilon_\perp - \sigma_\parallel / \sigma_\perp)$.

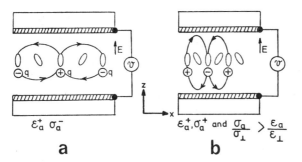

FIG. 30. Splay distortion for two kinds of materials. The $(\epsilon_-,\ \sigma_+)$ case is not dielectrically stable and gives rise to a Frederickcsz transition. It should be understood as: $(\Delta\epsilon \sim 0,\ \Delta\sigma > 0)$.

q_x much larger than $q_z = \pi/d$ and parallel to the plates. A necessary condition for instability to occur is that the destabilizing term $\alpha_3 q_x v_z$ be larger than the stabilizing term $\alpha_2 q_z v_x$. Using the incompressibility condition $q_x v_x + q_z v_z = 0$, this results in the inequality

$$q_x/q_z > (\alpha_2/\alpha_3)^{1/2}.$$

To help the formation of instabilities, one can minimize the stabilizing dielectric torque by choosing a material with a weak ϵ_a. The Orsay and Philips groups have observed instabilities of this type in materials of ϵ_a ranging from zero to 0.2. A splay square grid (Fig. 31a) at almost the same field for different thickness is observed down to zero frequency. Above a cutoff value $\sim\pi/d$ the wavevector of the grid follows a law of the type $q_x^{\,2} \sim \omega$. We relate it to a splay dielectric regime. At low frequency, localized butterfly wing-like instabilities appear which transform into "coffee beans" (see Fig. 31b). They appear at almost the same voltage for different thicknesses; we suggest that they may be identified with the flattened vortices of the conduction regime. At higher voltage, the corresponding flow becomes very rapid and seems to shrink in a small region of space ("coffee beans"), close to the plates. Note that the domain of existence of these "coffee beans" is a closed loop in a $V(\omega)$ diagram which may or may not intersect the dielectric threshold line (see Fig. 32). This is analogous to the conduction regime behavior in the planar geometry of $(\epsilon_a^{\,-},\ \sigma_a^{\,+})$ materials. Clearly we can compare these results to the Bénard problem in nematics; the situation $(\epsilon_a^{\,-},\ \sigma_a^{\,+})$ in planar materials is also obtained in the thermal case for $k_a^{\,+}$ (anisotropy $k_a = k_\| - k_\perp > 0$) (see Section III), whereas the case $(\epsilon_a^{\,+},\ \sigma_a^{\,-})$ for a homeotropic material compares with a nematic with $k_a^{\,+}$ heated from above.

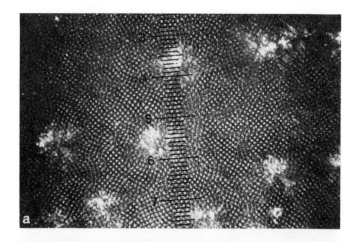

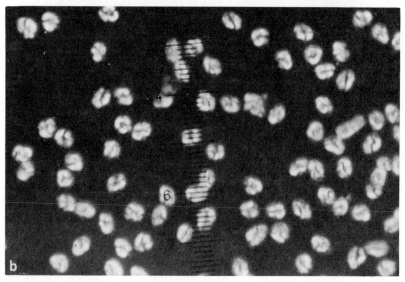

FIG. 31. (a) Homeotropic geometry. Grid pattern of a splay instability in a ($\Delta\epsilon \sim 0$, σ_+) material ($\omega/2\pi = 50$, $V = 170$ volts). (b) "Coffee beans" conduction-like instability in a $\Delta\epsilon$ ~ 0, σ_+ material ($\omega/2\pi = 4$kHz, $V = 90$ V).

A last kind of interesting materials are the nematics which exhibit a low dielectric relaxation frequency, down to the kHz range.[48] These materials have a positive dielectric anisotropy at low frequencies and a negative one at frequencies larger than the relaxation frequency (see for instance the review by de Jeu in this volume). This effect has been

[48] de Jeu, W. H., and Lathouwers, T. W., *Mol. Cryst. Liq. Cryst.* **26**, 255 (1974).

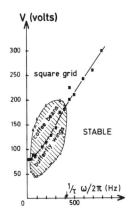

FIG. 32. A typical $V(\omega)$ diagram for a $(\Delta\epsilon \sim 0,\ \sigma_+)$ material $(CH_3O\ \phi C \equiv C\ \phi\ C_5H_{11})$. The splay grid pattern seems to be associated with a dielectric instability, conduction regime may correspond to the instability observed inside the closed-loop region (Orsay Group, private communication).

useful to visualize directly the relaxation frequency $\omega = 2\pi\tau^{-1}$ for $\epsilon_{\parallel}$; In the planar geometry, EHD instabilities (and the dynamic scattering mode) are only observed for $\omega > \omega_r$. For lower frequencies, the material undergoes a Freedericksz transition. An other possible application suggested by the Philips group was to use the change in dielectric anisotropy versus ω to restore rapidly an undistorted texture ("erasing" a display) in the DSM by shifting down the frequency of the applied field. In practice, the rapid thermal variation of ω_r complicates the realization of such devices. These ideas seem now obsolete, since almost all commercial displays using nematics work in the twisted wave guide mode rather than in the DS mode.[48a]

11. PRETRANSITIONAL PHENOMENA

Some interest has been aroused in recent years by the parallellism between the onset of linear instabilities and order–disorder phase transitions.[49] For this reason, "pretransitional" phenomena have been investigated; i.e., some properties of the system have been measured versus the applied field E, next to the threshold field E_c, in order to derive "power law" behavior of physical quantities in terms of the reduced variable $(E - E_c)/E_c$. Inelastic Rayleigh scattering is a convenient probe for the analysis of fluctuations close to an instability thresh-

[48a] Klingbiel, R. T., Genova, D. J., and Bücher, H. K., *Mol. Cryst. Liq. Cryst.* **27,** 1 (1974).

[49] See, for instance Haken, H., *Rev. Mod. Phys.* **47,** 67 (1975); Graham, R., in Lefèvre.[47]

old, provided that the medium is transparent and reasonably coupled to light waves. EHD instabilities in nematics are good candidates in this respect, because (a) the instabilities appear with a wavevector reasonably close to that of light waves, and (b) the cross section for light scattering by angular fluctuations of the director is quite large in nematics as compared with that in normal fluids.[50]

The behavior of the bend mode in the planar geometry of EHD is well described by Eq. (VI.1), since it is valid even in the limit of small amplitudes. The two modes of oscillation previously described also exist for the fluctuations of the system close to the instability threshold. Because of the source term of the form $E\psi$ and EQ in Eq. (VI.1), these modes are characterized by odd harmonic (ω, 3ω, 5ω, . . .) oscillations of one variable, coupled with even harmonic (0, 2ω, 4ω, . . .) oscillations of the other variable and vice-versa (see Fig. 34).

In the conduction regime, we look for a bend curvature solution of the form ψe^{st} (s close to zero near V_c). In the harmonic approximation (keeping only the Q dependence at the frequency ω), we find

$$s \sim -2(E_c - E)/E_c(1/T_E)$$

($1/T_E$ is the usual bend damping rate $K_3 q_x{}^2/\eta_B$).

The field variation of s is characteristic of a typical "mean field" type behavior (exponent 1). The damping time of the bend mode diverges and becomes infinite at the threshold, just because the destabilizing torque due to the coupling with space charges increases and cancels exactly the stabilizing torque. Such behavior is "classical" and has been observed in detail, in particular, using the Freederickzs transition instability.[13] Note that the range of field used ($1 < E_c/\Delta E_c < 30$) is probably too small to observe a nonclassical behavior. For large frequencies ($\omega\tau \gg 1$), there is no more gain mechanism for the conduction regime. The dielectric torque is stabilizing, and the damping rate increases just as

$$s \sim -(1/T_0 + \epsilon_a E^2/4\pi\eta_B).$$

This behavior has been checked by Martinand.[51]

In the dielectric regime, one has to look for solutions of the shape Qe^{st}, or $\psi e^{i\omega t}e^{st}$. One finds

$$s \sim -[(\xi^2 - 1)/\xi^2][(E_c - E)/E_c](1/\tau),$$

i.e., also a mean field dependence. The damped oscillations of ψ near threshold have been observed by the Orsay Group.[42] Along with the divergence of the damping time, one expects a divergence of the

[50] See Lekkerkerker, H. N. W., in Lefèvre.[47]
[51] Martinand, J. L., and Durand, G., Solid State Commun. 10, 815 (1972).

amplitude of the fluctuations, with a critical exponent $-1/2$. This last behavior has also been observed.[42] (Note that the same mean field exponents have also been obtained in the isotropic Bénard problem, where the 1/2 power of the dependance of the convective velocity has been checked with great accuracy.[14] See also the discussion of the dynamics of the thermal convection effects in nematics.)

The coherence length has been measured by the Rome group, and decreases as expected above threshold.[52] At the same time, rolls oscillate with a frequency linearly increasing with the field above threshold.[53] The length and number of disclination lines have been estimated by Galerne[54] in the dielectric regime. Further work is under way to understand the onset of disorder (DSM) above the "well-ordered" Williams instability.

12. DISCUSSION AND CONCLUSION

The model we have presented explains well the EHD of nematics, despite its limitations. Are these limitations very critical?

The first comes from the fact that the model deals only with ac excitation. A recent work of the Russian group[55,56] has demonstrated that the variable grating mode of Vistin[57] was due to a flexoelectric coupling, as suggested previously by Derzhanski.[58] A controversy remains for the nature of the dc excited EHD. In parallel with the Carr–Helfrich mechanism, proper to the nematic phase, the so-called Felici mechanism[59] can also lead to EDH, even in isotropic fluids. For dc, this mechanism is related to charge injection through the electrodes. Some experiments of the Grenoble group with a stabilizing magnetic field indicate that the Carr–Helfrich mechanism is also valid for dc, although other observations (mainly from the Russian Group) seem uncompatible with it.

Thus, some controversy remains. Note that the absolute ac conductivity does not enter the model, but in practice the lowest conductivity which allows the Carr–Helfrich mechanism is the one for which diffusion currents still remain negligible compared with conduction currents.

[52] Bartolino, R., Bertolotti, M., Scudieri, F. S., and Sette, D., *Appl. Opt.* **12**, 2917 (1973).
[53] Ribotta, R., Dunmur, J., and Durand, G., unpublished.
[54] Galerne, Y., private communication.
[55] Bobylev, Ju. P. and Pikin, S. A., *Zh. Eksp. Teor. Fiz.*, **72**, 369 (1977).
[56] Barnik, M. I., Blinov, L. M., Trufanov, A. N., and Umanski, B. A., *J. Phys. (Paris)* **39** (1978).
[57] Vistin, L. K., *Dokl. Akad. Nauk SSSR* **194**, 1318 (1970).
[58] Derzhanski, A. I., Petrov, A. G., Khinov, P., and Markovski, B. L., *Bulg. Phys. J.*, *I* **2**, 165 (1974).
[59] Felici, N., *Rev. Gen. Electr.* **78**, 717 (1969).

For the conduction regime, this requires that the Debye-screening length be smaller than simple thickness, i.e., for typical thin samples a maximum resistivity of the order 10^{10} or 10^{11} Ω·cm.

Another limitation of the model is the one-dimensional analysis. Note that this approximation is very good for the dielectric regime, where the system "adjusts" its instability wavevector q_x to a value much larger than d^{-1}. In the conduction regime, computer analyses[60,61] have also shown that the cutoff at $q_x = \pi/d$ is a very good approximation for the planar geometry.

The good agreement between the one- and two-dimensional models is probably due to the fact that in the planar geometry the hydrodynamic shear parallel to the plates gives a very weak torque on the molecules. For the homeotropic geometry the agreement is less obvious and two-dimensional analysis is needed, as already indicated in the thermal convection problem.

The two-dimensional computer analysis of the EHD problem made by Penz[62] is of interest. This author has computed the $E_c(q_x)$ threshold dependence on the wavevector q_x, which presents a minimum at $q_x \sim \pi/d$, as sketched on Fig. 33. This point gives the lowest threshold field of a mode of adjustable q_x, as in most experiments. If one could fix q_x at a different value by using parallel stripes as electrodes, one could in principle measure the $E_c(q_x)$ Penz curve. Unfortunately, the author derives the conclusion that, above threshold $(E > E_c)$, the system could oscillate with the two wavevectors q_{xa}, q_{xb} resulting in $E_c(q_{xa,b}) = E$. This claim is wrong and has been corrected in a subsequent paper.[59] Obviously, one single mode oscillates, with a wavevector between q_{xa} and q_{xb}. An estimate of the field dependence of q_x necessitates not only the knowledge of the wavevector dependence of the unsaturated gain, deduced from Fig. 33, but also an estimate of the q_x dependence of nonlinearities which give rise to saturation and allow the oscillation amplitude to stabilize. A guess of the q_x dependence of these terms for the one-dimensional model is given in Smith et al.[42] A detailed estimation of the nonlinearities occurring in this problem remains necessary to understand the properties of the EHD instabilities above threshold. Experiments are in progress to test the behavior of rolls in the conduction regime up to the dynamic scattering regime.

[60] Pikin, S. A., Sov. Phys.—JETP (Engl. Transl.) 33, 641 (1971); Pikin, S. A., and Shtolberg, A. A., Sov. Phys.—Crystallogr. (Engl. Transl.) 18, 283 (1973).
[61] Barnik, M. I., Blinov, L. M., Grebenkin, M. F., Pikin, S. A., and Chigrinov, V. G., Phys. Lett A 51, 175 (1975).
[62] Penz, P. A., Phys. Rev. 10, 1300 (1974).

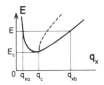

FIG. 33. The typical threshold dependence on the wavevector q_x computed with a two-dimensional analysis. The lowest threshold corresponding to a wavevector q_c is obtained for $E = E_c$. For larger electric fields E the instability is obtained in a band of wavevectors $q_{xa} < q < q_{xb}$. However, the observed variation of q above threshold (corresponding to a state of the fastest growth) inside these limits is different (broken line).

The last limitation of the model is the range of validity of the dielectric regime. As previously explained, diffusion currents can quench the space charge Q which builds up in the curvature ψ of the bend mode. As the optimum wavevector of the bend oscillation increases as $\omega^{1/2}$, there exists a maximum frequency ω_M where the bend period compares to the Debye–Hückel screening length, above which the Carr–Helfrich mechanism should loose its efficiency. A systematic study of the dielectric regime, for materials of differents ξ^2 (mainly with different dielectric anisotropy) has been made recently by the Russian group.[63] They found that, in the high-frequency limit, the "chevron" mode does not depend on the material anisotropy; the threshold voltage is found continuous at the nematic–isotropic transition temperature, with the voltage necessary to induce the high frequency Felici type of convective instability.[59] This kind of instability appears in isotropic liquids with a field threshold defined by the balance between stored dielectric energy $E_c^2/4\pi$ and the dissipation energy $\eta\omega$. This results in a threshold field $\sim\omega^{1/2}$ of the same order of magnitude as the dielectric regime threshold. Experiments are now running to clarify this point. Our present idea is that the isotropic Felici model can govern the high-frequency regime, but that the Carr–Helfrich mechanism explains well the dielectric regime below ω_M.

Appendix: Analog Simulation of Linear Instabilities

As pointed out in Guyon and Pieranski,[31] many instabilities in isotropic liquids, in liquid crystals, and even outside the field of physics can be described by a rather similar simple set of coupled equations. The system of two coupled linear differential equations given by Eqs. (V.7-1)

[63] Barnik, M. I., Blinov, L. M., Grebenkin, M. F., and Trufanov, A. N., *Mol. Cryst. Liq. Cryst.* **37**, 47 (1976).

and (V.7-2) is the simplest form, used through this study, of a more general form of equations. It has been applied to the description of hydrodynamic instabilities in nematics here as well as to surface undulations produced by rain flow!

Because of their general use, we will spend some time on these equations. Three different methods of solutions can be used:

1. An analytical solution has been obtained by Galerne[29] in the case of electrohydrodynamic equations in nematics as well as by Dubois-Violette,[28] and their results are discussed in the text.

2. A numerical integration has been done,[25] but does not give much new physical insight.

3. Finally, a simple electric circuit can simulate the problem. In the following we will discuss these *analog results*.

In order to avoid parasitic oscillations in the circuit, it is convenient to replace the differentiation operator by an integration. We replace the system of equations (V.7-1) and (V.7-2) by the equivalent one:

$$n_y = - \int \left(\frac{1}{T_y} n_y + As n_z \right) dt \qquad (A.1\text{-}1)$$

$$n_z = - \int \left(\frac{1}{T_z} n_z + Bs n_y \right) dt. \qquad (A.1\text{-}2)$$

The circuit which simulates this set of equations is represented in Fig. 34. The integrator has been realized by using the low cost IC operational amplifiers. The multiplication operation $s(t) \cdot n_{z,y}(t)$ is performed by two

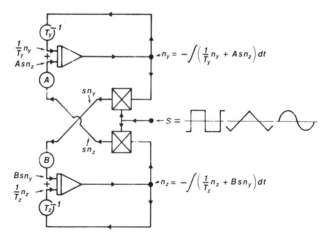

FIG. 34.   Circuit simulating set of equations (A.1-1) and (A.1-2).

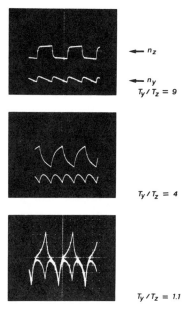

FIG. 35. Photographic displays of an oscilloscope, showing variation of $n_y(t)$ and $n_z(t)$ for different choices of the T_y and T_z, for a square wave signal.

standard four-quadrant integrated circuit multipliers. Finally, the multiplication by the same constant factor $(1/T_y, 1/T_z, B, A)$ is realized by means of the potentiometers.

The block diagram is very suggestive of the internal structure of the equations (A.1-1) and (A.1-2): It consists of two distinct loops for n_y or n_z with negative feedbacks. The positive feedback is realized by coupling these two loops through the use of multipliers which mix the n_z or n_y signals with $s(t)$. (In the hydrodynamic instabilities in nematics, $s(t)$ corresponds to the shear rate.)

Figure 35 gives some photographic displays of an oscilloscope giving the variation of $n_y(t)$ and $n_z(t)$ for different choices of the T_y and T_z and for a square-wave signal: When the two time constants are nearly equal (curves $T_y/T_z = 1.1$), the variations of n_y and n_z are very similar and practically identical over the first quarter of the period. However, n_z goes to zero more rapidly than n_y and crosses the null axis instead of just relaxing exponentially to zero as in the absence of coupling [this suggests that when $T_y/T_z = 1$, an instability cannot develop—cusp on Fig. 19 going to zero in the square wave case (curve C); see also Fig. 16]. On the other hand, when $T_y/T_z \gg 1$, n_y relaxes so slowly that it remains nearly constant as T_z oscillates around zero.

Threshold measurements can be done easily by changing the amplitude of s at constant T until the amplitude of the oscillations remains constant over several periods. The results agree with the numerical and analytical results but are less accurate. However, it is quite easy to adjust at will either the shape of the signal or the nature of the coupling (nonlinear one).

Lyotropic Liquid Crystals: Structures and Molecular Motions

JEAN CHARVOLIN

Laboratoire de Physique des Solides,* Université Paris Sud, Orsay, France

AND

ANNETTE TARDIEU

Centre de Génétique Moléculaire, C.N.R.S., Gif-sur-Yvette, France

I. Introduction

Lyotropic liquid crystals—which in fact exhibit both lyotropism and thermotropism—can be formed by amphiphilic molecules as different as lipids and copolymers. This chapter is concerned exclusively with lipids. Nevertheless, lipids encompass a large family of compounds from the

* Laboratoire associé au C.N.R.S.

209

simplest fatty acid salts and the ionic or nonionic surfactants to the more complex lipids of biological origin.

Early investigators of the mesomorphic state used to divide their interest between thermotropic (TLC) and lyotropic (LLC) liquid crystals.[1,2] Later, however, TLC and LLC studies followed separate paths because of distinctive, mostly experimental, features. Since the early days physicists have been able to apply their techniques and methods of investigation to TLC monodomains to explain a large number of macroscopic observations and to develop a most successful continuum theory by treating the molecules as rigid cylindrical rods. Accounts of these results are given in de Gennes[3] and in the part of this review devoted to TLC. A similar phenomenological approach to macroscopic behavior was more difficult to apply to LLC, as these systems do not easily grow large monodomains, contain two or three components, even much more in biological materials, and involve more flexible molecules. More particularly, the last two points preclude simple conceptual descriptions using one or two order parameters as in the case of TLC. Nevertheless, because of characteristics of their own, LLC rapidly attracted the interest of physical chemists and biologists. On the one hand the detergency and liposolubility of LLC were extensively studied, mainly because of technological applications. On the other hand, the occurrence of lipids in biological membranes led to the study of model systems such as lipid–water and lipid–protein–water; as a consequence the structural polymorphism of LLC is fairly well documented and many of the numerous phases may now be described in terms of molecular organization. Apart from their biological relevance, these structural studies also demonstrate new relationships between macroscopic order and molecular disorder.

The purpose of this chapter is to describe the present state of our knowledge of molecular organization and motion in LLC, with special emphasis on the complementary information obtained from X-ray structural analysis and from magnetic resonance (MR) studies of the microdynamic behavior. In the first part, we shall describe some aspects of the polymorphism; as the structures of the phases and their correlation with chemical composition are covered by several review articles,[4–6] the

[1] O. Lehmann, "Flüssige Kristalle." Engelmann, Liepzig, 1904.
[2] G. Friedel, *Ann. Phys. (Leipzig)* [4], **18**, 273 (1922).
[3] P. G. de Gennes, "The Physics of Liquid Crystals." Oxford Univ. Press (Clarendon), London and New York, 1974.
[4] V. Luzzati, *Biol. Membr.* **1**, 71 (1968).
[5] G. G. Shipley, *Biol. Membr.* **2**, 1 (1973).
[6] V. Luzzati and A. Tardieu, *Annu. Rev. Phys. Chem.* **25**, 79 (1974).

discussion will be restricted mainly to the interplay of order and disorder in the most commonly occurring phases. The second part will be devoted to more recent studies, concerning the characterization in space and time of the molecular disorder by MR techniques. Some comparisons will be made with corresponding observations in TLC.

II. Structural Polymorphism

1. GENERAL PROPERTIES OF THE LIPID MOLECULES

The amphiphilic character of lipids is due to the presence in the same molecule of a hydrophilic and a hydrophobic part, the two linked together by bonds sufficiently flexible to have rather independent behaviors. The hydrophobic part consists of one or two hydrocarbon chains, saturated or unsaturated (generally containing from 8 to 20 carbon atoms), while the hydrophilic groups are more heterogeneous. Some examples are given below.

Soaps: $[CH_3—(CH_2)_{n-2}—COO^-]_m \, X^{m+}$; X is an m valent cation.

Phospholipids of biological interest:

$$R_1—CO—OCH_2$$
$$|$$
$$R_2—CO—OCH$$
$$| \qquad O$$
$$CH_2—OPO \, X$$
$$O^-$$

R$_1$, R$_2$ are hydrocarbon chains

Phosphatidic acid: X = H

Lecithins: $X = CH_2—CH_2—N^+—(CH_3)_3$

Lysocompounds: H instead of R$_2$—CO

One property common to all lipids is the segregation of the hydrophobic and the hydrophilic moieties into distinct regions. In the presence of water, water and hydrocarbon are separated from each other by an interface covered by the polar groups of the lipid molecules. Yet, even in the anhydrous phases, the polar groups segregate out from the hydrocarbon region. In addition, the hydrocarbon chains of lipids are known to undergo order–disorder transitions as a function of temperature and/or water content.[4]

These two properties of lipids allow the formation of a wide variety of phases which display a high degree of long-range organization in one, two, or three dimensions, while remaining more or less disordered at the molecular level.

2. ANALYSIS OF THE X-RAY DIFFRACTION DATA

The structures presented in this article are all based upon X-ray diffraction studies. The first step in the structure analysis of a multicomponent system is to characterize the different phases, and to determine their range of existence, in other words to construct the phase diagrams. Several phase diagrams of lipid–water systems are shown in Fig. 1 (see also Shipley[5]).

Most often, the X-ray scattering experiments are performed on unoriented samples. Therefore, each phase is characterized by a family of reflections whose spacings and intensities are known. The spacings of the small-angle reflections, $s = 2 \sin \theta / \lambda \leq (10 \text{ Å})^{-1}$ specify the long-range organization—namely the symmetry and dimensions of the lattice. The reflections or diffuse bands in the high-angle region $s \geq (5 \text{ Å})^{-1}$ provide information on the conformation of the hydrocarbon chains.

In general, the structure analysis is based mainly upon the symmetry and dimensions of the lattice, the chemical composition, and the partial specific volumes, and leads to the determination of parameters such as the dimensions of the structure elements and the area available at the hydrocarbon–water interface; the intensity of the reflections and the electron density maps may provide valuable help.[4,7–10]

The justification of the structures described here can be found in the original publications.

3. GENERAL DESCRIPTION OF THE POLYMORPHISM AND CLASSIFICATION OF THE STRUCTURES

The classification used for TLC is inadequate for describing the large number of phases observed with lipid–water systems. Indeed nematic and cholesteric phases are never observed with lipids, and if some phases belong to the smectic class, phases ordered in more than one dimension are also frequently observed. Even the term "mesomorphic" introduced by Friedel[2] to specify states ordered in less than three dimensions is not well adapted to lipid–water phases as the distinction between one-, two-, and three-dimensional lattices does not appear to be of fundamental importance with lipids.

We prefer the classification introduced by Luzzati et al.,[4,6] which is

[7] V. Luzzati, A. Tardieu, and D. Taupin, J. Mol. Biol. **64**, 269 (1972).

[8] L. Mateu, V. Luzzati, Y. London, R. M. Gould, F. G. A. Vosseberg, and J. Olive, J. Mol. Biol. **75**, 697 (1973).

[9] Y. K. Levine, Ph.D. Thesis, University of London (1970).

[10] J. L. Ranck, L. Mateu, D. M. Sadler, A. Tardieu, T. Gulik-Krzywicki, and V. Luzzati, J. Mol. Biol. **85**, 249 (1974).

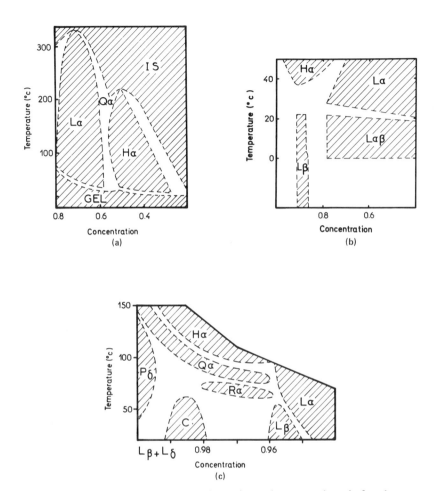

FIG. 1. Portions of phase diagrams. The regions where one phase is found pure are hatched. C is the weight concentration: lipid/(lipid + water). (a) System potassium palmitate–water. IS is the micellar isotropic solution, the other phases are described in the text. The potassium soaps having different chain length display the same behavior. With sodium soaps, the phase diagram is similar except for the cubic phase Qα which is absent and the Lβ gel which is replaced by the coagel (a crystal, water demixion). (b) System phosphatidic acid–water. The chain composition is heterogeneous. (c) Dry portion of the phase diagram of the system egg lecithin–water. The Lα phase can incorporate up to 45% water. For high hydration, the Lα phase containing 45% water is observed with water in excess. In these regions, the phase diagrams of synthetic lecithins are similar to that of egg lecithin. In the more hydrated region at low temperature, two phases wich chains in a mixed conformation, Pγ and Pαβ, are observed with egg lecithin[10] and two phases with chains in an ordered conformation, Pβ' and Lβ' are observed with synthetic lecithins[12] [see also in the updating list of references, Janiak *et al.* (1976)].

based upon the lattice space groups and the conformations of the hydrocarbon chains, completed by geometrical and topological considerations such as shape and content of the structure elements. In this classification the different phases are identified by Latin and Greek letters, which specify respectively, the lattice type and the organization of the chains.

For the most frequently observed lattices we use the following notation:

L for the one-dimensional lamellar lattice;
H for the two-dimensional hexagonal lattice (space group $p6m$);
P for the two-dimensional oblique or rectangular lattice (space groups $p2$ or cmm);
T, R, and Q for the three-dimensional tetragonal, rhombohedric, and cubic lattice of space group $I422$, $R3m$ and $Ia3d$,[4] respectively.

For the conformation of the hydrocarbon chains, we distinguish three main types:

α is the liquid-like conformation;
β, β', and δ are ordered ones;
$\alpha\beta$ and γ are mixed types.[6]

A more pictorial description is based upon the shape of the structure elements. Apart from micellar solutions where spherical particles can be found,[11] all the phases can be shown to consist of lamellar and rod-like elements. In lamellae containing phases, the structure elements may be infinite planar sheets packed in a one-dimensional lattice (phase L in Figs. 2a,b,c,f, 3a, 4), infinitely long ribbons of finite width or distorted sheets organized in two-dimensional lattices (Figs. 2d, 3b), or disks of finite size organized in a three-dimensional lattice.[4] In rod-containing phases, the structure elements are rods, all identical and crystallographically equivalent, either infinitely long or of finite length. The phase H (Fig. 3c) consists of infinitely long rods packed in a two-dimensional lattice. Other phases (T, R) consist of rods of finite length, linked three by three or four by four, forming two-dimensional hexagonal or square networks stacked into three-dimensional lattices.[4] Phase Q consists of rods joined three by three, forming two interwoven three-dimensional networks (Fig. 3d). In all the phases, with the exception of those formed by infinitely extended lamellae, the inner and outer volumes of the structure elements are topologically distinct. Therefore, two different situations can occur: The hydrocarbon chains may be inside and the

[11] F. Reiss-Husson and V. Luzzati, *J. Colloid Interface Sci.* **21**, 534 (1966).

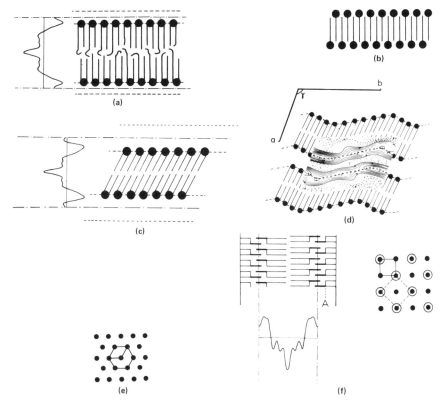

Fig. 2. Schematic representation of the molecular structure of phases with ordered chains. (a) Structure of the Lβ phase. The figure represents a section through a lamella. The black circles represent the polar groups, the straight lines represent the fully extended chains, the wriggles represent the chains in the liquid-like conformation. The chains are stiff and parallel, oriented perpendicular to the lamellae; a thin liquid layer is postulated in the center of the hydrocarbon layer, in order to take into account the length heterogeneity. On the left, electron density profile.[10] (b) Structure of the gel observed with a potassium soap. Here also the chains are stiff and parallel, and oriented perpendicular to the lamellae, but they are interdigiting. (c) Structure of the Lβ' phase. The chains are fully extended and tilted with respect to the normal to the lamella; here the end CH_3 groups of the two opposite monolayers are supposed to be in registry; the water layer is thicker than in the Lβ phase. On the left, electron density profile.[10] (d) Structure of the phase Pβ'. The figure represents a section of the two-dimensional periodic structure and shows the rippled lamellae. The insert is the electron density distribution with the negative contours dotted.[10] (e) Representation of the hexagonal organization of the hydrocarbon chains in a plane perpendicular to their axes, for the four preceding cases. (f) Structure of the Lδ phase. The polar groups are represented by heavy lines, and the hydrocarbon chains by light ones. The chains are coiled into helices with their axes perpendicular to the lamellae. The polar groups of two adjacent layers are in registry. On the left, electron density profile, on the right, section in a plane perpendicular to the hydrocarbon chains showing the tetragonal organization of the hydrocarbon chains and of the polar head groups. Filled circles represent the traces of the hydrocarbon chains, open circles represent the traces of the polar head groups.

polar medium outside (oil in water), or vice-versa (water in oil) (Fig. 3c,d).

Another possibility is to divide the phases into three main classes according to the conformation of the hydrocarbon chains: ordered, liquid-like, and mixed. Whenever the chains are ordered, the molecules can be considered as rigid rods, the structure elements can only be infinite sheets, and the polymorphism arises from the different organizations of the hydrocarbon chains.[10] In contrast, chains in a disordered conformation can fill volumes of various shapes; in this case, another type of polymorphism is encountered, corresponding to the shape and size of the structure elements.[4] Another source of polymorphism was discovered recently; when the hydrocarbon chains are heterogeneous, phases can be observed in which the chains segregate into microdomains with either an ordered or a disordered conformation. These phases are of the lamellar type. The "liquid" and "ordered" microdomains can be randomly distributed within the lamellae or periodically organized in two-dimensional lattices.[10]

4. STRUCTURES OF THE PHASES

We will now try to analyze in some detail the information obtained from X-ray diffraction studies of the local order and disorder in some of the phases. The results consist of information about the organization and conformation of the hydrocarbon chains and of the polar head groups, and on the correlations between neighboring molecules or structure elements. For the sake of clarity, we classify the phases according to the conformation of the chains. We will not consider here the crystalline phases which display a three-dimensional organization at the atomic level.

a. Lamellar Phases with Ordered Chains

The lamellar phases with ordered chains are observed at the low-water and low-temperature end of the phase diagrams (Fig. 1). The X-ray diagrams are characterized by the presence of sharp reflections in the high-angle region. The structure elements are symmetric bilayers. The hydrocarbon chains are stiff and all parallel organized according to two-dimensional lattices (Fig. 2).

A rectangular lattice was observed with Rb soap at low temperature,[4] a square one (side 4.8 Å) with dry lecithins (Lδ),[12,13] both with chains

[12] A. Tardieu, V. Luzzati, and F. C. Reman, J. Mol. Biol. 75, 711 (1973).
[13] V. Luzzati, T. Gulik-Krzywicki, and A. Tardieu, Nature (London) 218, 1031 (1968).

perpendicular to the lamellae, but the most common lattice is hexagonal with a side of about 4.8 Å. The chains may be perpendicular to the lamellae (β type),[12,14] or tilted (β' type) in two phases Lβ' and Pβ' observed with synthetic lecithins[12,15] (Fig. 2). The angle of tilt increases with the water content and is accompanied by a small distortion of the hexagonal lattice. In Lβ' the question of the interlamellar correlation of the tilt azimuth remains open.

In the β and β' conformations the thickness of the lipid layer corresponds to molecules with all the CH_2 groups in the extended trans–trans conformation; in the δ conformation,[12] the thickness of the lipid layer is smaller, indicating that the chains are possibly coiled into helices. It can thus be concluded that the two-dimensional ordering need not be accompanied by a full elongation of the chains. In these three types of conformations the symmetry of the lattice and the fact that the chains are crystallographically equivalent, indicates the presence of rotational disorder, either static or dynamic. When the lattice is rectangular the chain rotations may be partially frozen.

Proceeding with the description of the phases with chains in the β,β' conformations, two cases may be distinguished: Either the chains are interdigitating,[16] (Fig. 2b) or the CH_3 terminal groups are localized in the middle of the lamella (Fig. 2a,c,d,f), in a more or less disordered layer, according to the chain heterogeneity.[12,14]

As far as the organization of the hydrophilic head groups is concerned, no evidence has ever been obtained of any two-dimensional order when the chains are in the β and β' conformations. However, in the Lδ phase, the polar groups are organized according to a two-dimensional lattice having the same symmetry as the hydrocarbon chains[12] (Fig. 2f,g).

In the Pβ' phase, the lamellae are distorted[12] by wave-like ripples and the distortion propagates from layer to layer (Fig. 2d). Therefore the lattice is two-dimensional, one dimension being the lamellar repeat distance and the other the periodicity of the distortion which is of the order of hundreds of angstroms. In this type of phase strong correlations exist between the lamellae; these correlations are probably mediated by the polar head groups through the water layer. In this phase, as well as in the others, the correlations are long-range effects, which do not seem to involve correlations at the molecular level; indeed no indication of local three-dimensional order is observed.

[14] T. Gulik-Krzywicki, E. Rivas, and V. Luzzati, *J. Mol. Biol.* **27**, 303 (1967).
[15] Y. K. Levine and M. H. G. Wilkins, *Nature (London), New Biol.* **230**, 69 (1971).
[16] J. M. Vincent and A. E. Skoulios, *Acta Crystallogr.* **20**, 432 (1966).

b. Phases with Chains in the Liquid-Like Conformation: Type α

This type of conformation is highly disordered, like that of a liquid paraffin, but with the average of the chain orientations perpendicular to the lipid–water interface. This orientation is all the more pronounced as the area per polar group at the interface decreases.[17]

Although X-ray diffraction methods are of little help in establishing the structure of liquids, the X-ray study provides several arguments in favor of a liquid-like conformation:[6] (a) A diffuse band is observed in the X-ray diffraction diagrams around $s = (4.6 \text{ Å})^{-1}$ which is almost identical to that of a liquid paraffin, whereas for the β conformation a sharp reflection is observed at $(4.2 \text{ Å})^{-1}$. (b) In the lamellar phases, the thickness of the lipid leaflet decreases by about 30% when passing from the β to the α conformation ($\sim$60 and 40 Å in egg lecithin). (c) In lamellar phases the electron density profiles show a delocalization of the CH_3 groups when passing from the β to the α conformation.[10] (d) In the nonlamellar phases simple geometrical considerations indicate that the chains must be folded in a fairly irregular way in order to fill uniformly the oddly shaped volumes available to them (Fig. 3); it appears most unlikely that the conformation of the chains is profoundly different in the lamellar and the nonlamellar phases. (e) All the phases with the chains in the α conformation display a peculiar temperature effect analogous to that observed with rubber: The short dimension of the structure elements (thickness of the lipid bilayer in the lamellar phase, diameter of the rods in the hexagonal phase, etc.; see Fig. 3) decreases as the temperature is raised, with an unusually high linear thermal coefficient (of the order of $-10^{-3}/°C$).

i. Phases Observed in the Dry Region. With monovalent cation soaps, the phases observed are of the lamellar class. For each soap (K, Na, Li, Rb, Cs),[4,18,19] several two-dimensional phases are observed as a function of temperature. The structures of all these phases are similar, although their parameters differ: The polar groups are organized in infinitely long ribbons of finite width (Fig. 3b). One case was observed in which the polar groups are clustered in disks organized in a three-dimensional lattice.[4] At higher temperature, a one-dimensional lamellar phase is observed in which the polar groups are organized in planar sheets.

The phases obtained with divalent cation soaps (and most diacyl lipids) belong to the rod-like class. The polar groups are clustered in rods either of finite length and linked 3 by 3 or 4 by 4 to form three-

[17] Y. K. Levine, *Prog. Biophys. Mol. Biol.* **24**, 1 (1972).
[18] A. E. Skoulios and V. Luzzati, *Acta Crystallogr.* **14**, 278 (1961).
[19] B. Gallot and A. E. Skoulios, *Kolloid-z.* **209**, 164; **210**, 143 (1966).

dimensional lattices (phases Tα, Rα, Q$\alpha^{4,20-22}$) or infinitely long (phase Hα), organized in a two-dimensional hexagonal lattice. All these phases are of the type water in oil (Fig. 3).

Several arguments suggest that the degree of order of the polar head groups is quite high in all the phases, with the exception of the high-temperature lamellar phase of the monovalent cation soaps. (a) The temperature sequence of the phases is specific for the nature of the polar part. (b) In the ribbons and disks, the area per polar group is similar to that of the crystalline soaps, and independent of temperature. (c) The number of polar groups per rod length in each of the rod-like phases is specific for the cations and independent of temperature and of the length of the hydrocarbon chains.

ii. Phases Observed in the Hydrated Region. Three phases are commonly observed: Lα, Hα, and Qα (Fig. 3a,c,d). The lamellar phase Lα is the most widespread in lipid–water systems. The structure consists of planar lipid bilayers, all parallel and equidistant, separated by water layers; no other positional or rotational correlation appears to exist between the lamellae. The Hα and Qα phase have already been described; it may be noted that in phase Qα both the hydrocarbon and the polar medium are continuous and intertwined throughout the three-dimensional lattice.

The sequence of the phases observed with increasing water concentration appears to depend upon the number of hydrocarbon chains per polar head. With monovalent cation soaps the sequence is L, Q, H, isotropic solution (Fig. 1) with a few other phases occasionally between L and H. The water content of phases L and H varies in a fairly wide range, making it possible to study the effects of hydration on the size of the structure elements[4]; in these systems, phases H and Q are of the type oil in water. With diacylated lipids (a large number of these have been studied from chemically homogeneous synthetic compounds to highly heterogeneous extracts from membranes) the sequence of the phases observed is H, Q, L, with some other rod-like phases occasionally between H and L (see Fig. 1); H and Q are here of the type water in oil. When the diacyl lipids do not carry a net electrical charge the Lα phase reaches a maximum hydration; the hydration may be increased by adding a small amount of a lipid with a charged polar group to an uncharged lipid[23] and the hydration is very large if the lipid is charged. A

[20] V. Luzzati and P. A. Spegt, *Nature (London)* **215**, 701 (1967).

[21] V. Luzzati, A. Tardieu, and T. Gulik-Krzywicki, *Nature (London)* **217**, 1028 (1968).

[22] V. Luzzati, A. Tardieu, T. Gulik-Krzywicki, E. Rivas, and F. Reiss-Husson, *Nature (London)* **220**, 485 (1968).

[23] T. Gulik-Krzywicki, A. Tardieu, and V. Luzzati, *Mol. Cryst. Liq. Cryst.* **8**, 285 (1969).

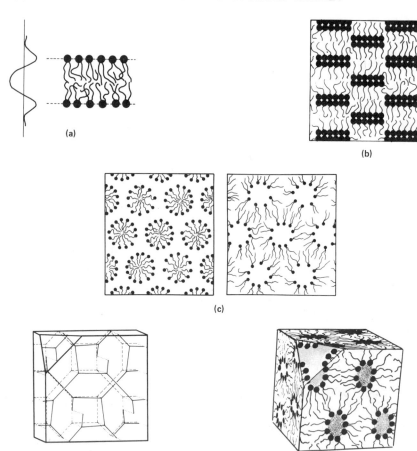

(a)

(b)

(c)

(d)

FIG. 3.   Schematic representation of the structure of phases with liquid chains. The polar groups are represented by dots, the hydrocarbon chains by wriggles. (a) and (b) are lamellar type phases, (c) and (d) are rod-like type phases. (a) Structure of the Lα phase. Section perpendicular to the lamellae. The electron density profile corresponds to a highly hydrated phase[7]; with lower water concentration the CH_3 groups may be more localized in the middle of the lamellae. (b) Structure of the two-dimensional centered rectangular phase of the anhydrous sodium soap. The loci of the polar groups are infinitely long ribbons of finite width.[18] (c) Structure of the Hα phase. The structure consists in infinitely long rods organized in a two-dimensional hexagonal lattice. Here is shown a section perpendicular to the rods. The left side corresponds to the type oil in water, as observed with monovalent soaps or lysolecithin; the right side corresponds to the type water in oil, as observed with diacyl phospholipids or divalent cation soaps. (d) Structure of the Qα phase. The cell is centered cubic, space group $Ia3d$. The structure consists of rods of finite length, all identical and crystallographically equivalent, joined three by three to form two interwoven three-dimensional networks. The left frame represents one unit cell, with the position of the axes of the rods. The right frame shows the distribution of water and hydrocarbon for a structure of the type water in oil, as observed in egg lecithin (Fig. 1).

general observation is that in each lipid system some phases may be absent, but the sequence of the phases is never reversed, and as the lipid concentration decreases at constant temperature, the area per polar group increases or remains constant, but never decreases, even if phase boundaries are crossed.

It may be noted that in lipids the separation between the thermotropic and the lyotropic polymorphism is rather loose. Indeed the same types of phases, Lα, Hα, Qα, etc., are observed in the dry and hydrated regions; as a rule the phases are of the type oil in water in the presence of water, and vice-versa in the dry, or almost dry, regions. The differences mainly concern the variety of phases observed, generally greater in the dry region and more strongly dependent upon the chemical composition of the head groups, which also appear more organized in the dry phases.

All the phases with the chains in the α conformation display a remarkable association of a highly developed long-range order with an extreme short-range disorder, to the extent that the structures may be visualized as a distribution of a polar and a hydrocarbon continuum, separated from each other by the hydrophilic groups of the lipid molecules.

iii. Energy Considerations. In an attempt to rationalize the structure of the different phases the contribution of several forces can be taken into account. (a) The repulsion between hydrocarbon and water, which leads to a segregation of the polar and apolar media and to the formation of an interface. This repulsion tends to decrease the area per chain at the interface. (b) The electrostatic interactions. A thermodynamic model taking into account these two types of forces has been put forward by Parsegian;[24] yet it should be noted that the nature and structure of the phases observed with ionic and neutral lipids often are quite similar. (c) The entropic contribution of the disordered conformations of the chains.[4] Thickening of the hydrocarbon leaflet (at constant volume) decreases the area available to one hydrophilic group; thus a surface tension acting against an increase of the mean area A per polar head at the interface is equivalent to a force stretching the hydrocarbon chains in a direction perpendicular to the interface. Each chain is thus analogous to a disordered linear polymer, stretched by an external force against thermal motion. (d) Steric factors related to the relative bulkiness of the polar and apolar moieties. Making the polar groups bulkier with respect to the hydrocarbon moiety has the effect of promoting phases with high surface/volume ratios and with distributions of the type oil in water and vice-versa.[4]

[24] A. Parsegian, *Science* **156**, 939 (1967).

Moreover, the very existence of planar and rod-like elements of finite width or length suggests the presence of local strains. These strains are probably due to the presence of both a fairly ordered distribution of the polar ends of the lipid molecules (quasicrystalline with dry phases and area limited by hydration in hydrated ones) and to a limited flexibility of the chains. This may lead to a situation in which the area per polar group is either too large or too small for the chain (or chains) anchored to that group. In this case, strains develop and accumulate as the size of the elements increases, and are released at the edge of the ribbons or at the junctions of several rods.

iv. Local Curvature. It may be interesting to consider how the local molecular ordering is characteristic of the different structures. One obvious point is the local curvature of the structure elements. If we consider the sequence of the phases observed in different systems with increasing water content, the curvature of the interface of the hydrocarbon regions varies in a fairly regular way, passing from concave in hexagonal structures of the type water in oil to planar and then convex in hexagonal phases of the type oil in water. Even the cubic phases fit well into this description as they present an intermediate situation in which both concave and convex interfaces are present. Such a variation of curvature appears to provide a convenient way to increase the area per polar group with increasing water content, while keeping a satisfactory packing of the hydrocarbon chains. Winsor[25] has stressed the relationship between curvature and the stability of the different phases.

c. Vesicles

In a large excess of water, diacylated phospholipids spontaneously form closed multilayers (liposomes), which after extensive sonication give rise to the formation of isolated lipid vesicles. These vesicles consist of a spherical lipid bilayer ($\sim$300 Å in diameter) whose properties are close to those of the bilayers of an Lα phase.[17] Such a system is not in thermodynamic equilibrium although it may be stable over a period of days. The structural characterization of the vesicles is rather loose and the packing of lipids in such distorted bilayers is not well understood, but useful information about their local structure and behavior has been obtained from EPR experiments.[26,27]

[25] P. A. Winsor, *Liq. Cryst. Plast. Cryst.* **1**, 199 (1974).

[26] R. B. Kornberg and H. M. McConnell, *Biochemistry* **10**, 1111 (1971).

[27] C. Taupin and H. M. McConnell, *Mitochondria Biomembr., Fed. Eur. Biochem. Soc., Meet., 8th,* Vol. 28, p. 219 (1972).

d. Phases with Mixed Conformations

Whenever the chains are chemically homogeneous, the β to α transition involves two phases, one with all the chains in the ordered, the other in the disordered conformation. With diacyl lipids with heterogeneous chains, the chains most often undergo a segregation during the α to β transition, and phases are observed in which the longest and most saturated chains are probably organized in β microdomains, the other chains remaining in the α conformation.[6,10,28] It appears that the conformations of the α and β domains are similar to those observed in the Lα and Lβ phases. The size of the domains may be estimated to be of the order of hundreds of angstroms. In some low-water phases the domains may become organized in two-dimensional lattices, the dimensions of the lattice corresponding respectively to the lamellar repeat distance and to the size of the domains. One example, the phase Pγ is shown in Fig. 4b. Such a lattice indicates the presence of strong correlations between lamellae, probably mediated by the polar head groups. In the phase L$\alpha\beta$ (Fig. 4a), where the layers of water between the lipid leaflets are sufficiently thick to hinder the correlations between lamellae, each lamella becomes a random mosaic of the two types of domains.[10]

It may be noted that the Lα → L$\alpha\beta$ and Lα → Pγ transitions involve the segregation of the lipid molecules, and thus diffusion phenomena at fairly large distances.[29]

5. DISCUSSION AND COMPARISON WITH THERMOTROPIC LIQUID CRYSTALS

One of the most remarkable characteristics of lipids is their ability to combine in one phase a periodically ordered long-range organization (in one, two, or three dimensions) and a highly disordered short-range conformation. We have emphasized the liquid-like disorder of the α conformation, which is the most widespread of all the conformations. We should stress the sharp distinction between the α and the other partly ordered conformations: The presently available X-ray diffraction evidence is against the existence of any intermediate conformation. The other conformations are at least partly ordered, namely the chains are stiff and parallel, and organized, with rotational disorder, according to

[28] V. Luzzati, A. Tardieu, T. Gulik-Krzywicki, L. Mateu, J. L. Ranck, E. Shechter, M. Chabre, and F. Caron, *Mitochondria Biomembr., Fed. Eur. Biochem. Soc., Meet., 8th,* Vol. 28, p. 173 (1972).
[29] Y. Dupont, A. Gabriel, M. Chabre, T. Gulik-Krzywicki, and E. Shechter, *Nature (London)* **238,** 331 (1972).

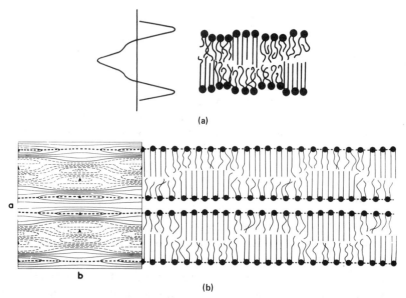

FIG. 4. Schematic representation of phases with hydrocarbon chains in a mixed conformation. (a) Schematic view of the structure of the lipid lamellae of the Lαβ phase; the lamellae can be visualized as disordered mosaics of α and β domains. (b) Phase Pγ of the system egg lecithin–water. The lattice is two-dimensional centered rectangular, space group *cmm*. Each monolayer is formed by an ordered mosaic of α and β domains, each domain being infinitely long and of finite width. The symmetry plane is in the water layer, and monolayers of the same conformation are associated through their polar faces. In the electron density map the negative contours are dotted.[7]

two-dimensional lattices. Nevertheless, none of the phases displays local three-dimensional ordering of the molecules; this allows these phases to be differentiated from "plastic crystals." Finally, the occurrence in lipid–water systems of several types of structure elements—lamellae or rods, of finite or infinite size—may be stressed once more.

Bearing these properties in mind, we can recall some of the analogies and differences which may be established between LLC obtained with lipids and TLC obtained with other compounds. From the chemical standpoint, the lipids commonly studied possess long hydrocarbon chains and appear to be rather flexible molecules; on the other hand, most of the compounds forming TLC consist of bulky, rigid, and elongated aromatic parts, linked to rather short hydrocarbon chains. This difference in rigidity of the molecule is probably related to the differences observed in the type of polymorphism. Indeed phases like nematic, cholesteric and smectic C are never observed with lipids, whereas phases with rod-like structure elements or with lamellae of finite size do not occur in TLC. The only phases observed both with

TLC and LLC are the smectic phases. A correspondence has been established between the phases $L\beta$ or $L\beta'$ and smectic B or H, because of the occurrence in both of a two-dimensional hexagonal lattice within the lamellae, and between the phase $L\alpha$ and smectic A, because of the lack of order within the lamella, and correlations between lamellae. However, if we take into consideration the conformation of the hydrocarbon part in TLC and LLC, some differences become apparent. In the phases $L\beta$ and $L\beta'$, the hexagonal lattice is formed by the hydrocarbon chains (cross section ~20 Å^2), while in smectic B and A the dimensions of the lattice are fixed by the aromatic part of the molecules (cross section ~25 Å^2). Furthermore, in LLC the $L\beta$ to $L\alpha$ transition corresponds to an order–disorder transition of the chains; in TLC instead it may be inferred from NMR, ESR, X-ray, and dilatometric measurements[30–32] that the order–disorder transition of the hydrocarbon part takes place at the solid–smectic B transition, and the conformation of the hydrocarbon part is very similar in smectics B, C, and A. Another difference corresponds to the existence of longitudinal correlations between molecules of adjacent layers in smectic B phases of short-chain compounds.[33] Such correlations are never observed with lipids.

Thus, until now the polymorphism, as well as the type of local disorder, appears to present different characteristics in TLC and LLC. Nevertheless, it is perhaps reasonable to suppose that changing the balance between the hydrocarbon and the aromatic part in TLC, either by increasing the chain length or changing the types of aromatic core, could lead to a polymorphism similar to that of lipid–water systems.

It should be noted, as a final technical comment on the characterization of local disorder by X-ray diffraction techniques, that while the description of the phases with ordered hydrocarbon chains may be fairly accurate, the technique is of little help for the study of the type of short-range disorder observed in the α conformation. It will be the purpose of the second part to analyze it in terms of anisotropy, diffusion, and deformations of the molecules.

III. Local Behavior of Lipids

6. DIFFERENT APPROACHES

An evident correlation exists between the disordered conformation (type α) of the hydrocarbon chains of the lipid molecules and the

[30] B. Deloche, J. Charvolin, L. Liebert, and L. Strzelecki, *J. Phys. (Paris)* **36**, C1-21 (1975).
[31] F. Poldy, M. Dvolaitsky, and C. Taupin, *J. Phys. (Paris)* **36**, C1-27 (1975).
[32] D. Guillon and A. E. Skoulios, *C.R. Hebd. Seances Acad. Sci., Ser. C* **278**, 389 (1974).
[33] J. Doucet, A. M. Levelut, and M. Lambert, *Mol. Cryst. Liq. Cryst.* **24**, 317 (1973).

diversified geometry of the structure elements. This has prompted the investigation of lipid–water systems with techniques sensitive to local fields and fluctuations in order to characterize the molecular disorder and particularly the evolution of the disorder all along a lipid chain. Among these techniques the magnetic resonance ones, nuclear (NMR) and electronic (EPR), have provided very fruitful information because of their selectivity and the suitability of their characteristic frequencies to those of molecular motions in lyotropic liquid crystals (LLC). Infrared, Raman, and fluorescence[34–36] spectroscopies are also being applied in this field; they take advantage of their sensitivity to the nature and the structure of the molecular environment, but their quantitative interpretation in terms of chain dynamics is not straightforward. Finally, any detailed model developed from the results of the previous investigations has to take into account the thermodynamic data,[37] which will not be presented here.

After giving a brief summary of the type of information which can be obtained from magnetic resonance studies we shall present and discuss some of their most characteristic results. Our presentation will be a very limited review of the abundant literature published during the last 10 years in this domain. We want to consider here the microdynamic behavior of lyotropic liquid crystals in relation to their structure. For this purpose we have chosen to analyze some typical experiments made with chemically simple monoacyl lipids (presenting isomerizations in the chain only) in the well determined structures described previously. We have discarded most of the important work in which the system under investigation is mainly considered for its analogy with biomembranes; most often in such cases the molecules are diacyl phospholipids (presenting isomerizations in the chains as well as in the polar heads), organized in vesicles or liposomes whose structures are not quantitatively characterized yet. We suggest that the interested reader consult the well documented reviews by Johanson and Lindman[38] and Tiddy,[39] in which are recorded and discussed all those works otherwise dispersed in very diverse publications. Also, we shall not consider the use of LLC as solvents for structural studies of solute molecules.[105]

[34] B. J. Bulkin and N. Krishnamachari, *J. Am. Chem. Soc.* **95,** 1109 (1972).

[35] K. Larsson, *Chem. Phys. Lipids* **10,** 165 (1973).

[36] A. S. Waggoner and L. Stryer, *Proc. Natl. Acad. Sci. U.S.A.* **67,** 579 (1970).

[37] M. C. Phillips, R. M. Williams, and D. Chapman, *Chem. Phys. Lipids* **3,** 234 (1969).

[38] A. Johansson and B. Lindman, *Liq. Cryst. Plast. Cryst.* **2,** 192 (1974).

[39] G. J. T. Tiddy, "NMR of Liquid Crystals and Micellar Solutions," Chem. Soc. Spec. Period. Rep., Vol. 4. Chem. Soc., London, 1975.

7. RELEVANT ASPECTS OF MAGNETIC RESONANCE[40]

Nuclear spins and paramagnetic spins are local probes which can be very sensitive to the structure and the fluctuations of their environment. As a matter of fact the energy levels of the spins in the presence of an external magnetic field depend not only upon the value of this field (Zeeman Hamiltonian $\mathcal{H}_z$) but also upon the numerous interactions of the spins with their magnetic or electric surroundings (Hamiltonian $\mathcal{H}_i$). When the interactions are functions of spatial parameters, the spin system is coupled to the degrees of freedom of the lattice through $\mathcal{H}_i$ which is time dependent when motions occur. The position of the resonance and its lineshape could in principle be calculated from the time-dependent Hamiltonian

$$\mathcal{H} = \mathcal{H}_z + \mathcal{H}_i(t).$$

a. Some Spin-Lattice Interactions

Let us consider the most frequent $\mathcal{H}_i$ encountered in liquid-crystal studies. The interactions are tensorial and their values depend upon the orientation of the tensor in a frame of coordinates related to the external magnetic field which is the axis of spin quantization.

In proton magnetic resonance (PMR) experiments the magnetic dipolar interactions between neighboring protons are predominant and depend not only upon the orientation of the proton–proton axis in the field but also upon the distance between the two protons; a typical value of this coupling in frequency units is about 10^4 Hz. With such an interaction it should be possible to study the motions of a pair of methylene protons on a hydrocarbon chain. Unfortunately this coupling is not at all selective because the protons of the pair are also coupled to the other neighboring protons; it is thus difficult to resolve methylene groups of the same chain or of different chains. More selective interactions have therefore been investigated.

For the NMR studies of the C^{13} nuclei naturally present on the chain skeleton two magnetic interactions have to be considered: One is the coupling with the orbital of the binding electrons which gives rise to the so called "chemical" shift of the resonance; the other is the dipolar coupling with the methylene protons. The first is very sensitive to the location of the C^{13} nuclei along the chain and the chemical shifts are distributed over a few 10^{-2} Hz; this coupling allows differentiation between methylenes but, being mainly a scalar isotropic coupling, it is in

[40] C. P. Poole, Jr., and H. A. Farach, "Relaxation in Magnetic Resonance." Academic Press, New York, 1971.

first order insensitive to motions (the anisotropic part of this coupling is not yet accurately known). The second is sensitive to the amplitude and rate of the methylene motions, but in anisotropic systems such as liquid crystals the nonaveraged part of the dipolar interaction broadens the C^{13} lines to such an extent that the C^{13} resolution is lost. One can get rid of the residual dipolar broadening by decoupling the H^1 and C^{13} resonances;[41] nevertheless this complicates the experiment and its interpretation. C^{13} magnetic resonance presently finds its full application in the study of isotropic systems, as for instance liquid alkanes[42] or lipids in vesicles.[43]

Specific information concerning the local behavior of anisotropic lipid systems has been obtained from EPR studies of spin-labeled molecules.[44] Here a paramagnetic nitroxide radical is attached at a definite position along the chain; its resonance is mainly dominated by the hyperfine coupling of the unpaired electron with the nitrogen nuclear moment, giving a typical value of the coupling of about 10^8 Hz. This probe is very sensitive to the local behavior; however, the fact that the spin-labeled molecule is a foreign probe in the structure can complicate the interpretation.

The difficulties inherent with the previous methods have recently promoted the NMR study of deuterated molecules (DMR) despite the burden of isotopic substitution. The interaction considered here is the electric quadrupolar interaction which, by coupling the deuteron quadrupole with the electric field gradient of the electronic structure of the deuterium bond, gives a doublet structure to the DMR line. A typical value of the coupling in the CD bond is about 10^5 Hz. Since in the first approximation the dipolar coupling between neighboring deuterons is negligible because of their low magnetic moment, the behavior of each methylene along a chain can then be studied.

EPR and NMR have, of course, their own fields of application but, when applied in similar systems, comparison of their results can also bring useful complementary information, as will be illustrated later on.

b. Motions and Lineshapes

In the presence of motion the total Hamiltonian is more conveniently written

$$\mathcal{H} = \mathcal{H}_z + \langle \mathcal{H}_i(t) \rangle + (\mathcal{H}_i(t) - \langle \mathcal{H}_i(t) \rangle).$$

[41] A. Pines, D. J. Ruben, and S. Allison, *Phys. Rev. Lett.* **33**, 1002 (1974).

[42] Y. K. Levine, N. J. M. Birdsall, A. G. Lee, J. C. Metcalfe, P. Partington, and G. C. K. Roberts, *J. Chem. Phys.* **60**, 2890 (1974).

[43] J. C. Metcalfe, N. J. M. Birdsall, and A. G. Lee, *Mitochondria Biomembr., Fed. Eur. Biochem. Soc., Meet., 8th*, Vol. 28, p. 197 (1972).

[44] H. M. McConnell and B. Gaffney, *Q. Rev. Biophys.* **3**, 91 (1970).

The time-dependent interaction Hamiltonian $\mathcal{H}_i(t)$ has been decomposed into two terms. The first is its time-averaged value, which is nonzero if the motion is anisotropic. The second contains the time dependence and fluctuates with a zero time average. Both terms contribute to the absorption lineshape, although in rather different ways.

The first static term displaces the Zeeman levels, either giving a structure to the resonance line or broadening it, depending on whether this term has defined values or distributed ones. In the latter case, which is predominant in dipolar solids, the linewidth characterizes the width of the distribution, and the lineshape is then approximately Gaussian and its width is usually defined by the second moment (M_2). Due to this distribution the spins do not precess at the same frequency and the inverse linewidth is also a measure of the mean time for the spins to get out of phase.

The second fluctuating term acts more subtly according to the frequency of the motion of interest (hereafter τ^{-1}). The fluctuation frequency has to be compared with the spin dephasing rate associated with the second term. This dephasing rate can be written as the square of the mean quadratic value of the modulated interaction $\langle(\mathcal{H}_i(t) - \langle\mathcal{H}_i(t)\rangle)^2\rangle$ (hereafter called h^2). Thus, if the fluctuation is slow, i.e., if

$$\tau^2 h^2 > 1,$$

no averaging is observed and the second term contributes to the linewidth as the first does; but if the fluctuation becomes more rapid, i.e., if

$$\cdot\,\tau^2 h^2 < 1,$$

the spins see only the mean value of the fluctuating term, which is equal to zero; the line narrows accordingly. This fluctuating term then intervenes in the linewidth through its low frequency components which can induce transitions conserving the magnetic energy but limiting the lifetime of a spin state. This contribution to the linewidth is weak and most often visible only when the first static term is zero, that is to say in isotropic fluid media. In this extreme case the lineshape is Lorentzian and its width is due to relaxation effects; the width is a measure of the inverse of the transverse relaxation time (T_2).

It is clear that the dynamic information contained in the fluctuating term cannot be extracted easily from the inspection of lineshapes except in the latter extreme case. To get more detailed information it is necessary to look in detail at the relaxation processes.

c. Motions and Nuclear Spin-Lattice Relaxation

The fluctuating Hamiltonian, which has a wide frequency spectrum, can induce transitions between different Zeeman energy levels (i.e., without keeping this energy constant). Thus, after a perturbation of the spin population on the Zeeman levels, the exponential return to thermal equilibrium is characteristic of the efficiency of the spin-lattice coupling, and its rate equal to the spectral density of the fluctuating Hamiltonian at the Zeeman frequency. For instance, the spectral density of a fluctuating magnetic field having an exponential correlation function is

$$J(\omega) = |h^2| \, 2\tau/(1 + \omega^2\tau^2),$$

where h^2 is the mean quadratic strength of the fluctuating field and τ the characteristic time of its variations. The spin-lattice relaxation rates can be measured at zero frequency (transverse relaxation T_2^{-1} in liquids), at the low frequency of the dipolar fields (dipolar relaxation T_D^{-1} in dipolar solids), in the 10^4 Hz range (rotating frame relaxation $T_{1\rho}^{-1}$), and in the 10^7 Hz range (longitudinal relaxation T_1^{-1}). These measurements can thus provide information about correlation functions in a rather wide dynamic range, from a few Hz to about 10^8 Hz. (The above formula for the relaxation time is valid for fast or slow motion, although the derivation of the formula is different. Special consideration is necessary for motions of intermediate rate for which $\tau^2h^2 \sim 1$.)

d. Nature of the Information Obtained with Magnetic Resonance

Magnetic resonance techniques appear well fitted for describing the molecular behavior in relation to the structure of the phase: The orientation or the reorientations of a particular chemical group of the molecule can be investigated. However, some experimental features have to be kept in mind when interpreting magnetic resonance experiments. In linewidth studies a motion can be detected only when averaging effects appear, that is, if the characteristic frequency of the motion (τ^{-1}) is higher than the characteristic frequency of the magnetic resonance interaction under consideration. In DMR a motion will be detected if $\tau^{-1} > 10^5$ Hz, in EPR if $\tau^{-1} > 10^8$ Hz; the same disorder can then be seen as dynamic in DMR and static in EPR if $10^5 < \tau^{-1} < 10^8$ Hz. This complementary aspect of the two experiments can be of interest. The measurement of nuclear relaxation as a function of frequency also makes it possible to investigate motions slower than the motional narrowing limit (analogous EPR measurements would be technologically much more complex) and, as already stated, the dynamical range covered can extend from a few Hz to about 10^8 Hz.

All this information concerning the behavior of one particular methylene, or nitroxyde radical, has a very local character. The search for information about correlated motions of neighboring molecules would require the study of the relative motions of their methylene groups. This

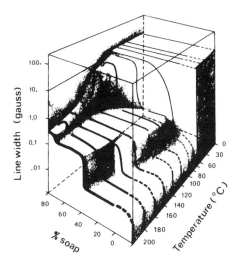

FIG. 5.   PMR linewidth of sodium palmitate in D_2O as a function of temperature and concentration. From Lawson and Flautt.[46] (Reprinted with permission from the Journal of Physical Chemistry. Copyright by the American Chemical Society.)

could in principle be obtained from the intermolecular contribution to the linewidth; however, it is much smaller than the intramolecular one and the expected effect from collective molecular motions in lyotropics is a very weak one. In this respect collective motions are probably better investigated with light-scattering techniques than with NMR, as it may be inferred from thermotropic studies where collective motions in smectic phases have been shown to have wavelengths extending up to the optical range.

8.   EVIDENCES FOR A DYNAMICAL DISORDER

We start the discussion by presenting the exhaustive PMR linewidth study of the sodium palmitate–water system by Lawson and Flaut.[45] The linewidths of palmitate protons are shown in Fig. 5 as a function of temperature and water content. Three primary levels of linewidth are apparent and can be associated with the phases of the system. The upper level ($\sim$10 G) comes from the crystalline phases. Here the structural data give the relative positions of the protons of the chains; the dipolar lineshapes can be calculated *a priori*, and the agreement with the experimental data is good if free rotation of the end methyl groups around their symmetry axis is admitted.[46] The lower level ($\sim$0.01 G)

[45] K. D. Lawson and T. J. Flautt, *Mol. Cryst.* **1**, 241 (1966); *J. Phys. Chem.* **72**, 2066 (1968).

[46] K. D. Lawson and T. J. Flautt, *J. Phys. Chem.* **69**, 4256 (1965).

corresponds to the fluid micellar zone. The lines are very narrow, all dipolar interactions are averaged out by isotropic motions—overall micellar rotation or/and translational molecular diffusion in the micelle. The intermediate level ($\sim$0.1 G) corresponds to the liquid-crystalline phases: the anhydrous phases, with which we shall not be concerned,[47] and the hydrated lamellar and hexagonal phases. Here the linewidths are about two orders of magnitude smaller than in the crystalline phases, demonstrating the existence of motions, but, as they are still larger than in the micellar phase, these motions preserve some residual dipolar interactions and can be assumed to be anisotropic. It appears, however, that either the experiment or the chosen parameter (the half-height linewidth) is not sensitive enough to differentiate between lamellar and hexagonal phases.

We shall first define the nature of these motions by simply studying NMR and EPR lineshapes for a few particularly illustrative experiments in the well-known lamellar, cubic, and hexagonal phases; then we shall determine and discuss their characteristic times.

a. Lamellar Phase, Lα

Clear qualitative ideas concerning the behavior of the hydrocarbon region of the bilayer have been provided by the EPR of labeled fatty acid[48a] or labeled phospholipids dissolved in the phase.

The EPR spectra demonstrate the existence of anisotropic motions, and a continuous decrease of the order parameter is observed when the label is displaced from the polar head to the methyl end, as shown in Fig. 6.[48] This suggests that the chains are rotating around their long axis and that their segments are less and less constrained when going from the interface into the hydrocarbon medium. In the Lβ phase, where the chains are rigid, the order parameter is close to unity and almost independent of the location of the label along the chain.

This method is of great efficiency in characterizing the disorder; however, particular care in the interpretation has to be taken around the transitions, when segregation effects can occur: Investigation of the Lαβ phase (Fig. 4) of the system phosphatidic acid–water has shown that this

---

[47] J. A. Ripmeester and B. A. Dunell, *Can. J. Chem.* **49**, 731 (1971).
[48] J. Seelig, N. Limacher, and P. Bader, *J. Am. Chem. Soc.* **94**, 6364 (1972).
[48a] General formula:

$$CH_3\!\!-\!\!(CH_2)_n\!\!-\!\!-\!\!C\!\!-\!\!-\!\!(CH_2)_m\!\!-\!\!CO_2H$$

$$O \qquad N\!\overset{\cdot}{\rightarrow}O$$

The degree of order of the spin label is defined by $S = \frac{1}{2}\langle 3\cos^2\theta - 1\rangle$, where θ is the angle between the nitrogen $2p\pi$ orbital and the normal to the bilayer plane.

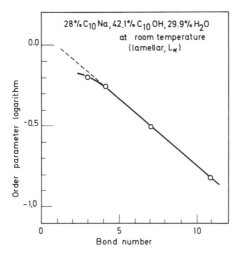

FIG. 6.   Logarithmic plot of the order parameter S_3 of a nitroxide spin label where the spin label is displaced along the chain, from the polar head, in a ternary system, soap–alcohol–water. From Seelig.[48b] (Reprinted with permission from the Journal of the American Chemical Society. Copyright by the American Chemical Society.)

system can be characterized differently according to the position of the label on the chain.[49] When the nitroxide is close to the polar head the labeled molecule is incorporated indifferently in the α or β domains and the phase is characterized as L$\alpha\beta$, whereas when the nitroxyde is far from the polar head, it is expelled from the β domains and the phase is seen as Lα.

With PMR the system is investigated without perturbation, through the protons of its molecules. However, the weak resolution of this method makes tedious any attempt for setting out a qualitative "disorder profile" in the bilayer, such as the one obtained from EPR. This can be seen in Fig. 7, which represents the PMR line for an unoriented sample of the lamellar phase; the lineshape has no fine structure assignable to the different methylene groups. Although the study of lamellar samples oriented between glass plates brings some resolution improvements,[50] the very nature of a proton spin system in an anisotropic medium introduces a major limitation. The spin dipolar interactions are not expected to be totally averaged by the anisotropic motions and we are then dealing with a coupled proton spin system instead of individual protons (or methylenes). This explanation was suggested by G. Tiddy for an oriented ternary lamellar system, the linewidth of which was

[48b] J. Seelig, *J. Am. Chem. Soc.* **92**, 3881 (1970).
[49] F. Caron, L. Mateu, P. Rigny, and R. Azerad, *J. Mol. Biol.* **85**, 279 (1970).
[50] C. Dijkema and M. J. C. Berendsen, *J. Magn. Reson.* **14**, 251 (1974).

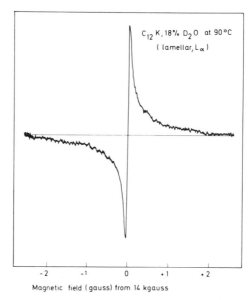

C_{12} K, 18% D_2O at 90 °C
(lamellar, L_α)

-2   -1   0   +1   +2
Magnetic field (gauss) from 14 kgauss

FIG. 7.   Derivative of the PMR spectrum of paraffin protons in a nonoriented lamellar sample.

dependent upon its orientation in the field,[51] and was confirmed by studying potassium laurate–water and lecithin–water[52] systems with pulsed NMR techniques developed for the studies of dipolar solids. Thus the lineshape of an unoriented sample contains two contributions: the homogeneous one of the coupled spins for each monodomain, and the inhomogeneous broadening introduced by the isotropic angular distribution of monodomains in the sample; the interpretation of the spectrum is thus not straightforward.[53–55] To go further with magnetic resonance, it was necessary to work with decoupled spins (to get rid of complex problems of spin dynamics[53]) in oriented samples (to get rid of the angular inhomogeneous broadening). This was achieved recently by the use of deuterated molecules.

A DMR spectrum obtained with an oriented sample of deuterated potassium laurate–water[56] in the phase $L\alpha$ is shown in Fig. 8. The

[51] G. J. T. Tiddy, *Nature (London), Phys. Sci.* **230**, 136 (1971).

[52] J. Charvolin and P. Rigny, *Nature (London), New Biol.* **237**, 127 (1972).

[53] M. Bloom, *Pulsed Nucl. Magn. Reson. Spin Dyn. Solids, Proc. Spec. Colloq. Ampere, 1st, 1973*; M. Bloom, E. E. Burnell, S. B. W. Roeder, and M. I. Valič, *J. Chem. Phys.* **66**, 3012 (1976).

[54] C. M. A. Seiter and S. I. Chan, *J. Am. Chem. Soc.* **95**, 7541 (1973).

[55] M. Wennerström, *Chem. Phys. Lett.* **18**, 41 (1973).

[56] J. Charvolin, P. Manneville, and B. Deloche, *Chem. Phys. Lett.* **23**, 345 (1973).

effective quadrupolar coupling constants, extracted from the doublet structure of the DMR line[56a] are much smaller than those of a static C—D bond, which are 167 kHz.[57] These residual quadrupolar splittings, and their variation with the orientation of the magnetic field,[56] directly shows that molecular motions with uniaxial symmetry take place and average a large part of the static coupling. As these splittings are distinct, the amplitude of the motions must vary along the chain. The narrowness of the lines also suggests that the motions are faster than the modulated interactions, i.e., τ is smaller than about 10^{-5} sec. Each line was attributed to methyl, methylenes, or groups of methylenes, assuming that the farther the group is from the polar head the stronger is the averaging effect. This identification was confirmed by Seelig and Niederberger, working with step-by-step deuterated molecules in the decanol–sodium decanoate–water system.[58] An interesting feature of the spectrum is that six methylenes close to the polar head have about the same degree of order; this characteristic was also encountered in the ternary system of Seelig and Neiderberger[58] and with lecithin,[59] and is not seen

---

[56a] In the presence of a quadrupolar coupling, the NMR line of a deuteron is split into a symmetric doublet. With an axial electric field gradient (e.f.g.) the doublet spacing is given, in frequency units, by

$$\Delta\nu = (3/4)(e^2qQ/h)(3\cos^2\theta - 1), \tag{1}$$

where e^2qQ/h is the static quadrupolar coupling constant, θ is the angle between the e.f.g. axis (in the case discussed here the C—D bond axis) and the magnetic field. Motions of this axis modulate θ and a time average is to be taken if the motion frequencies are larger than the static interaction. If the motion is anisotropic and has an axis of symmetry, (1) becomes

$$\Delta\nu = \frac{3}{4}\frac{e^2qQ}{h}\left\langle\frac{3\cos^2\theta' - 1}{2}\right\rangle(3\cos^2\phi - 1); \tag{2}$$

θ' and ϕ are the angles made by the symmetry axis with, respectively, the C—D bond and the magnetic field. An effective quadrupolar coupling constant is measured, which is

$$\frac{e^2qQ}{h}\left\langle\frac{3\cos^2\theta' : 1}{2}\right\rangle,$$

and

$$\left\langle\frac{3\cos^2\theta' - 1}{2}\right\rangle$$

can be defined as the order parameter, S, of the C—D bond relative to the axis of symmetry.

[57] L. J. Burnett and B. M. Muller, *J. Chem. Phys.* **55**, 5829 (1971).
[58] J. Seelig and W. Niederberger, *Biochemistry* **13**, 1585 (1974).

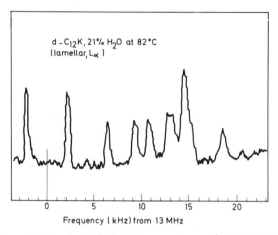

$d - C_{12}K, 21\% H_2O$ at 82°C
(lamellar, L_α)

Frequency (kHz) from 13 MHz

FIG. 8.   DMR spectrum of paraffinic deuterons in an oriented lamellar sample. The external magnetic field is at $\pi/2$ from the normal to the lamella. Only half the spectrum, which is symmetric around the zero central frequency, is shown. Each line is characteristic of a methylene, or a group of methylenes, with a definite degree of motional averaging. From Charvolin et al.[56] (Copyright by North-Holland Publishing Company. Used with permission.)

in spin label results. Later on we shall compare EPR and DMR results while analyzing the chain behavior in terms of deformations.

Other molecular motions have to be considered as well: translational diffusion, flip-flop in the bilayer, exchange between bilayers. Flip-flop and exchange can be estimated to be too infrequent to affect NMR and EPR lines.[26,60] On the other hand the translational diffusion frequency is rather large as suggested by proton spin-echo measurements in the anhydrous $L\alpha$ phase of sodium palmitate[61] (diffusion coefficient $D \simeq 5 \times 10^{-7}$ cm²·sec⁻¹) and EPR experiments in lecithin bilayers[62] ($D \simeq 2 \times 10^{-8}$ cm² sec⁻¹). These orders of magnitude for translational diffusion in the bilayers confirm their rather fluid character.

b. Cubic Phase, Q_α

Only PMR experiments have been done on this phase, which is characterized by very narrow linewidths, analogous to those of the micellar solution.[45] In both cases the lines are so narrow that the shapes of the spectra are dominated by the weak chemical shifts of the protons,

[59] A. Seelig and J. Seelig, *Biochemistry* **13**, 4839 (1974).
[60] V. F. Bystrov, N. I. Dubrovina, L. I. Barsukov, and L. D. Bergelson, *Chem. Phys. Lipids* **6**, 343 (1971).
[61] R. Blinč, V. Dimič, J. Pirš, M. Vilfăn, and I. Zupančič, *Mol. Cryst. Liq. Cryst.* **14**, 97 (1971).
[62] P. Devaux and H. M. McConnell, *J. Am. Chem. Soc.* **94**, 4475 (1972).

as shown in Fig. 9. Thus the $Q\alpha$ phase, which is highly viscous, in contrast with the fluid micellar solution, appears as a liquid through PMR investigation. This behavior has to be explained by the existence of motions averaging all the dipolar interactions between protons. If we remember that in the $Q\alpha$ phase the hydrocarbon region consists of rods of finite length connected three by three and is continuous through the three-dimensional lattice (Fig. 3d) it becomes clear that rapid diffusion of the molecules in such a tubular structure will be equivalent to an isotropic motion even if the whole space is not accessible to the molecules. We shall show later on that this translational diffusion throughout the sample was effectively measured, thus confirming the connected rod structure.

c. Hexagonal Phase, $H\alpha$

In Fig. 5 the linewidth in the hexagonal phase appears similar to that in the lamellar one. This is quite surprising considering the structural differences between the two phases (Fig. 3a,c). Suspecting that the

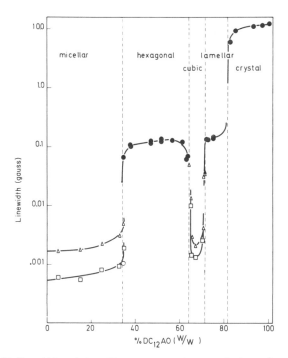

FIG. 9. PMR linewidths' of the different phases of the nonionic surfactant dimethyldodecylamine oxide in D_2O at 30°C. ($\square$) NCH_3 protons; ($\triangle$) methylene protons. From Lawson and Flautt.[45] (Copyright by Gordon and Breach. Used with permission.)

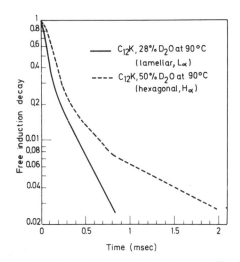

Time (msec)

FIG. 10. Comparison of the PMR free induction decay (or Fourier transform of the absorption spectrum) of paraffin protons in lamellar and hexagonal nonoriented samples of potassium laurate–D_2O. From Charvolin and Rigny.[63] (Copyright by the American Institute of Physics. Used with permission.)

chosen parameter, the half-height linewidth, was too insensitive Lawsson and Flautt compared the ratio of the linewidths measured at half and one-eighth height[45] and found some differences which suggested that the line was narrower in the hexagonal phase. In order to extract all the information contained in the lineshape its Fourier transform (or free induction decay) was approximated to exponential shape functions.[63] The results are shown in Fig. 10 and demonstrate that the decay time is indeed longer for $H\alpha$ than for $L\alpha$, or in other words, that the lines are narrower and the residual dipolar interactions smaller in $H\alpha$. As with the $Q\alpha$ phase, this behavior may be explained by an averaging effect of the translational diffusion. However, in the $H\alpha$ phase, unlike the $Q\alpha$ phase, the diffusion takes place on infinitely long cylinders so that not all the orientations of the molecules in space are equally allowed; only the component of the diffusion in a plane perpendicular to the cylinder axis leads to averaging, which cannot therefore be complete. Such a motion would reduce the dipolar interactions by a factor of 2 compared to the $L\alpha$ case and this is in agreement with the experimental data. Apart from diffusion a weak contribution to the reduction of dipolar interactions can be due to the increase of area per polar group when passing from $L\alpha$ and $H\alpha$.

[63] J. Charvolin and P. Rigny, *J. Chem. Phys.* **58**, 3999 (1973).

Obviously the translational diffusion around a cylinder takes a short time compared with the characteristic time of PMR experiments since averaging effects are always observed. With EPR experiments these effects are sometimes observed[64] and sometimes not.[49,65] In the latter case the molecules seem to be randomly distributed around the cylinder axis. These experiments might suggest that the mean time for a molecule to reorient up-down around the cylinder axis is of the order of 10^{-8} sec. Thus an estimate of the diffusion coefficient would be about 10^{-7} or 10^{-6} $cm^2 \cdot sec^{-1}$, values not too far from those given previously for the $L\alpha$ phase.

d. Qualitative Description of Lipid Behavior

Magnetic resonance gives a dynamical picture of the disorder detected by X-ray diffraction studies. The lipid molecules change their shape by internal motions, isomeric rotations, flexions, and torsions. The existence of a soap–water interface and the presence of neighboring chains prevent the intramolecular reorientations from being isotropic. The molecules also diffuse laterally in the structures, keeping their polar groups fixed on the interface. This translational diffusion appears rather rapid, whatever the nature of the phase. The effect of this diffusion on the NMR spectra depends upon the curvature of the interface: In rod structures the diffusion on the curved interface strongly affects the NMR spectra through the orientations of the intramolecular vectors in the external field, whereas it is without effect when taking place on the plane interface of the lamellar structure. Finally it must be noticed that while NMR may make a clear distinction between anisotropic and isotropic phases, it cannot be used to determine the type of structure elements of an anisotropic phase (lamellae or rods) with certitude.

e. Some Remarks about Water and Ions

DMR studies of water in the hydrated phases are also highly sensitive to the anisotropy of the structure.[66,67] As shown in Fig. 11 for the potassium laurate–D_2O system, a D_2O quadrupolar splitting is observed in the anisotropic phases $L\alpha$ and $H\alpha$, which disappears in the isotropic phase $Q\alpha$. This is reminiscent of the sudden narrowing of the paraffinic PMR spectra upon going from anisotropic to isotropic structures (Fig. 9). Most likely, as for the soap molecules, water translational diffusion

[64] J. Seelig and H. Limager, Mol. Cryst. Liq. Cryst. 25, 105 (1974).
[65] J. M. Boggs and J. C. Hsia, Proc. Natl. Acad. Sci. U. S. A. 70, 1406 (1973).
[66] K. D. Lawson and T. J. Flautt, J. Phys. Chem. 72, 2058 (1968).
[67] J. Charvolin and P. Rigny, J. Phys. (Paris) 30, C4-76 (1969).

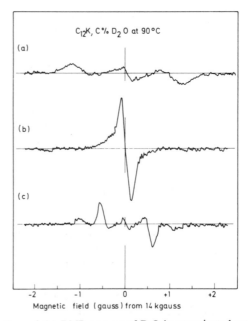

FIG. 11.   Derivatives of the DMR spectra of D_2O in nonoriented samples of potassium laurate–D_2O. The central line observed in the lamellar and hexagonal phases is most likely a DMR artifact [double quantum transition, see M. Wennerstrom *et al.*, *J. Magn. Reson.* **13**, 348 (1974)]. (a) Lamellar, Lα, $c = 27$; (b) cubic, Qα, $c = 36$; (c) hexagonal, Hα, $c = 46$. From Charvolin and Rigny.[67] (Copyright by Comission des Publications Françaises de Physique. Used with permission.)

occurs and the averaging effect is complete in the Qα phase but only partial in the Hα. The residual quadrupolar splitting observed in the Lα phase reveals the influence of the interface on the water behavior. However, for high hydration levels the aqueous medium is certainly not homogeneous. Some water molecules may interact strongly with the interface while others, in the center of the water layer, are far from its influence: exchanges in the water layer introduce an averaging effect whose magnitude is difficult to estimate as the relative importances of the water populations are not known. As far as the lipid–water interactions are concerned, it is also difficult to discriminate between the influences of numerous parameters: The water molecules can interact with the polar heads through hydrogen bonding or ion solvation; their overall rotation is then slightly anisotropic and the quadrupolar coupling constants of the deuteron chemical bonds are only partially averaged. The same deuterons also experience external field gradients due to the charge distribution around the polar heads. Finally, deuteron chemical exchange between water molecules and some polar heads[68,69] may introduce other averaging effects.

In order to isolate some of these parameters, systematic studies varying the chemical nature of the polar head, the hydration level, the ionic strength, and the charge density at the interface are at present under way. Particularly data on the electric field gradients in the aqueous medium, without hydrogen bonding contributions, have been obtained from the decrease, of the quadrupolar broadenings or splittings of various positive and negative ions with increasing water content.[69] Since our aim in this review is mainly the description of the lipid medium, we have given here only the general trend of the NMR studies of the aqueous medium. These studies are considerably detailed in the already quoted review by Johansson and Lindman.[38] Development of macroscopic investigations such as conductivity measurements,[70] and of local ones such as infrared spectroscopy of water and polar head bonds[71] should help the interpretations. Finally, it is obvious that the understanding of the aqueous medium is complementary to that of the lipid medium; the water molecules at the interface certainly reflect in some way the behavior of the lipids. Evidence of such a correlation has been suggested by preliminary relaxation studies of D_2O in potassium laurate,[72] where some aspects of the dynamic behavior of the chains are visible through the water behavior.

9. Analysis of Lipid Motions

Having described the detection of lipid motions in relation to the macroscopic symmetry of the structures, we shall now analyze quantitatively each motion in relation to the local structural parameters, particularly the mean area per polar head.

a. Translational Diffusion

We have suggested previously that the liquid-like behavior of the $Q\alpha$ phase, seen through PMR lineshape studies, might be interpreted as resulting from the total averaging of dipolar interactions by the transla-

[68] A. Johansson and T. Drakenberg, *Mol. Cryst. Liq. Cryst.* **14**, 23 (1971).

[69] G. Lindblom, N. O. Persson, and B. Lindman, in "Chemie, Physikalische Chemie and Anwendungstechnik der Grenzflächenaktiven Stoffe" Vol. II, p. 939. Carl Hansen Verlag, München 1973.

[70] J. François, *J. Phys. (Paris)* **30**, C4-83 (1969).

[71] J. François, *Kolloid Z. & Z. Polym.* **251**, 594 (1973).

[72] J. Charvolin and P. Rigny, *Chem. Phys. Lett.* **18**, 515 (1973).

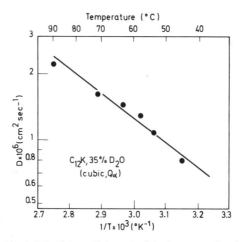

FIG. 12.   Translational diffusion coefficient D of the laurate molecules in a cubic sample as a function of the inverse temperature. From Charvolin and Rigny.[63] (Copyright by the American Institute of Physics. Used with permission.)

tional diffusion of the molecules all around the rods. In order to test this hypothesis we took advantage of the narrowness of the proton spectral lines of $Q\alpha$ to apply the NMR spin-echo methods commonly used to measure self-diffusion coefficients in isotropic liquids.[40] In this method, an external magnetic field gradient is applied; when the molecules diffuse they "see" a varying external magnetic field, and an additional transverse relaxation mechanism, characteristic of the diffusion, is introduced. In liquids, where the intrinsic relaxation times are long, this process can be efficient with commonly available field gradients of a few tens of gauss per centimeter; typically diffusion over distances of about 1 μm during a few 10^{-2} sec can be measured, i.e., diffusion coefficients $D \simeq 10^{-7}$ cm²·sec⁻¹. On the other hand, in solids, where the transverse relaxation time, which is dominated by dipolar interactions, is very short, this type of measurement would need too large field gradients. In the $Q\alpha$ phase of potassium laurate–water the D value measured this way is 2×10^{-6} cm²·sec⁻¹ at 90°C.[73] The temperature variation of D is shown in Fig. 12; the activation energy of the diffusion process is about 5.7 kcal mole⁻¹. Similar studies have been performed on the $Q\alpha$ phase of the dodecyltrimethyl–ammonium chloride–water system.[74] Such values can be compared with the diffusion coefficients of bulk water at 25°C and of liquid paraffins,[75] which are of the order of 10^{-5} cm²·sec⁻¹. Thus, the

[73] J. Charvolin and P. Rigny, *J. Magn. Reson.* **4**, 40 (1971).
[74] T. Bull and B. Lindman, *Mol. Cryst. Liq. Cryst.* **28**, 155 (1974).
[75] D. C. Douglass and D. McCall, *J. Phys. Chem.* **62**, 1102 (1958).

molecules flow very rapidly in the rod structure and this microscopic behavior is in striking contrast to the high macroscopic viscosity of the Qα phase. Moreover, as the measured diffusion takes place over distances of about 1 μm, i.e., the dimension of one monodomain, these experiments confirm the structural model of an hydrocarbon continuum in the Qα phase.[22]

We have previously shown evidence, from EPR experiments, of spin-label diffusion in anisotropic lamellar structures. Unfortunately, in this phase it is difficult to get as direct evidence as that on the Qα phase discussed above, owing to the existence of residual dipolar interactions which preclude the formation of "liquid-like" spin echo. An attempt to circumvent this difficulty was the similar study of a stimulated echo which suggested also a diffusion coefficient of about 10^{-6} cm²·sec⁻¹.[76] Recent improvements in the applicability of the spin-echo method will certainly advance these investigations in the near future.[77] Finally, although we have no quantitative data for the diffusion in the Hα phase, it appears certain now that the translational diffusion is responsible for the previously discussed narrowing of the PMR line in this phase (Fig. 10). Despite the limited number of conclusive experiments on the anisotropic phases, all the data converge in favor of the existence of a translational diffusion in all phases with disordered chains, whatever their structure is.

In simple soap–water systems the diffusion coefficient D is rather large, around 10^{-6} cm²·sec⁻¹; such a D value corresponds to local diffusive behavior close to that of a liquid of large molecules. In fact, using the well-known relation

$$\langle l^2 \rangle \propto D\tau,$$

where $\langle l^2 \rangle$ is the mean square value of the path covered in time τ one can estimate to be above 10^{-9} sec the time it takes for one molecule to move past by another. We stress here some features of the diffusion when it occurs over distances of the order of, or larger than, the characteristic dimensions of the structure elements: In the lamellar structure we are dealing with a two-dimensional diffusion,[78] while in the cubic structure the diffusion takes place in three dimensions. In the latter case, however, the process is highly unusual since only a part of the space is accessible, the molecules being confined to the tubular structure of the hydrocarbon medium; the definition of D, and its usage have certainly to be revised accordingly. Nevertheless we can have

[76] R. T. Roberts, *Nature (London)* **242**, 348 (1973).

[77] R. Blinč, J. Pirš, and T. Zupančič, *Phys. Rev. Lett.* **30**, 546 (1973).

[78] R. Naqui, *Chem. Phys. Lett.* **28**, 280 (1974).

confidence in the orders of magnitude of D and we can consider two expected effects of this diffusion in relation to relaxation in NMR. In the high-frequency range, around 10^9 Hz, a PMR relaxation is to be seen whatever the structure is when the intermolecular interactions are modulated by the diffusion of a few angstroms. The orientation of the intramolecular vectors is hardly affected by this elemental jump; to get an important effect the intramolecular vectors have to reorient appreciably in the external field. This occurs in the rod structures when the molecules reorient by diffusion around the rod axis over distances comparable with the rod diameter. A frequency of about 10^7 Hz can be estimated for such a displacement, to which will correspond, in this lower frequency range, an extra relaxation effect detectable in rod structures only.

b. Chain Deformations

The different methylene groups of a deuterated chain can be distinguished (see Fig. 8) according to differences in the motional averaging of the static quadrupolar coupling of their C—D bonds. In a lamellar phase, where no long-range effect of the translational diffusion is to be expected, this averaging is due to chain deformations and swinging.[56] In the lamellar phase of potassium laurate these motions have uniaxial symmetry around the normal to the interface in times shorter than about 10^{-5} sec; as discussed in footnote 56a the measured quadrupolar splittings are then proportional to the order parameters of the C—D bonds with respect to the normal to the interface. The potassium laurate methylene and methyl order parameters extracted from Fig. 8 are shown in Fig. 13 for three values of the mean area A per polar head at the interface.[79] Such curves are faithful representations of the chain flexibility in bilayers of defined A.

If we consider first the curve corresponding to the lowest value of A ($A = 32.6$ Å^2) a striking feature is that, with the exception of the first and the last four groups, all the other methylenes have nearly the same order parameters (see also Fig. 8). This behavior differs from that of an isolated chain, fixed at one end, where each segment is expected to have a greater orientational freedom than the preceding one and the order decreases continuously with a characteristic persistence length of about three or four segments.[80,80a,81] However, for a chain in a bilayer the steric

[79] B. Mely, J. Charvolin, and P. Keller, *Chem. Phys. Lipids* **15**, 161 (1975).

[80] Y. K. Levine, P. Partington, and G. C. K. Roberts, *Mol. Phys.* **25**, 497 (1973).

[80a] As represented in Fig. 14, a chain subgroup has three rotational isomers due to rotations around the C—C bonds;[81] then a free chain with n segments has about 3_n rotational isomers.

[81] P. Flory, "Statistical Mechanics of Chain Molecules." Wiley (Interscience), New York, 1969.

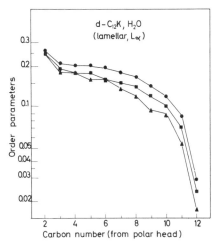

FIG. 13. C—D bond order parameter with respect to the lamellae normal as the mean area A per polar head at the interface varies, in an oriented lamellar sample of potassium laurate–water (see Fig. 8). A ($Å^2$): (●) 32.6; (■) 35; (▲) 39.1.

repulsions between neighboring chains should also be considered. Recently, de Gennes has considered the situation where the mean spacing between chains, measured normally to the chain elongation axis, remains uniform in the regions of the bilayers without chain ends; his general treatment shows that the steric interactions constrain the chains in such a way that the orientational disorder cannot increase along the chain as it would for an isolated chain.[82] The decrease of the order parameters for the last few segments is then explained by the presence of many chain ends in the central region of the bilayer, which makes the steric hindrance much less stringent. If this interpretation is correct, then the fact that this drop involves the last three segments of the chain implies that the chain ends are not confined to a place in the exact center of the bilayer, but rather that they are distributed over a region of sizable thickness. In turn, this implies that the deformations of the chains are not restricted to small-angle torsions or flexions of the bonds, but also involve isomeric rotations around the C—C bonds which can move the chain end appreciably. A detailed description of the accessible conformations would be an overwhelming task; a few very specific models have been developed, however, considering only the simplest isomers. For example, the occurrence of isolated gauche conformations (t, g, t) with equal probabilities all along the chain would yield a continuous decrease of the order of the methylenes along the chain and

[82] P. G. de Gennes, *Phys. Lett. A* **47**, 123 (1974).

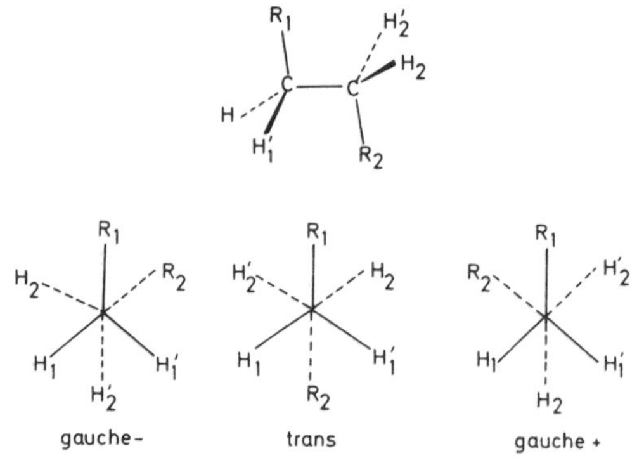

gauche-   trans   gauche +

FIG. 14.   A chain subgroup and its three rotational isomers. The two gauche (g^- or g^+) have higher energies ($E \sim 500$ cal·mole^{-1}) than the trans one (t).

is ruled out by the DMR measurements. Physically such a deformation is highly improbable, except perhaps near the methyl end, for it ignores the steric repulsions between neighboring chains. Seelig[58] considers only defects, such as "kinks," which do not change the overall direction of the chain (a "kink"[83] combines two gauche isomerizations of different signs; it can be written t, g^+, t, g^-, t or t, g^-, t, g^+, t. Trans sequences on both sides of it stay parallel to each other; they are only laterally shifted). Such a defect, migrating rapidly back and forth along the chain, decreases equally the order of all the methylene groups; the order parameter of the "plateau" in Fig. 13 would correspond, in this model, to a probability $\simeq 0.8$ for a link to be in a trans state. Of course similar local defects keeping the overall direction of the chains perpendicular to the bilayer could be considered as well.[84] Such models, which introduce intermolecular cooperativity in an arbitrary way by eliminating presupposed unfavorable conformations, skip the interesting aspect of possible collective deformations of lipids in bilayers.

Several approaches to the "many-chain" problem have been proposed: detailed two-chain interaction,[85] exact two-dimensional lattice models,[86] and a self-consistent molecular field approximation.[87] In this last one, Marcelja extends to lyotropic liquid crystals some of the ideas he has developed for thermotropic systems.[88] More specifically, the

[83] W. Pechhold, *Kolloid Z. & Z. Polym.* **228**, 1 (1968).
[84] E. Dubois-Violette, F. Geny, L. Monnerie, and O. Parodi, *J. Chim. Phys.* **66**, 1865 (1969).
[85] P. Bothorel, J. Belle, and B. Lemaire, *Chem. Phys. Lipids* **12**, 96 (1974).
[86] J. F. Nagle, *J. Chem. Phys.* **58**, 252 (1973).
[87] S. Marcelja, *Biochim. Biophys. Acta* **367**, 165 (1974).
[88] S. Marcelja, *J. Chem. Phys.* **60**, 3599 (1974).

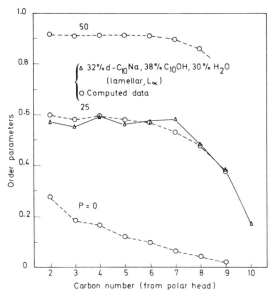

FIG. 15. Order parameters, in lamellar phase at 28°C, all along a decanoate chain (the definition of the order parameter here is twice that of Fig. 13). Dashed lines: calculation for different values of the lateral pressure P (in dyn/cm), solid line: experimental. Data from Ref. 88. From Marcelja.[87] (Copyright Elsevier/North-Holland Biomedical Press. Used with permission.)

packing of the lipid polar heads on the interface is introduced through a lateral pressure term which is estimated from surface pressure measurements in monolayers; then statistical averages are evaluated by summation over all conformations of a chain in the field due to neighboring molecules. The calculated behavior of the order of the chain links agrees well with experimental results, as shown in Fig. 15. A palmitate chain (C_{16}) would then have four links in gauche conformations, in agreement with earlier thermodynamic calculations.[89] This theory also gives reasonable agreement with the measured thermodynamic properties of the Lα $\rightarrow$ Lβ transition, where all the chains take a rigid, all trans conformation.

It is of interest to point out that spin-label experiments show a monotonic decrease of the order from head to tail (see Fig. 6 and Seelig *et al.*[48]), in contrast with the previously discussed "plateau" obtained from deuteron experiments. Although it is not possible at the moment to discriminate against perturbing effects from the nitroxide label, it is nevertheless interesting to attempt a comparison of EPR and DMR data

[89] J. F. Nagle, *Proc. Natl. Acad. Sci. U. S. A.* **70**, 1443 (1974).
[90] B. Gaffney and H. M. McConnell, *J. Magn. Reson.* **16**, 1 (1974).

in terms of lipid behavior only. Recently, Gaffney and McConnell[90] suggested that one could reconcile the two types of experiments by taking into account their different time scales. In their model, local fast motions, with frequencies larger than about 10^8 Hz, are responsible for the monotonic decrease of the spin-label order parameters and each methylene occupies a larger effective volume than the preceding one. The region close to the interface is then a weak density region and a plausible solution for this packing problem can be a collective bending of the chains: Their elongation axis, normal to the bilayer in the tail region, is tilted in the vicinity of the interface.[91] This model accounts for the EPR spectra if one assumes a static bending angle or, to say it in an other way, if the bilayer is assumed to have local biaxiality for times longer than 10^{-8} sec. The failure of DMR experiments to detect any biaxiality in potassium laurate and lecithin bilayers[56,59] could be explained by the fact that the lifetime of this biaxiality falls in between 10^{-5} and 10^{-8} sec. This tilt angle, affecting the first methylenes, would then introduce a reducing factor in the expressions for their DMR splittings, transforming into a plateau the monotonic decrease of their order caused by fast motions. In this model, in contrast to that of de Gennes, the mean lateral spacing is not constant through the bilayers. As already stated, this assumption is mostly supported by the need of a bending angle as a fitting parameter for the EPR spectra.[91] Nevertheless, it remains a description of the behavior of labeled chains which, in the absence of direct conclusive experiments, can still be suspected to be somewhat different from that of nonlabeled molecules as demonstrated by recent experiments in the thermotropic field.[31]

Returning to Fig. 13, we can analyze now the evolution of the methylene order parameters when the mean area per polar head A increases, as a function of the temperature or water content, as shown in Fig. 16.[19] These variations of the thermodynamic conditions of the system thus imply an increase of the chain disorder. The data of Fig. 13 support this idea; the C—D bond order parameters decrease when A increases, confirming the use of A as a global characterization of the microscopic disorder.[79] However, the shapes of the decay are changing, showing that the disorder does not increase equally for all the methylenes. For the first four methylenes in particular the disorder does not increase for $A \geq 35$ Å^2. This last result suggests a stiffness effect, certainly related to the anchoring of the lipid molecule at the interface. Anchoring influences could also be invoked to understand the behavior of the first methylene in a binary system compared to that in a ternary system. In the soap–water case (Fig. 13) the first methylene order parameter is

[91] B. Gaffney and H. M. McConnell, *Proc. Natl. Acad. Sci. U. S. A.* **68**, 1274 (1971).

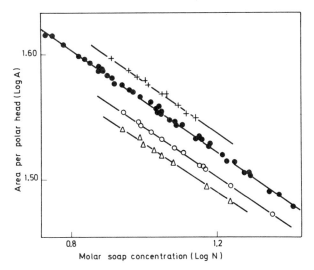

FIG. 16. Variation of the mean area A per polar head with temperature and concentration in the lamellar Lα phases of potassium soaps. A does not depend upon chain length from C_8 to C_{22}. (+) 104°C; (●) 86°C; (○) 65°C; (△) 45°C. From Gallot and Skoulios.[91a] (Copyright by Dietrich Steinkopff Verlag. Used with permission.)

significantly larger than that of others in the "plateau" region, whereas no difference is observed in the soap–alcohol–water case (Fig. 15).

c. Characteristic Times

Orders of magnitudes of the characteristic times estimated from lineshape studies were given in the previous paragraphs. More accurate determinations can be obtained from relaxation rate measurements (T_1^{-1} and $T_{1\rho}^{-1}$) which, in principle, give access to the spectral densities of the motions. In this respect deuteron relaxation is very promising because of the selectivity of DMR spectra. Unfortunately, relaxation experiments on deuterated molecules are just beginning and cannot yet be analyzed. We can, however, present and discuss some results derived from proton relaxation experiments.

Figure 17 shows the evolution towards equilibrium of the Zeeman energy of paraffinic protons in potassium laurate lamellae. The departure from exponentiality at long times suggests some distribution of T_1 relaxation times. Since, as we have shown, the spin states are connected through the residual dipolar couplings between methylenes, the T_1 relaxation times of motionally different methylenes are homogenized through the so-called "spin-diffusion" process[40] (experiments in lecithin

[91a] B. Gallot and A. E. Skoulios, *Kolliod-z.* **208**, 37 (1966).

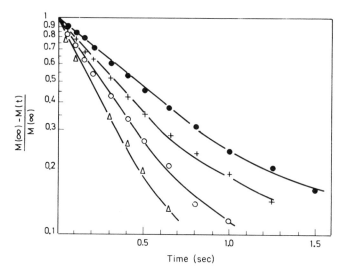

FIG. 17. Recovery of the Zeeman energy of the paraffin protons after a $\pi/2$ pulse excitation at 30 MHz, in a lamellar nonoriented sample. (●) 90°C; (+) 70°C; (○) 58°C; (△) 43°C.

bilayers[92,93] suggest, for instance, that spin diffusion is rapid and that the relaxation occurs via common energy sinks which could be protons in the polar heads or close to them). If such is the case in potassium laurate the nonexponentiality observed in Fig. 17 might result from the T_1 distribution due to the isotropic angular distribution of the lamellae in the unoriented sample. Thus it appears that the proton relaxation does not provide easily interpretable information about the details of chain motions. However, the frequency dependence of this relaxation, by giving access to the spectral densities of the motions, can still provide useful, although macroscopic, dynamic information.

Variations of the proton relaxation rates with frequency have been investigated in the Lα, Qα, and Hα phases of the potassium laurate–D$_2$O system.[63] The frequency ranges extend from a few kHz to 70 kHz using $T_{1\rho}$ techniques and from a few MHz to 60 MHz using ordinary T_1 techniques.[94] We can distinguish three frequency regions where important relaxation effects occur.

In the high-frequency region the relaxation process appears independently of the nature of the phase: The relaxation time goes through a

[92] J. T. Daycock, A. Darke, and D. Chapman, *Chem. Phys. Lipids* **6**, 205 (1971).

[93] G. W. Feigenson and S. I. Chan, *J. Am. Chem. Soc.* **96**, 1312 (1974).

[94] T. C. Farrar and E. D. Becker, "Introduction to Pulse and Fourier Transform NMR Methods." Academic Press, New York, 1971.

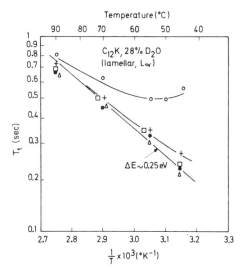

FIG. 18. Arrhenius plots of the longitudinal relaxation times T_1 of the paraffin protons at different frequencies for a lamellar nonoriented sample. Frequency, MHz: (○) 60; (+) 30; (●) 12; (□) 10; (△) 7. From Charvolin and Rigny.[63] (Copyright by the American Institute of Physics. Used with permission.)

minimum at 60 MHz and 55°C (Figs. 18 and 19) indicating the existence of a motion having this characteristic frequency at 55°C. The independence of T_1 on the nature of the phase was also noticed in more complex phospholipid systems.[95] This suggests that the motion considered here, being insensitive to the curvature of the interface, is a purely local, rapid motion. The temperature shift of the T_1 minimum gives an estimate of the activation energy of the motion, $\Delta E \simeq 6$ kcal/mole. The values of the characteristic time ($\tau \simeq 10^{-9}$ sec) and activation energy ($\Delta E \simeq 6$ kcal/mole) suggest that the relaxation mechanisms could be translational diffusion, rapid isomerizations in the chains, and/or rotational diffusion of the molecule around its elongation axis.

In the medium-frequency region, from 10 to 30 MHz, differences between the phases are apparent. The relaxation rate is frequency independent in the lamellar phase Lα, while it goes on increasing in the cubic phase Qα when the frequency decreases, as shown in Figs. 18 and 19. Thus a relaxation mechanism, inoperative in Lα, acts in Qα at medium frequency (~10 MHz) and its effect is more visible at high temperature. Obviously such a mechanism can be related to the long-range effect of the diffusion which, in the isotropic rod structure, totally averages the dipolar interactions left by the rapid motions discussed

[95] D. Chapman and N. J. Salsbury, *Trans. Faraday Soc.* **62**, 2607 (1966).

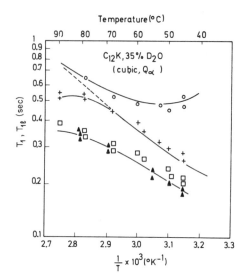

FIG. 19. Arrhenius plots of the relaxation times T_1 and $T_{1\rho}$ of the paraffin protons in a cubic sample. T_1: (O) 60 MHz; (+) 30 MHz; (□) 10 MHz; $T_{1\rho}$: (▲) 8–80 kHz. From Charvolin and Rigny.[63] (Copyright by the American Institute of Physics. Used with permission.)

above. The characteristic time of 10^{-8} sec proposed previously for the reorientation of the molecular elongation axis, by diffusion along a curved interface, agrees well with the observation of Fig. 19.

Finally, in the low-frequency region, up to 80 kHz, great differences occur between the Lα and Qα phases, as can be seen by comparing Figs. 19 and 20. The relaxation rate is now frequency-independent in Qα, whereas it increases abruptly in Lα. If the situation is clear from Qα, where the averaging out of the dipolar interactions by the diffusion along the interface prevents the manifestation of any relaxation mechanism at lower frequencies, the analysis of the results for Lα is not at all straightforward. Here the relaxation is measured at frequencies comparable with that of the nonaveraged dipolar interactions, and in such a case, as discussed by Goldman[96] for low-frequency relaxation by a rapidly fluctuating local field, T_D^{-1} is expected to be larger than T_1^{-1}. Such a behavior can be seen on Fig. 20; however, the temperature dependence of the $T_{1\rho}^{-1}$ curve differs much from that of T_1^{-1}. The rapid motions responsible for the high-frequency relaxation (T_1^{-1}) are then not the only cause for the increase of the relaxation rate in the very low-frequency range, which could be also related to the presence of slow motions with characteristic times around 10^{-6} sec. Therefore, the chain

[96] M. Goldman, "Spin Temperature and Nuclear Magnetic Resonance in Solids." Univ. Press (Clarendon), London and New York, 1970.

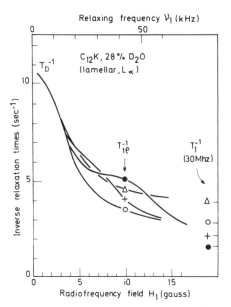

FIG. 20. Relaxation rate $T_{1\rho}^{-1}$ of the paraffin protons as function of the radio-frequency field strength (or relaxing frequency) for a nonoriented lamellar sample at different temperatures. Temperature, °C: (●) 90; (+) 70; (○) 55; (△) 44. From Charvolin and Rigny.[63] (Copyright by the American Institute of Physics. Used with permission.)

deformations appear to have distributed correlation times. The slowest could be a long-wavelength deformation of the molecule, e.g., some kind of pendulum motion. We cannot extract accurate determinations for the fastest motions; they are certainly short-wavelength deformations, affecting a few methylenes on a chain, such as the "kinks" and other defects discussed previously. A theoretical treatment[97] based on a Rouse analysis, familiar in the field of polymer physics,[98] estimates their characteristic times to be about 10^{-8} or 10^{-9} sec; these could interfere with translational and rotational diffusion in the high-frequency relaxation range. Note that, since each motion modulates several interactions of different origins, it is practically hopeless to try to interpret the values of the modulated interactions extracted from relaxation data. This is why we relied only on frequency and temperature dependences of the relaxation rates and never on their absolute values.[99]

The variations of the relaxation rates as a function of frequency in the

[97] G. Agren, *J. Phys. (Paris)* **33**, 887 (1972).
[98] P. E. Rouse, *J. Chem. Phys.* **21**, 1272 (1965).
[99] D. F. S. Natush and R. M. Newman, *J. Magn. Reson.* **4**, 358 (1971). (Another example of the frequency dependence of the PMR relaxation in LLC.)

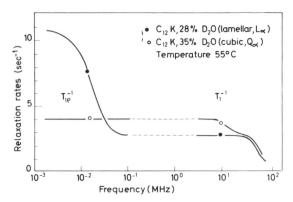

FIG. 21.   Schematic representation of the paraffin proton relaxation rates as functions of the relaxing frequencies. No measurements have been made in the 0.1–1 MHz range, in between the domains of the $T_{1\rho}$ and T_1 techniques. The cubic and lamellar phase are compared. This figure sums up the three preceding ones.

Lα and Qα phases are presented in a schematic way in Fig. 21. Each cutoff frequency on the figure corresponds to one of the previously discussed characteristic times. The phase Hα was not included, but it nevertheless provides a good test for the consistency of this description of lipid dynamics. The situation presented by the Hα phase is intermediate between those of Lα and Qα. Since the translational diffusion along a cylindrical interface is not isotropic this motion can modulate only one-half of the intramolecular interactions and the rest are modulated by the chain deformations. Relaxation effects from both motions are then seen simultaneously in Hα,[100] whereas they exclude each other in Lα and Qα. Relaxation through fast reorientation of the chain elongation axis by lateral diffusion also has been proposed recently as a possible mechanism for understanding the PMR lineshapes of small lecithin vesicles,[101] the tumbling of which is too slow to account for the sharpness of the lines.

10.   Discussion

The magnetic resonance studies presented above may appear somewhat dispersed and fragmentary when compared to the systematic aspects of the structural ones. Only a few phase diagrams of simple systems (dimethyldodecylamine oxide, sodium palmitate, potassium

[100] J. Charvolin, Thèse de Doctorat d'Etat, Orsay (1972).
[101] M. Bloom, E. E. Burnell, M. I. Valič, and G. Weeks, *Chem. Phys. Lipids* **14,** 107 (1975).

laurate) have been investigated through lineshape and relaxation studies in the region with disordered chains. Otherwise most of the results come from studies of various phospholipids in structures of lamellar type. Comparing all these NMR results on chain behavior it appears that they depend much more upon the structure of the phases than upon the chemical nature of the compound.

We have analyzed these results as the superposition of the effects of local molecular dynamics and of long-range diffusion along the interface. While the latter depends strongly upon the geometry of the phase, there are only a few differences between the phases as far as the former is concerned. This common microscopic behavior can be described in terms of lipid translational diffusion and deformations: The molecules diffuse in the structures with their polar heads on the interface and they change their shape by flexions, torsions, or isomeric rotations. The chain deformations appear strongly limited by the interface and by the presence of neighboring molecules. As a matter of fact, one can distinguish three regions in the bilayer when going from the interface towards the center. The three or four first segments of the chain see a stiffness effect due to the anchoring of the polar head in water and to the proximity of the interface. Beyond this first zone the segments get more disordered, but steric constraints, imposed by neighboring molecules, limit the number of accessible conformations. Finally, the disorder is important in the central region of the bilayer where the presence of chain ends partially removes these steric constraints. The dynamical parameters for the microscopic behavior of the potassium laurate are summarized in Table I; these figures might change within about one order of magnitude with the nature of the compound. The small values of the characteristic times reveal the rather fluid behavior of the hydrocarbon

TABLE I. ORDERS OF MAGNITUDE OF CHARACTERISTIC TIMES AT 90°C AND ACTIVATION ENERGIES OF THE CHAIN MOTIONS IN THE POTASSIUM LAURATE–WATER SYSTEM

| Motions | Deformations of the chains (a) and rotations around their elongation axis (b), orientational fluctuations of this axis (c) | | Translational diffusion $D = 2 \times 10^{-6}$ cm²·sec⁻¹ |
	Slowest [most likely (c)]	Other [(a) and (b)]	
Correlation time (sec)	$\sim 10^{-6}$	Distribution $< 10^{-6}$	for a local jump $\sim 10^{-9}$
Activation energy	apparent ~ 0	No accurate determination but $\neq 0$	~ 6 kcal/mole

medium. An interesting feature is the distribution of the deformations times up to a few microseconds, showing the existence of slow motions of, or within, the bilayer. In this respect one has to compare the times of the slowest deformation motion (10^{-6} sec) and of the local diffusive jump (10^{-9} sec). If the pendulum motions of the molecules, which are certainly their slowest deformation modes, were uncorrelated, the molecules would have to wait about 10^{-6} sec between two diffusive jumps. The rapid diffusion can then be understood if the slow motions are collective motions in the lamellae. In this respect the slow motions could depend upon the macroscopic structure (lamellar or cylindrical for instance), whereas the fast motions are qualitatively independent of it and governed essentially by the density of the medium.

IV. Final Comment

Lyotropic phases formed by lipids with chains in the α conformation can be pictured, in a first approximation, as entanglements of two immiscible liquids of diversified topologies. The molecules, water as well as lipids, flow in the structure elements with diffusion times which are not far from those observed in viscous liquids; the lipid–water interface constraints, of course, the lateral diffusion and introduces local anisotropy in aqueous and lipid media. The curvature of this interface does not seem to have any drastic influence on the purely local behavior of the lipids (diffusive jumps, rapid isomerizations), but can be suspected to be of importance when the long-range behavior of the lipid medium (collective reorientations of the local anisotropy axis) is considered. Such an analysis of the lipid disorder justifies the recent phenomenological approaches to the elastic and hydrodynamic[102,103] properties of LLC and also suggests experimental investigations of these properties through light-scattering studies[104,105] applied until now in TLC only.

ACKNOWLEDGMENTS

The authors would like to thank V. Luzzati and P. Rigny for many helpful and stimulating discussions, Prof. P. G. de Gennes for his constant interest, and Prof. M. Bloom and F. Poldy for critical reading of the manuscript.

[102] W. Helfrich, *Phys. Lett. A* **43**, 409 (1973).
[103] F. Brochard and P. G. de Gennes, *Liq. Cryst., Proc. Int. Conf., 1973* (1975).
[104] L. Powers and N. A. Clark, *Proc. Natl. Acad. Sci. U. S. A.* **72**, 840 (1975).
[105] C. L. Khetrapal, A. C. Kunwar, A. S. Tracey, and P. Diehl, in "NMR Basic Principles and Progress" (P. Diehl, E. Fluck, and R. Kosfeld, eds.), Vol. 9. Springer-Verlag, Berlin and New York, 1975.

NOTE ADDED IN PROOF

Since this review was completed the development of the NMR studies of deuterated molecules foreshadows considerable improvements in our knowledge of the molecular behavior in liquid crystals. The references below present the tendencies of this development in lyotropic liquid crystals.

Descriptions of disordered media of saturated chains,
B. Mely, J. Charvolin, and P. Keller, *Chem. Phys. Lip.* **15**, 161 (1976);
of unsaturated chains,
A. Seelig and J. Seelig, *Biochemistry* **16**, 45 (1977);
of an ordered medium of saturated chains,
B. Mely and J. Charvolin, *Chem. Phys. Lip.* **19**, 43 (1977).
Studies of the interfacial regions of soap systems,
K. Abdolall, M. I. Valic, and E. E. Burnell, *Chem. Phys. Lip.* (in press);
of lipid systems,
J. Seelig, H. Gally, and R. Wohlgemuth, *Biochim. Biophys. Acta* **467**, 109 (1977).
Investigations of the organisation of biological membranes,
G. W. Stockton, K. G. Johnson, K. W. Budler, A. T. Tulloch, Y. Boulanger, I. CP. Smith, J. M. Davies, and M. Bloom, *Nature (London)* **269**, 267 (1977).

Concerning the structural work:
Nature of the thermal pretransition of synthetic phospholipids: Dimyristoyl and Dipalmitoyl-lecithin,
M. J. Janiak, D. M. Small, and G. G. Shipley, *Biochemistry,* **15**, 4575 (1976).

SOLID STATE PHYSICS, SUPPLEMENT 14

Liquid Crystals and Their Analogs in Biological Systems

Y. BOULIGAND

E.P.H.E. et Centre de Cytologie Experimentale, C.N.R.S., Ivry-sur-Seine, France

> The aspect of molecular pattern which seems to have been most underestimated in the consideration of biological phenomena is that found in liquid crystal.
> (Needham, 1942)

I. Introduction

The future importance of liquid crystals in the life sciences has been pointed out from the earliest research on this particular state of matter.[1,2] Several biologists have published comments on the possible significance of liquid crystals in the organization of cells and tissues. The embryologist Needham, quoted above, compared the successive steps in the progressive determination of the different axes of a limb in a vertebrate embryo with the various phase transitions between isotropic and crystalline states in a mesogenic compound. We will discuss below this hypothesis. Frey-Wyssling[3] devoted several paragraphs of his classic book *Submicroscopic Morphology of Protoplasm* to the question of liquid crystals. A great number of tissues and cell organelles show an organization which is related to the molecular structure of the different mesomorphic liquids such as smectics, nematics, and cholesterics. Myelinic systems in organisms are made of stacked cell membranes and are close analogs of smectic liquids or lyotropic lamellar mesophases.[4-6] Analogs of nematics and cholesterics have been observed in the organic matrix of several skeletal tissues.[7-14] Desoxyribonucleic acid can form a cholesteric network in certain chromosomes.[11,12] It seems, accordingly, that Lehmann's and Needham's predictions were fully justified.

We would like to make an attempt to render the biological terminology used in this paper accessible to nonbiologists. Cells are fundamental components of living organisms. Cells show in many cases an autonomous way of life (protozoa, cells in culture). There is a basic organization common to cells which consists of a nucleus limited by a double membrane, fenestrated with pores, and containing the genetic material or chromatin (Fig. 1). The chromatin is more or less condensed at a given time and is a complex of desoxyribonucleic acid (DNA) and basic proteins. The cytoplasm surrounding the nucleus consists mainly of an aqueous medium with salts, sugars, proteins, and other macromolecules

[1] O. Lehmann, "Flüssige Kristalle und die Theorien des Lebens," *J. Ambr. Barth,* Leipzig, 1908.

[2] H. Kelker, *Mol. Cryst. Liq. Cryst.* **21,** 1 (1973).

[3] A. Frey-Wyssling, "Submicroscopic Morphology of Protoplasm." Elsevier, Amsterdam (1953).

[4] B. B. Geren, *Exp. Cell Res.* **7,** 558 (1954).

[5] D. Chapman, *Ann. N.Y. Acad. Sci.* **137,** 745 (1968).

[6] J. D. Robertson, *In* "Handbook of Molecular Cytology" (A. Lima-de-Faria, ed.), p. 1403. North-Holland Publ., Amsterdam, 1969.

[7] Y. Bouligand, *C.R. Hebd. Seances Acad. Sci.* **261,** 3665 and 4864, (1965).

[8] Y. Bouligand, *J. Microsc. (Paris)* **5,** 34a (1966).

[9] Y. Bouligand, *Electron Microsc., Proc. Int. Congr., 6th, 1966* Vol. 2, p. 577 (1966).

[10] Y. Bouligand, *Mem. Mus. Nat. Hist. Nat., Ser. A* **40,** 189 (1966).

[11] Y. Bouligand, *J. Microsc. (Paris)* **6,** 41a (1967).

[12] Y. Bouligand, *J. Phys. (Paris)* **30,** C4-90 (1969).

[13] Y. Bouligand, *Electron Microsc., Proc. Int. Congr., 7th, 1970* Vol. 3, p. 105 (1971).

[14] Y. Bouligand, *J. Microsc. (Paris)* **11,** 441 (1971).

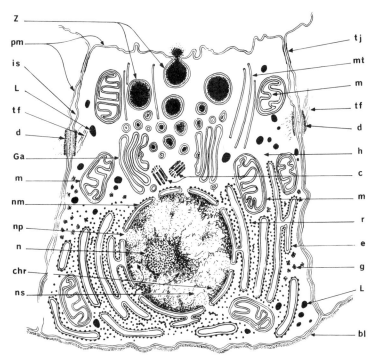

FIG. 1. Schematic representation of a pancreatic acinar cell. This type of cell is close to the ideal generalized cell: All the fundamental organelles are found enclosed within it. These cells produce the pancreatic enzymes involved in digestion (amylases allowing the hydrolysis of starch and other polysaccharides, lipase cleaving lipids into glycerol and fatty acids, and several proteolytic enzymes; carboxypeptidase breaking the peptide bond of small polypeptides, trypsin and chymotrypsin which cut heavier proteins, but in an inactive form). All the represented cell machinery is involved in enzyme synthesis. bl: basement lamina; c: centrioles; chr: chromatin; d: desmosome; e: rough endoplasmic reticulum = ergastoplasm (note the presence of numerous ribosomes attached to the surface of this reticulum); g: glycogen; Ga: Golgi apparatus; h: hyaloplasm = fraction of the cytoplasm extending between the organelles, being an isotropic sol in general (sometimes a gel); is: intercellular space; L: lipidic droplet; m: mitochondrion with its internal partitions or cristae; mt: microtubule; n: nucleolus, a dense region of the nucleus which is rich in RNA and has its own DNA forming a fibrous network (the granular fraction is made of precursors of ribosomes); nm: nuclear membrane (double); np: nuclear pore; ns: nuclear sap (medium free between chromatin and nucleolus); pm: plasma membrane (bilayer); r: ribosome; tf: tonofilaments attached to desmosomes forming a cytoskeleton with microtubules; tj: tight junction (electrophysiological role); Z: secretion vacuole of zymogen (a complex of enzymes); r: ribosomes.

in solution and various suspended organelles. One has structural elements as filaments and tubes of various dimensions, which are impli-
cated both in the static and dynamic aspects of the cytoplasm. This
latter contains a discrete number of organelles involved in definite
functions (see Fig. 1): energy production (mitochondria), protein synthesis, packaging and export (endoplasmic reticulum, Golgi apparatus,
secretory vacuoles), orientation of cell division (centrioles); various
forms of intercell attachments or junctions (desmosomes, tight-junc-
tions). For a more complete description of cells and organelles, see
classic books such as Du Praw[15] (1968) and Lima-de-Faria.[16]

II. Historical Summary

Each important advance in the physics of liquid crystals has had a
noticeable biological counterpart. This summary will examine alternately
physical and biological observations.

1. SOAPS, MYELIN, AND CHOLESTEROL DERIVATIVES

Liquid crystals are birefringent liquid phases of organic compounds
and examples such as those given by soaps were observed more than
100 years ago. The myelin forms given by lecithins and different extracts
from white matter present in the vertebrate brain have also been known
since 1850. Variously coiled tubes made of concentric lamellae appear
when such materials come into contact with water (Fig. 2). The myelinic
systems are given by amphiphilic molecules with two different parts: one
hydrophilic and one hydrophobic; they are often birefringent and liquid.
It was only during the last decades of the past century that the liquid-
crystal concept was established by a botanist, Reinitzer, and a physicist,
Lehmann. The first of these two authors tried to purify cholesterol
benzoate.[17] The crystals of this substance melt at 145°C and give a turbid
liquid which becomes clear only at 179°C. This turbidity is suspicious in
a pure substance. Reinitzer and Lehmann thought that the turbidity
could be related to the presence of impurities. Reinitzer observed the
birefringence of the turbid liquid, its rotatory power, and, in certain
conditions, the diffusion of superb physical colors. Later, Lehmann with
the collaboration of several chemists, studied a great number of these
substances, mainly cholesterol derivatives showing liquid phases. Many

[15] E. J. DuPraw, "Cell and Molecular Biology." Academic Press, New York, 1968.
[16] A. Lima-de-Faria (ed.), "Handbook of Molecular Cytology." North-Holland Publ.,
   Amsterdam, 1969.
[17] O. Lehmann, Z. Physiol. Chem. 4, 462 (1889).

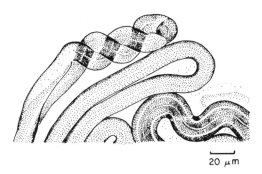

20 μm

FIG. 2.   Curved and coiled myelinic tubes given by sodium oleate and water, after a micrograph by Nageotte.[45]

of these latter are strongly birefringent when at rest between slide and coverslip.

2.   OPTICAL PROPERTIES OF LIQUID CRYSTALS AND INSECT CUTICLE

The extraordinary optical properties of liquid crystals and, in particular, cholesterics have been interpreted by Mauguin[18] with reference to a certain model given by Poincaré. This theory was completed by de Vries.[19] A. A. Michelson, the physicist known for his studies of the velocity of light, observed the reflection of circulary polarized light by iridescent cuticles of certain beetles.[20] Later, Gaubert[21] and Mathieu and Farragi[22,23] indicated the identity of the optical properties of the body-wall of these insects and those of cholesteric liquid crystals.

3.   GEOMETRY AND TOPOLOGY IN LIQUID CRYSTALS AND BIOLOGICAL MATERIALS

From 1910 to 1923 Friedel and Grandjean discovered the main geometrical and topological properties of liquid crystals and defined on this basis the smectic, nematic, and cholesteric organizations.[24,25] Their work does not contain any reference to biology, but several of their new concepts are of great importance for certain biological analogs. In

[18] C. Mauguin, *Bull. Soc. Fr. Mineral.* **34**, 6 and 71 (1911).
[19] M. de Vries, *Acta Crystallogr.* **4**, 219 (1951).
[20] A. A. Michelson, *Philos. Mag.* **21**, 219 (1911).
[21] P. Gaubert, *C.R. Hebd. Seances Acad. Sci.* **179**, 1148 (1924).
[22] J.-P. Mathieu and N. Farragi, *C.R. Hebd. Seances Acad. Sci.* **205**, 1378 (1937).
[23] J.-P. Mathieu and N. Farragi, *Bull. Soc. Fr. Mineral.* **61**, 174 (1938).
[24] G. Friedel and F. Grandjean, *Bull. Soc. Fr. Mineral.* **33**, 192 and 409 (1910).
[25] G. Friedel, *C.R. Hebd. Seances Acad. Sci.* **176**, 475 (1923).

developing certain works of these authors, we have come upon Möbius strips,[26,27] a situation related to the concept of the Möbius crystal first discussed by Frank.[28,29] We observed that Möbius singularities also exist in biological materials and, particularly, in certain chromosomes in which they could have a functional significance.[30,31]

Focal conics and related curves have been described in detail by Friedel and Grandjean.[24] Such distortions exist in several biological materials.[12,30-34] One also has to note the beautiful drawings made by Grandjean of smectic rodlets or "bâtonnets" displaying a remarkably symmetrical lattice of focal conics (see Friedel and Grandjean[24]) and seeming to draw their origin from the contemplation of certain flower organs. We must note that Grandjean was not only a brilliant physicist who first introduced the molecular field theory in liquid crystals[35] but was also a great zoologist who wrote important papers on the mechanisms of biological evolution.[36,37]

4. STRUCTURES OF LAMELLAR LIQUID CRYSTALS

Researches on the molecular structure of liquid crystals were undertaken after the First World War. The lamellar structure of smectics was confirmed by X-ray studies (de Broglie and Friedel[38,39]). Each smectic layer is a liquid film whose molecules are generally normal to the layer. The molecules move within one layer and can be compared to people in a very crowded party, with everyone standing or walking about, and trying to speak to everyone else. This gives a picture of the plane-restricted diffusion within the smectic layers.

Liquid films given by soaps added to water lead to a stepped structure which has been analyzed by Perrin.[40] This author observed molecular mono- and bilayers and indicated the analogy with liquid crystals.

[26] Y. Bouligand, J. Phys. (Paris) 35, 215 (1974).
[27] Y. Bouligand, J. Phys. (Paris) 35, 959 (1974).
[28] F. C. Frank, Philos. Mag. [7] 42, 809 (1951).
[29] F. C. Frank, Discuss. Faraday Soc. 25, 19 (1958).
[30] Y. Bouligand, J. Phys. (Paris) 36, C1-173 and 331 (1975).
[31] Y. Bouligand, J. Biochim. 57, vii (1975).
[32] Y. Bouligand, Tissue & Cell 4, 189 (1972).
[33] Y. Bouligand, J. Phys. (Paris) 33, 525 (1972).
[34] Y. Bouligand, J. Phys. (Paris) 33, 715 (1972).
[35] F. Grandjean, C.R. Hebd. Seances Acad. Sci. 164, 280, 431, and 636 (1917).
[36] F. Grandjean, Arch. Sci. 10, 477 (1957).
[37] F. Grandjean, Acarologia 14, 454 (1972).
[38] M. de Broglie and E. Friedel, C.R. Hebd. Seances Acad. Sci. 176, 738 (1923).
[39] M. de Broglie and E. Friedel, C.R. Hebd. Seances Acad. Sci. 180, 269 (1925).
[40] J. Perrin, Ann. Phys. (Paris) [9] 10, 160 (1918).

Grandjean[41] confirmed this point of view by the description of stepped drops given by smectics. The works of Langmuir[42] and Perrin[40] suggested that the fatty acids were vertical in the monomolecular films spread on water, their orientation being due to the presence of the hydrophilic group at one extremity of the molecule.

The X-ray diffraction patterns of crystallized fatty acids revealed the alternating orientation of the paraffinic chains and the polar ends (de Broglie and Trillat[43]). These studies led to the commonly accepted conception of bilayers in soaps and in many lyotropic systems. Later, Friedel[44] and Nageotte[45] gave descriptions of the liquid-crystal phases in myelin extracted from the brain or in lecithins. They observed the fluid motions and various textures which are classic in smectics. The interpretations of these two authors often differed. Nageotte probably observed a B phase (or the β configuration, see Section III,8). Note that the B phases have been described from X-ray data by Lawrence and Rawlins.[46]

5. BIOLOGICAL MEMBRANES

It seems that the liquid-crystal concept must be applied to different biological membranes. The chemistry of the plasma membrane was first approached by the study of red blood cells. Gorter and Grendel[47] showed that the extracted lipoids can be spread on water and give a film whose surface is twice the external area of the whole population of erythrocytes. The shape of red cells in normal blood is constant and their number is easily determined. Drawing their conception from Langmuir's ideas, the authors suggested that this film is made of a monomolecular layer of vertical molecules with the hydrophobic poles pointing upwards. It seems likely that the red cell membrane is a bilayer with the hydrophilic poles pointing outwards, with adjunction of proteins. These results have been confirmed by Waugh and Schmitt[48] by the measurement of the thickness of the superimposed bilayers using an interferential method introduced by Perrin.[40] The number of bilayers is easily found by direct observation of their limits in the stepped structure and by comparison with figures given by red cell membranes.

[41] F. Grandjean, C.R. Hebd. Seances Acad. Sci. **166**, 165 (1918).

[42] I. Langmuir, J. Am. Chem. Soc. **39**, 1848 (1917).

[43] L. de Broglie and J.-J. Trillat, C.R. Hebd. Seances Acad. Sci. **180**, 1845 (1925).

[44] G. Friedel, C.R. Hebd. Seances Acad. Sci. **185**, 330 and 1237 (1927).

[45] J. Nageotte, Actual. Sci. Ind., 431–434 (1936).

[46] A. S. C. Lawrence and F. I. G. Rawlins, Sci. Prog. **28**, 339 (1933).

[47] E. Gorter and R. Grendel, J. Exp. Med. **41**, 439 (1925).

[48] D. F. Waugh and F. O. Schmitt, Cold Spring Harbor Symp. Quant. Biol. **8**, 233 (1940).

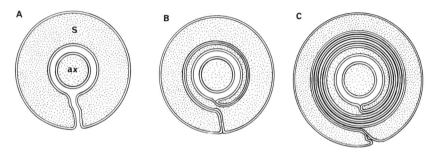

Fɪɢ. 3.   Cross sections showing the origin of myelin around an axon (ax). Axons are cylindrical expansions of nerve cells involved in the transmission of impulses. The axons are often surrounded by certain cells (called Schwann cells: S). The plasma membrane of these latter develops extensively and forms a spiralized coating, as can be seen from the three steps represented in A, B, C; redrawn after Robertson.[6]

A second approach to the physical and chemical properties of the plasma membrane was suggested later by the first electron microscope investigations on myelin morphogenesis around the axons of certain nerve cells (Geren[4]). The membrane of the adjacent cells (called "Schwann cells") develops extensively and forms a spiralized coating as can be seen from Fig. 3. The resulting myelin is thus a stacking of plasma membranes. It was already known from X-ray diffraction patterns, birefringence studies, and chemical investigations that myelin was made of concentric bilayers with intercalated proteins, but this was the first observation of the continuity between these bilayers and the plasma membrane. It follows that our knowledge of myelin properties is a great contribution to that of the cell membrane. In particular, the fluid character of the myelin was known to many early researchers (review in Frey-Wyssling[3]) and it must apply to the cell membrane.

Diffusion in the plane of the plasma membrane was suggested very early by cytologists. The liquid character of the plasma membrane appears when cells are studied with the help of a micromanipulator (Chambers[49]). Fluidity has to be invoked to explain the formation of vesicles by exo- and endocytosis. Exocytosis corresponds to the release of secretion from cells within bilayer-limited vesicles. In endocytosis extracellular materials are taken up within invaginations of the cell membrane. Bennett[50] has been led to assume the existence of fluid streams in the plasma membrane to interpret his electron micrographs by endocytosis.

It should be noted that the plasma membrane deformations cannot be

[49] R. Chambers, Ann. Physiol. Physiochim. Biol. 6, 233 (1930).
[50] S. Bennett, J. Biophys. Biochem. Cytol. 2, Suppl. 2, 99 (1956).

explained by the concept of elasticity alone. A thin rubber membrane is elastic, but is not liquid. In it, two neighboring molecules of polymer remain very close, even in the case of a strong deformation. In contrast, a ceaseless diffusion occurs in a liquid film. The density of molecules remains constant but the neighborhood of any particular molecule is rapidly changed. Owing to its liquid character the plasma membrane forms only closed surfaces. A free edge would be unstable. A hole can exist in a rubber sheet and will deform according to the distribution of tensions. A stable hole is inconceivable in a film of soap and water or in a plasma membrane. This latter appears to be a two-dimensional liquid insoluble in the surrounding aqueous phases. It has been impossible to observe the successive steps of the exo- and endocytosis which need a brief molecular rearrangement at points of separation and fusion.

The fluid character of the plasma membranes was known very early and seems to have been forgotten by most biochemists and physiologists. It has been rediscovered today by the sophisticated methods of molecular biology, as can be appreciated from recent papers on the plasma membrane regarded as a "fluid mosaic" (Singer and Nicolson;[51] Edidin[52]).

III. Smectics and Related Systems in Cells and Tissues

6. THE ELECTRON MICROSCOPIC INVESTIGATION OF BIOLOGICAL MATERIALS

The ultrastructural organization of cells and tissues is that which has been revealed mainly by electron microscopic examination. The usual form of this process is to take a fresh specimen and to fix it by immersion in an aqueous solution of a substance which insolubilizes the major biochemical compounds and also stops numerous enzyme activities. OsO_4 is one of the often used substances in fixation (see general books such as Brachet and Mirsky;[53] de Robertis et al.;[54] Lima-de-Faria[16]). OsO_4 links unsaturated compounds, such as many lipids, and makes them insoluble in organic solvents. After fixation, the sample is dehydrated in alcohol washes of increasing concentration and is progressively penetrated by a resin such as epon or araldite, which finally is

[51] S. J. Singer and G. Nicolson, *Science* **175**, 720 (1972).
[52] M. Eddin, *Symp. Soc. Exp. Biol.* **28**, 1 (1974).
[53] J. Brachet and A. E. Mirsky, "The Cell," Vols. 1–6. Academic Press, New York, 1959–1964.
[54] E. P. D. de Robertis, W. W. Novinski, and F. A. Saez, "General Cytology," 3rd ed. Saunders, Philadelphia, 1963.

allowed to polymerize. Thin sections are obtained with an ultramicrotome. Contrast is improved by floating sections on solutions of heavy metallic compounds. Many artifacts can originate from such methods. The best way to discuss the results is to compare the pictures obtained with various fixing agents and to study the material by the freeze-etching technique: The fresh material is rapidly frozen and cut at −100°C. After a brief time of water sublimation, which makes the fine structures stand out in relief, a shadowed replica is made.

7. BIOLOGICAL BILAYERS

Cells are bounded by a closed thin membrane called a "plasma membrane" which appears to consist of two electron-dense layers separated by a uniform clear zone (see Robertson[6]). In certain cases, the outer layer is slightly denser than the inner one. The cell membrane shows an inner side in contact with the cytoplasm and an outer side in contact with the extracellular medium. One often sees a granular component called "glycocalyx" or "cell-coat" lying on the outer side of the cell membrane. When two cells are in contact, the extracellular medium is reduced to a narrow zone separating the two plasma membranes.

The ultrastructural aspect and the proposed chemical nature of the cell coat are highly variable. In contrast, the cell membrane ultrastructure is very constant. The system of two dense layers separated by a uniform clear zone is also found in the membranes bounding the nucleus and different cytoplasmic organelles: mitochondria, endoplasmic reticulum, Golgi apparatus, etc. This bilayer organization has been called the "unit membrane" and is very general, despite a high degree of variability in chemical composition. Among the major components one finds phospholipids (such as lecithin, phosphatidyl-ethanolamine, phosphatidyl-serine, mono- and diphosphatidyl-glycerol, phosphatidyl-inositol), cholesterol, extremely varied proteins, and small amounts of polysaccharides. Cholesterol is absent in the bacterial membrane and in the internal cristae (infolded membranes) of mitochondria (see Chapman[55]).

It has been shown that the intermediate clear layer corresponds to the paraffinic chains of the phospholipids and that the two dense layers are given both by the hydrophilic extremities and proteins. The interpretation of these figures is based on the comparison of natural membranes and models. For instance, various phospholipids can be extracted from cell membrane. After addition of water, the material is fixed and prepared for electron microscopy (review in Robertson[6]). Thin sections

---

[55] D. Chapman, *In* "Membranes and Ion Transport" (E. E. Bittar, ed.), Vol. 1, p. 24. Wiley, New York, 1970.

show unit membranes lying roughly parallel. Their average distance increases with the degree of hydration. The varying interval due to water accumulation between the successive bilayers of constant thickness can be estimated from X-ray diagrams. Such models show a regularly decreasing birefringence when water is progressively added. The birefringence is positive in the absence of water and is due to the orientation of phospholipids. The water has a refractive index different from the mean index of bilayers. The addition of water introduces accordingly a negative form of birefringence (see Frey-Wyssling[3]) and this latter can mask the positive intrinsic birefringence of phospholipids. The sign of the birefringence observed changes when the bilayers separate with a sufficient addition of water.

Surprisingly, the distribution of proteins in natural lamellar systems, such as the nerve myelin, keeps a periodic distribution after extraction of lipids. The proteins form equidistant layers showing a periodicity very close to that observed before extraction. These results are confirmed by the fact that the radially positive birefringence (oriented normally to the layers) is reversed in the nerve myelin to a radial negative birefringence when the lipids are extracted (see Robertson[6]). Such studies indicate that most proteins are floating on the polar ends of the phospholipids. More recent work suggests that a small part of the proteins can be more or less-deeply embedded in the thickness of the phospholipid layer and certain of them could span the entire thickness of the bilayer (see Singer and Nicolson[51]). The proteins are probably oriented by the phospholipidic environment and the distribution of polar and nonpolar residues must be involved in this process. More generally, it is well known that molecules in solution in a mesomorphic phase are oriented by the medium and this fact has been used in magnetic resonance studies (Jain et al.[56]). Some of the proteins of the plasma membranes are enzymes involved in selective permeability and play a great role in the active transport of certain ions.

A schematic diagram of the arrangement of unit membranes in the cell, its nucleus, and its organelles is given in Fig. 1. This picture shows a section of a system which must be regarded as three-dimensional and which is made of closed and more or less folded and corrugated surfaces. Connections occur between the nuclear envelope, the endoplasmic reticulum, and the cell membrane.

Series of superimposed bilayers are frequent in biological materials. Superb examples can be found in the atlas of biological untrastructures published by Porter and Bonneville[57], and, for instance, in the myelin

---

[56] P. L. Jain, H. A. Moses, H. O. Lee, and R. D. Spence, *Phys. Rev.* **92,** 844 (1953).

[57] K. R. Porter and M. A. Bonnevile, "An Introduction to the Fine Structure of Cells and Tissues." Lea and Febiger, Philadelphia, 1964.

sheath around certain nerve fibers, which is described in Fig. 2. Each cone or each rod in visual cells of vertebrates is made of a considerable stack of flat vesicles. Note that the parallel flat vesicles are replaced by a hexagonal packing of bilayered tubes in the compound eye of insects (see Wolken[57a]). The chloroplasts of plant cells also are complex systems of lamellae. These structures are birefringent and show the unit-membrane arrangement (two dense layers separated by a clear zone) when observed in good conditions at the highest magnifications of the electron microscope.

The Golgi apparatus and the rough endoplasmic reticulum (Fig. 1) are other noteworthy examples of superimposed flat sacs limited by bilayers. Certain mitochondria show a parallel alignment of their internal cristae. Other mitochondria (mainly in protozoa) show tubular cristae. The external plasma membrane of cells can form hexagonal packing of cylindrical microvilli. All these arrangements are not artifacts and are observed after fixation, dehydration, and embedding, or by freeze-etching.

8. MODELS

The biophysics of lipidic associations has been studied extensively. Water–lipid systems have been prepared from brain, mitochondria, chloroplasts, and other organelles. They have been compared with phases given by pure synthetic phospholipids or their mixtures. The ternary or quaternary systems given by water, lecithin, bile salts, and more complex associations also have been analyzed by wide- and small-angle X-ray diffraction patterns and by polarizing microscopy. These models have been described mainly by the groups of Luzzati,[58-61] Chapman,[5,55,62,63] and Small et al.[64-66]

[57a] J. J. Wolken, "Photoreceptors and Evolution." Academic Press, New York, 1975.
[58] V. Luzzati, In "Biological Membranes" (D. Chapman, ed.), Vol. 1, p. 71. Academic Press, New York, 1968.
[59] V. Luzzati and F. Husson, J. Cell Biol. 12, 207 (1962).
[60] V. Luzzati, H. Mustacchi, A. E. Skoulios, and F. Husson, Acta Crystallogr. 13, 660 (1960).
[61] V. Luzzati, A. Tardieu, and D. Taupin, J. Mol. Biol. 64, 269 (1968).
[62] D. Chapman, "The Structure of Lipids." Methuen, London, 1965.
[63] D. Chapman, Liq. Cryst. Plast. Cryst. 1, 288 (1975).
[64] D. M. Small, M. C. Bourges, and D. G. Dervichian, Biochim. Biophys. Acta 125, 563 (1966).
[65] D. M. Small, M. C. Bourges, and D. G. Dervichian, Biochim. Biophys. Acta 137, 157 (1967).
[66] D. M. Small, M. C. Bourges, and D. G. Dervichian, Biochim. Biophys. Acta 144, 189 (1967).

The paraffinic chains undergo a transition between a stiff conformation (with parallel alignment, hexagonal packing, rotational disorder) and a highly disordered conformation (respectively called type β and type α according to a notation introduced by Luzzati). This reversible transition depends on temperature and on the chemical composition of the mixture. The β arrangement of the hydrocarbon chains is very similar to that of the molecules of a smectic B with no correlated ordered layers. The phospholipid molecules of the water–lipid systems give lamellae (indefinite bilayers), indefinite rods of constant diameter, rodlets of definite length and diameter, or spheres of equal radius. (In the absence of water, one also obtains indefinite ribbons of constant thickness and width, or disks.) For the structural units, the polar groups may be either inside or outside and, accordingly, two opposite topologies appear: "oil in water" and "water in oil." The different structural units are associated in phases showing a periodicity in one, two, or three dimensions (stacked lamellae, hexagonal packing of rods, two- or three-dimensional networks of jointed rods).

In lamellar systems given by mixtures of two synthetic and slightly different lecithins, intermediary phases between α and β types appear and a probable segregation of molecules leads to a large-scale periodic organization (Ranck et al.[67]).

The water–lipid systems are not all mesomorphic in the sense given by Friedel,[68] that is to say: birefringent and liquid without any two- or three-dimensional periodicity (usually smectics show only a one-dimensional periodicity). In water–lipid systems, phospholipids and water can diffuse even if the lattices show a two- or a three-dimensional periodicity. For a more complete discussion of the order in lyotropic phases, see the chapter by Charvolin and Tardieu in this volume.

9. The Liquid Character and the Morphology of Membranes

The various shapes of membranes (see Fig. 1) and different phenomena, such as exocytosis and endocytosis, are related to their fluid character. We have seen that the unit membranes, which are bilayers with added proteins and glycoproteins, can form flat sacs and tubes, often showing parallel accumulation or hexagonal stacking. These situations are very different from the preceding lamellar and hexagonal phases studied in water–lipid models (with "oil" in lamella and "water" in between). However, transitions from lamellae to rods or spheres

[67] J. L. Ranck, L. Mateu, D. M. Sadler, A. Tardieu, T. Gulik-Krzywicki, and V. Luzzati, J. Mol. Biol. 85, 249 (1974).
[68] G. Friedel, Ann. Phys. (Paris) [10] 18, 273 (1922).

observed in models are an interesting approach to certain problems of morphogenesis and have a bearing on the question of the exo- and endocytosis. One must keep in mind that proteins play a great role in membrane and probably in morphological questions. The fusion of a vesicle with a plasma membrane is facilitated by the presence of a ring of some globular proteins (eight in many cases, see Satir[69]). Similar rings also have been observed around nuclear pores, joining the two external and internal bilayers of the nuclear envelope in cells. It seems likely that such proteins are local modifiers of the elastic constants and introduce a spontaneous curvature.

Recent studies show that the fluid character of the membrane is necessary to interpret the variations of shape in red cells. The passage from a biconcave disk to an echinoid morphology (Fig. 4) changes the surface shape and this is inconceivable by pure elasticity. In the course of this transformation, the bilayers keep their constant thickness, which is evident from electron microscopy, and one must admit the existence of diffusion in the plane of the plasma membrane. The study of the transition between normal biconcave shape and sphere has led Helfrich and Deuling[70] to assume that the membrane behaves as a two-dimensional liquid and shows a spontaneous curvature opposite to the curvature of the sphere. This second characteristic is related to the asymmetry of the plasma membrane. The adjacent layers of proteins differ on the two sides of the phospholipidic bilayer. Such differences are also visible in myelin (Fig. 2) and the electron-dense levels are alternately lighter or darker. Another beautiful illustration of the fluid character of the red cell membrane is the interpretation of the ceaseless flickering observed in dark field or phase-contrast microscopy (Brochard and Lennon[71]). Other modes of oscillations are visible at the surface of several sorts of cells like fibroblasts, certain leucocytes, etc., and would need a comparable study (see Policard and Baud;[72] Bloom and Fawcett[73]).

10.   OTHER SMECTIC ANALOGS

Smectics and their analogs are very numerous in cells and tissues and our review of this area is incomplete. We will end this section with

[69] B. Satir, *Symp. Soc. Exp. Biol.* **28**, 399 (1974).

[69a] M. Bessis, "Living Blood Cells and Their Ultrastructure." Springer-Verlag, Berlin and New York, 1973.

[70] W. Helfrich and H. J. Deuling, *J. Phys. (Paris)* **36**, C1-327 (1975).

[71] F. Brochard and J. F. Lennon, *J. Phys. (Paris)* **36**, 1035 (1975).

[72] A. Policard and C. A. Baud, "Les Structures Inframicroscopiques Normales et Pathologiques des Cellules et des Tissus." Masson, Paris, 1958.

[73] W. Bloom and D. W. Fawcett, "A Textbook of Histology," 8th ed. Saunders, Philadelphia.

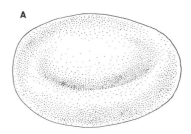

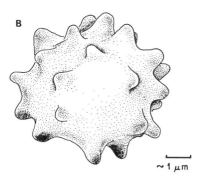

~ 1 μm

Fig. 4. In certain media, the normal biconcave shape of red cells (discocytes) (A) in the blood can transform into a spiculated form called echinocyte (B), without any variation of volume and cell area (Bessis[69a]).

examples very far from membranes and their derivatives. Certain viruses show a cylindrical shape with a definite length. In some electron micrographs, in section, one can observe superb smectic stackings of these viruses (Warmke and Christie;[74] Miličič et al.[75]). Other preparations correspond to a nematic configuration (Gourret[76]). Before these electron microscopic investigations, smectic and nematic arrangements had been recognized by Bernal and Fankuchen[77] in these viruses. Another three-dimensional arrangement similar to a plywood has also been observed (Herold and Munz[78]).

The superficial grease secreted on certain insect cuticles seems to be a smectic liquid composed of amphiphilic molecules and behaves as an

[74] H. E. Warmke and R. G. Christie, *Virology* **32**, 534 (1967).
[75] D. Miličič, Z. Stefanac, N. Juretić, and M. Wrischer, *Virology* **35**, 356 (1968).
[76] J.-P. Gourret, private communications (1975).
[77] J. D. Bernal and I. N. Fankuchen, *Nature (London)* **139**, 923 (1937).
[78] F. Herold and K. Munz, *J. Gen. Virol.* **1**, 375 (1967).

electret. Beament[79,80] has shown that a 6 V potential applied across the cuticle of the cockroach gave a persistent voltage decreasing exponentially from 4 to 0.2 V in 10 minutes.

IV. Nematics, Cholesterics and Their Analogs

11. NEMATICS AND CHOLESTERIC INCLUSIONS IN TISSUES

Several papers have been devoted by Stewart[80a] to the existence of nematic or slightly cholesteric droplets in the cells of different tissues and organs. Mesomorphic spherulites have been observed in the corpus luteum of the human ovary, which produces the progesterone involved in the maintenance of the embryo in the uterus. These droplets contain free and esterified cholesterol as main components. Triglycerides and phospholipids are also present. Other steroids form spherulites in the adrenal cortex. This endocrine gland is known for its production of the cortical steroids involved in many aspects of metabolism. Mesomorphic spherulites are often found in the walls of arteries in certain arterioscleroses. As cited above, nematic and cholesteric liquid crystals have been recognized by polarizing microscopy in certain viral accumulations in plant cells (Bernal and Fankuchen;[77] Wilson and Tollin[81]).

12. STACKED ROWS OF PARALLEL ARCS IN FIBROUS SYSTEMS

For some years we have been interested in the fact that thin sections of different animal tissues, of various plant cell walls, and of certain chromosomes show stacked rows of parallel arcs, which are strikingly similar, despite the differences in the chemical nature and in the function of these biological materials. This arrangement also has been observed in pathological structures (hemoglobin in sickle cell anemia, viral tumors, etc.). We have published a tentative list of these materials[32–34] which is not yet complete.

13. THE EXAMPLE OF THE CRAB BODY WALL

The organic matrix of the crab body wall is a noteworthy example. The organic network remains after decalcification and is made of chitin (a polymer of acetyl-glucosamine) and proteins. The material appears to be made of superimposed lamellae, which are parallel to the surface of

[79] J. W. L. Beament, *Biol. Rev. Cambridge Philos. Soc.* **36**, 281 (1961).
[80] J. W. L. Beament, *Adv. Insect Physiol.* **2**, 67 (1964).
[80a] G. T. Stewart, *Mol. Cryst. Liq. Cryst.* **1**, 563 (1966).
[81] H. R. Wilson and P. Tollin, *J. Ultrastruct. Res.* **33**, 550 (1970).

the carapace (Fig. 5). We define the plane of this latter as being horizontal. Each lamella contains parallel arcs when observed in oblique sections (Fig. 5). In contrast, in vertical section (i.e., normal to the cuticle plane) the arcs disappear but the stratification persists (Fig. 6). One sees alternate layers of fibrils in longitudinal view and in cross section. As the plane of the section approaches the plane of the layers, the arcs appear larger. In all oblique sections which cut through the entire thickness of the material, the number of lamellae remains constant; each lamella contains one row of parallel arcs.

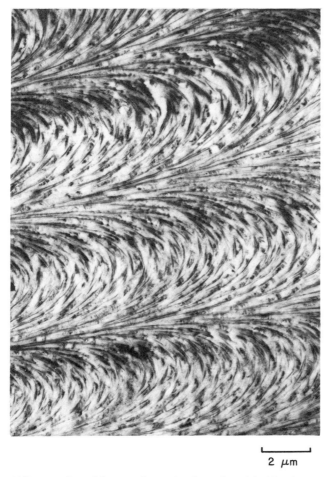

2 μm

FIG. 5.   Oblique section of the organic matrix of a crab cuticle (*Carcinus maenas*). The material has been decalcified. The fibrils made of chitin and proteins draw staked rows of parallel arcs.

As it obviously appears in section, the arced figures are superimposable by all horizontal translations. Since this property holds for any of the
oblique sections, it follows that the fibril direction is constant within any
given horizontal plane. This means that the fibril direction in any plane
depends solely on its level in the thickness of the cuticle.

If one considers two oblique sections which are opposite and symmetric to each other with respect to a single vertical axis (Fig. 7), the

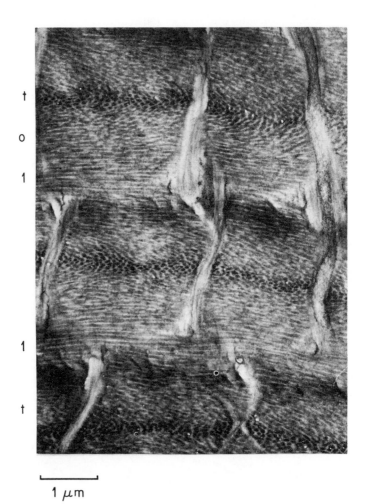

1 μm

FIG. 6.   Vertical section of the organic matrix of a crab cuticle (*Carcinus maenas*).
Vertical means that the section plane is normal to the general direction of the cuticle.
Fibrils lie alternately in the section plane (l) and perpendicularly to it (fibrils cut
transversely: t). The fibrils lying at the intermediary levels are cut obliquely (o).

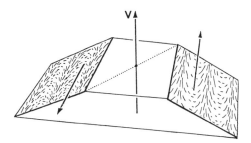

FIG. 7.   Two oblique sections symmetrical about a vertical axis (V). The concavities of the bow-shaped patterns are opposite in the two sections.

observations show that the rows of arcs in each section are equivalent, except that the concavity is opposite. This means that the two oblique sections are superimposable by a rotation by 180° about the vertical axis. It follows that within a single horizontal plane, the fibrils simultaneously are parallel, and are superimposable by a 180° rotation about any vertical axis. One can easily verify that only horizontal or vertical orientations are compatible with these two constraints. As a uniform vertical distribution of fibrils does not give, in section, any arced configuration, the fibrils must be horizontal.

The oblique sections which approach the horizontal allow a measurement of the changing horizontal fibril direction in successive levels. This measurement needs only a very slight correction for the obliqueness of the section plane. This can be omitted in a first approximation. The fibril direction at each level has been represented in a dial (Fig. 8). Each dial corresponds to a particular level in the thickness of the cuticle. The directions observed on the successive dials appear to rotate in a continuous way. In other words, it follows that our system is comparable to a piece of plywood where the grain of the different layers rotates by a constant small angle from layer to layer. This is different from a common plywood, where the grains in any two successive layers lie at right angles.

The basic properties revealed in the sections are illustrated in the pyramidal model of Fig. 9, which shows steps of rotation by a constant angle of 30°. A series of rectangles, where parallel and equidistant straight lines have been drawn, is arranged in the form of a pyramid trunk. Series of bow-shaped patterns appear on the oblique sides. The arcs show opposite concavities on the right and on the left. It seems clear that a cholesteric organization occurs in the organic matrix of the crab body wall.

The twisted fibrous materials are not arranged with the exactness of

the model. We have to consider the average directions of fibrils in small
volumes (the biggest dimension being, for instance, one-tenth of the
thickness of one-half helicoidal pitch). A certain degree of disorder is
defined by the deviation of the fibrils from the average direction in each
sample volume. The model does not show any fibril interweaving

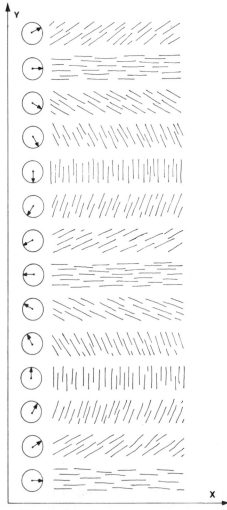

FIG. 8.   Oblique section very near the horizontal direction. The arced patterns are very
wide. The observed directions correspond with a negligible correction to the exact
direction they have in the horizontal planes. Each direction is represented in a dial; the
direction is regularly rotating when the horizontal level is changed. The Ox axis is
horizontal; the Oy axis is a line of steepest slope.

FIG. 9. Pyramidal trunk made of a series of rectangles of a definite thickness. Parallel and equidistant straight lines have been drawn on each rectangle. Their direction changes by an angle of 30° from one level to the following one. The sense of rotation has been chosen left-handed, as it is found in all the biological materials we have studied.

between different horizontal levels, but such vertical connections could occur in the fibrous system.

Stacked series of parallel arcs can be observed between the glass and cover slip with cholesterics whose half-helicoidal pitch varies between 10 and 15 μm, the thickness of the liquid-crystal slab being of the same order. The Brownian motion is directly visible in the phase-contrast microscope, and if the cholesteric axis is tilted with respect to the optical axis of the microscope the ceaseless movements of transitory molecular swarms appear to outline the parallel arcs. Series of arcs or "commas" also have been observed under different experimental conditions by Friedel[68] and Rault and Cladis.[82] Systems of spirals due to distortions are seen both in the sections of the organic matrix in the crab cuticle and in preparations of cholesteric liquids. Their origin is identical (Bouligand[32-34]).

14. PSEUDOMORPHOSES OF CHOLESTERICS AND NEMATICS

Similar series of parallel arcs are common in numerous arthropod cuticles and, particularly, in insects whose cuticle reflects circularly polarized light and shows a strong rotatory power as do cholesteric liquids. Note that the colors of insects do not depend on temperature.

[82] J. Rault and P. E. Cladis, *Mol. Cryst. Liq. Cryst.* **15,** 1 (1971).

Recent work in this field will be found in Bouligand,[12] Neville and Caveney,[83] Caveney,[84] and Neville[85]).

Such hard materials are naturally polymerized liquid crystals. Synthetic cholesterics also have been polymerized (Strzelecki and Liébert[86]). They exhibit in thin section the geometrical and optical properties of liquid crystals, but are not fluid (Bouligand et al.[87]). One can define an order parameter in these systems describing the degree of alignment in a small volume around a given point.

The helicoidal pitch of the crab cuticle keeps a constant value at a definite depth, but shows strong variations in the course of the cuticular secretion cycle between two successive molts. In locusts, a nematic pseudomorphosis forms during the day, whereas a cholesteric "plywood" is normally deposited during the night.[85] A thin section of the elongated tibia of the hind leg of a locust shows twisted lamellated zones in alternation with layers of preferred microfibrillar orientation. It seems that a twisting substance is secreted by the epidermis with a varying concentration during the cuticle cycle. The helicoidal pitch is stabilized by a rapid polymerization of the freshly deposited material. The abrupt variations of twist could not occur if the cuticle was mesomorphic in the bulk a long time after secretion.

In the crustacean cuticle, electron micrographs show the successive steps of the fibril differentiation starting from small subunits, as can be seen in Fig. 10. It seems likely that the secretion passes through a very dense and brief colloidal state. Very thin filaments (or segments) are first deposited in the vicinity of the cytoplasmic membrane of the epidermal cells. The subunits appear from small thickened and electron-dense areas of this membrane. They are progressively gathered into larger units and form a fibrous network. Each fibril is a bundle of filaments and numerous exchanges occur here and there. Such systems were studied by Ostwald.[88] His work on the colloidal mesophases has been extended by several papers on birefringent gels, mainly by Hermans,[89,90] Frey-Wyssling has reviewed these in his book.[3] We reproduce in Fig. 11 a schematic diagram showing the progressive formation of microcrystals

[83] A. C. Neville and S. Caveney, Biol. Rev. Cambridge Philos. Soc. 44, 531 (1969).
[84] S. Caveney, Proc. R. Soc. London, Ser. B 178, 205 (1971).
[85] A. C. Neville, "Zoophysiology and Ecology," Vol. 4/5. Springer-Verlag, Berlin and New York, 1975.
[86] L. Strzelecki and L. Liébert, Bull. Soc. Chim. F. 597, 605 (1973).
[87] Y. Bouligand, P. E. Cladis, L. Liébert, and L. Strzelecki, Mol. Cryst. Liq. Cryst. 25, 233 (1973).
[88] W. Ostwald, Z. Kristallogr., Kristallgeom., Kristallphys., Kristallchem. 79, 222 (1931).
[89] P. H. Hermans, Kolloid-Z. 83, 71 (1938).
[90] P. H. Hermans, Kolloid-Z. 97, 231 (1941).

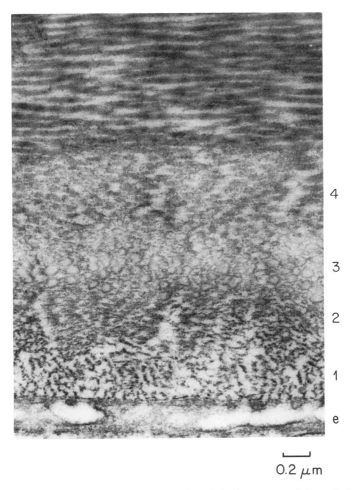

0.2 μm

FIG. 10. Electron micrograph of a thin section of the integument of the crab *Carcinus maenas* at the limit between the epidermal cytoplasm (e) and the freshly deposited cuticle. The organic matrix of the body wall is made of fibrils (chitin and proteins) and will later be mineralized (calcite). One distinguishes four main steps leading to the fibril differentiation (1–4).

in an oriented polymer solution. Such systems cease to be liquid and transform into gels when the concentration is progressively increased. Crystalline bundles of parallel polymers also appear in the course of the consolidation of the cuticle in arthropods. The first deposited cuticular material is a dense gel which is birefringent. This colloidal state allows certain twisting forces (van der Waals) to come into play.

These considerations of cuticle deposition suggest a process of self-

A   B   C

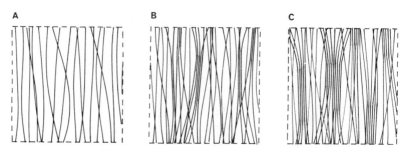

FIG. 11.   Microcrystallization of an anisotropic polymer solution (after Frey-Wyssling[3]).
An increase of the polymer concentration leads to the crystalline packing of chains, which,
however, emerge in the liquid phase and pass from one crystal to another one. This
process transforms an anisotropic sol(A)(fluid) into a birefringent gel(C)(nonfluid).

assembly whose mechanism would be related to the growth of liquid
crystals.

One finds equivalents of nematics and cholesterics in the organic
matrix of numerous other skeletal tissues (reviewed by Bouligand[32]).
It also seems that a brief passage through a nematic or a cholesteric
phase is rare. The first secretion is often an oriented gel. However, there
are some possible exceptions. In the formation of hairs, the condensation of keratin can involve liquid birefringent droplets (Birbeck and
Mercer[91]). Note that many secretions are cholesteric in a first liquid step
and transform into a hard crystallized material, different from a cholesteric pseudomorphosis. This is the case in the oothecal secretion of the
praying mantis (Kenchington and Flower,[92] Neville and Luke[93]). Certain
fresh silks secreted by bumblebees seem to me to be cholesteric, as can
be seen from Fig. 6 in a paper of Rudall.[94]

15. CHROMOSOMES

The chromosomes of certain unicellular organisms (dinoflagellates)
show a mass of fibrils which have been shown to be DNA and form
stacked rows of parallel arcs (Fig. 12). These latter are strikingly similar
to those observed in the organic matrix of the crab body wall. The
observed patterns correspond to a cholesteric layering of the DNA
(Bouligand,[7] Bouligand et al.[95]). The half helicoïdal pitch varies between

[91] M. S. C. Birbeck and E. H. Mercer, J. Biophys. Biochem. Cytol. 3, 203 (1957).
[92] W. Kenchington and N. E. Flower, J. Microsc. (Oxford) 89, 263 (1969).
[93] A. C. Neville and B. M. Luke, J. Cell Sci. 8, 93 (1971).
[94] K. M. Rudall, In "Aspects of Insects Biochemistry" (T. W. Goodwin, ed.), Biochem.
     Soc. Symp. 25, 83. Academic Press, New York, 1965.
[95] Y. Bouligand, M.-O. Soyer, and S. Puiseux-Dao, Chromosoma 24, 251 (1968).

1000 and 1500 Å. The chromosomes are more or less elongated but they may also be spherical. The cholesteric layers lie normal to the longitudinal axis of the chromosome.

If the chromosome is spherical, the diameter perpendicular to the layers is taken to represent the longitudinal axis. The explanation of the arced patterns observed in section is provided by the geometry of a stack of disks where parallel and equidistant lines have been drawn (Fig. 13). These disks form a series of uniformly rotated layers. The intersections of two oblique planes show the origin of the repeated arced patterns in these chromosomes. The filaments are not so well ordered in chromosomes as in the model (as can be seen from Fig. 12). In each

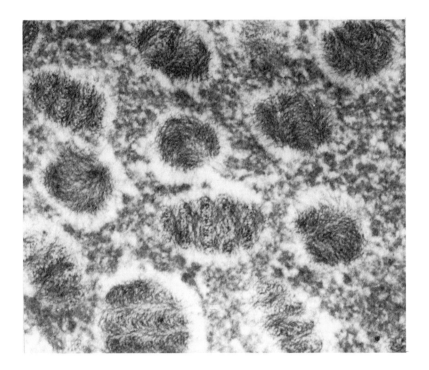

0.2 μm

Fig. 12.   Electron micrograph of a section of dinoflagellates (unicellular algae) showing chromosomes in the form of elongated ellipsoids of revolution. The stacked rows of parallel arcs are more or less visible according to the direction of the long axis of the chromosomes with respect to the section plane. Certain sections are longitudinal and show alternately filaments in longitudinal and in transversal section.

small volume of the chromosome one can define the average direction of DNA and observe the deviation of individual or groups of filaments with respect to this mean direction. The order parameter is smaller than unity.

Similar stacked rows of parallel arcs have been documented in bacterial chromosomes, mainly by Giesbrecht,[96,97] Ryter and Piéchaud,[98] Ryter,[99] Grund,[100] and J.-P. Gourret (private communications, 1975). The dimensions of these chromosomes are small and do not allow, in general, more than two or three rows of parallel arcs. Some beautiful micrographs of such bacterial chromosomes have been obtained by Gourret (Fig. 14). These series of parallel arcs also correspond to the cholesteric packing of DNA. In bacteria the chromosome DNA filament is known to be very long and generally ring-shaped. Its length in *Escherichia coli* is 1.2 mm. This filament is often unique, but there are two after the duplication occurring before cell division. This DNA filament is folded back on itself many times in order to fit in the chromosome which is very small, usually less than 1 μm in its longest dimension.

The most visible part of the dinoflagellate chromosome in thin section represents only that fraction of the total DNA which is densely packed. It is probable that peripheral loops project into the surrounding medium and form a more dilute phase (Bouligand et al.,[95] Soyer and Haapala[101]). Such peripheral loops must also exist in bacterias, as is suggested by certain autoradiographic studies (Caro,[102] Franklin and Granboulan[103]). The loops in dinoflagellates are probably more or less extended according to the physiological state of the cell.

The simplest model which can be given of the folding of the DNA filament in the dinoflagellate chromosome is shown in Fig. 15. One must note that this model is speculative, the exact circuit of the DNA filament being impossible to establish within the present state of our preparation and observation techniques. On the other hand, the model of Fig. 13, which concerns only the distribution of the average directions of DNA in the chromosome, is not hypothetical and is a geometrical reconstruction

[96] P. Giesbrecht, *Zentralbl. Bakteriol., Parasitenkd., Infektionskr. Hyg., Abt. 1: Orig.* **83**, 1 (1961).
[97] P. Giesbrecht, *Zentralbl. Bakteriol., Parasitenkd., Infektionskr. Hyg., Abt. 1: Orig.* . **187**, 452 (1962).
[98] A. Ryter and M. Piechaud, *Ann. Inst. Pasteur, Paris* **105**, 1071 (1963).
[99] A. Ryter, *Colloq. Int. C.N.R.S.* **163**, (1965).
[100] S. Grund, *Zentralbl. Veterinaermed., Reiche B* **21**, 73 (1974).
[101] M.-O. Soyer, and O. K. Haapala, *J. Microsc. (Paris)* **19**, 137 (1974).
[102] G. Caro, *Symp. Int. Soc. Cell Biol.* **1**, 228 (1962).
[103] R. M. Franklin and N. Granboulan, *J. Mol. Biol.* **14**, 623 (1965).

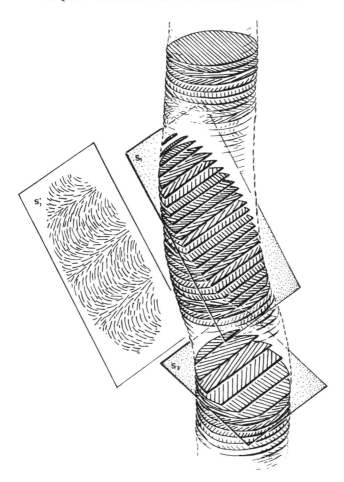

FIG. 13. Certain dinoflagellates have very elongated and cylindrical chromosomes. This model explains the origin of the arced patterns. A series of stacked disks with parallel and equidistant lines shows a regular twist with a left-handed orientation as has been proved to occur in these chromosomes. Two section planes at different angles relative to the long axis are shown.

deduced from the consideration of various serial sections. The existence of a certain degree of disorder has been neglected in the two models.

The dinoflagellate chromosome shows a distribution of directors similar to that present in a droplet or a rodlet of MBBA (methoxy-benzilidene-butylaniline, a classical nematic) twisted by addition of a cholesterol derivative. Such droplets are obtained in suspension in the isotropic liquid by decreasing the temperature. The difference is that in the chromosome the DNA forms a very long filament, a Watson and

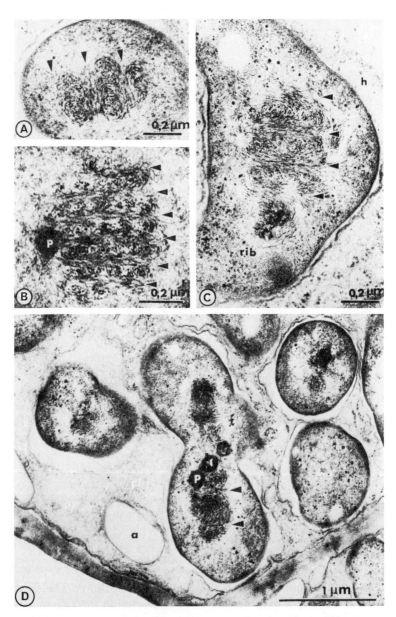

Fig. 14.   Electron micrograph of bacteria in thin section showing the fibrous network of DNA in the chromosomes. These bacteria belong to the genus Rhizobium and are involved in the use of atmospheric nitrogen for biosynthesis. They live in lateral nodules of the roots of the legume *Sarothammus scoparius* L. The DNA filaments are seen alternately longitudinally (arrows) and in section. A, B, C: cutting plane almost parallel to the cholesteric axis; D: oblique section in a dividing cell; a: starch grain; h: host cell; P: polyphosphate grain; rib: ribosomes. (Courtesy of Dr. J.-P. Gourret.)

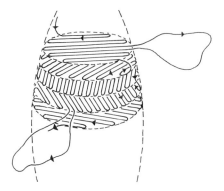

FIG. 15. Hypothetical path of the DNA double helix in a dinoflagellate chromosome. The successive planes of backfolding of the filament are not differentiated in reality. Their representation here in discrete layers is done to facilitate the drawing. Some loops extend in the external medium.

Crick double helix, folded back on itself to form the small chromosome. In contrast, the MBBA droplet is made of individual small molecules. The geometry of the director distribution is, however, similar.[32-34]

Several authors have observed cholesteric mesophases given by DNA in concentrated aqueous solutions (Robinson,[104] Bouligand,[12] Lerman[105]). The chromosomes of dinoflagellates and bacteria correspond to very concentrated solutions of DNA and are gels. They resemble the mesomorphic droplets floating in an isotropic medium. It is worth remembering that cells engaged in periodic divisions show alternate condensation and dilution of the chromosomes. These latter are visible in the condensed stage during the actual process of division and invisible in the dilute stage, which lasts for the rest of the cell cycle. The comparison with mesomorphic germs growing from the isotropic phase indicates that the reversible condensation of chromosomes could be related to the mechanism of the transition between a mesomorphic and an isotropic phase.

Flory[106] has described the transitions and phase equilibria involving polymer chains in ordered configurations. An equilibrium can occur between isotropic dilute solutions of the polymer and the anisotropic solution phases. The chains can emerge from the anisotropic phase and enter the isotropic phase. As we have seen above, the same situation occurs in chromosomes with the DNA lateral loops. The question arises

[104] C. Robinson, *Tetrahedron* **13**, 219 (1961).
[105] L. S. Lerman, *Cold Spring Harbor Symp. Quant. Biol.* **38**, 59 (1973).
[106] P. J. Flory, *J. Polym. Sci.* **49**, 105 (1961).

as to whether all the DNA filaments which approach the interface do in fact cross it and form loops in the more dilute surrounding medium.

A similar problem has been treated by Flory[107] for the case of synthetic polymers and he found that only a certain percentage of chains emerge into the isotropic medium. The marked difference in DNA density between the condensed regions of chromosomes and the surrounding medium points to a similar conclusion for this material. In this case, a great proportion of DNA filaments of the dense region must be folded back when they meet the interfacial zone, as has been schematized in Fig. 15. As the DNA double helix offers a certain rigididy, one must expect a nonnegligible radius of curvature. This situation could explain certain distortions often visible in the vicinity of the interface. Another question is whether the two attachment points are very close, in the interfacial zone, or far apart. It is not yet possible really to answer this question. The first hypothesis appears to be more plausible by reference to other kinds of chromosomes showing loops attached to very restricted zones of the longitudinal axis.[108]

We have proposed above a comparison between the reversible condensation of chromosomes and the transition between a mesophase and the isotropic liquid. The external loops can be more or less extended. Their expansion and retraction could be simply controlled by the variation of certain concentrations leading to a phase transition. A comparable change of size of droplets occurs in the case of simple mesophases such as those given by MBBA and cholesterol benzoate. Droplet sizes can be decreased simply by the addition of a solvent, e.g., toluene; evaporation of the solvent leads to an enlargement of the mesomorphic droplets. The use of a solvent is thus a simple way to displace the equilibrium between the amorphous liquid and the mesophase. Returning to the question of chromosomes, Chambers[109] showed a long time ago that "shrinkage or condensation of chromosomes upon injury to the nucleus is a general phenomenon. For example, in the living grasshopper spermatocyte the nucleus, typically in the prophase stage, is optically homogeneous. When the nucleus is pricked, chromosome filaments appear. These progressively thicken until they are transformed from characteristic early prophase chromosomes into chromosomes resembling those of the metaphase, stage" (Chambers[109]). One must note that the chromosomes studied by this author do not show the system of arced patterns described above. In this material, the orientations of DNA seem to be masked by the presence of certain proteins

[107] P. J. Flory, *J. Am. Chem. Soc.* **84**, 2857 (1962).
[108] H. G. Callan, *Int. Rev. Cytol.* **15**, 1 (1963).
[109] R. Chambers, *Cellule* **35**, 107 (1924).

(histones). Such is the case of chromosomes in the great majority of organisms. However, the chromosome condensation can still be related to a phase transition. Certain concentrations in the nucleus must be modified by injury to the nuclear envelope and probably by cytoplasmic contents rushing into the nucleus. It is remarkable that in these materials the natural condensation of chromosomes is preceded by the disappearance of the nuclear membrane.

The value of these comparisons between chromosome condensation and phase transitions is, as yet, difficult to estimate. Chromosomes are condensed phases of DNA covered with proteins; other chemical species are present. Certain aspects of polymer crystallography ought to apply to these materials, particularly those concerning the problem of packing. The examples cited above illustrate this viewpoint. However, one must keep in mind the profound differences separating on one hand the very simple system formed by a cholesteric droplet (or rodlet) floating in the isotropic phase, given by one pure substance or a binary mixture and, on the other hand, the chromosome with its genetic information, its replication properties, and its complicated biochemistry.

V. Mesomorphism and Morphogenesis

The morphogenesis of living organisms is one of the great problems in biology. The egg divides into several cells, which are similar at first and then progressively differentiate. Cell differentiation is studied mainly by the methods of genetics and biochemistry. It seems, however, that other approaches will be necessary to explain the geometrical and topological phenomena appearing in the course of embryonic development. One can recognize in normal development parts of the embryo which correspond to the future organs. The classical method of studying this problem is to soak minute pieces of agar in vital stains (e.g., neutral red, Nile blue), dyes which are tolerated by living tissues, and to hold them against one region of the embryo for a short time. The dye is retained for a long time by cells with which it was in contact and also by their progeny. Thus, the presumptive rudiments of eye, anterior, and posterior limbs can be delineated. The organ districts become determined at a definite stage. This means that before this stage the excision of the organ district does not prevent the organ from developing normally out of the surrounding tissues. Once the organ district has been determined, its excision leads to the absence of the organ, although the rest of the embryo continues to develop. The partial excision of a limb district, a short while after its determination, results in a limb which is normal but smaller than usual. A determined district is subdivided into several presumptive rudiments of the various parts of the future organ. These subdistricts are determined in turn.

The progressive stages of embryonic determination include the setting up of axes of polarity. This process has been admirably illustrated by the experiments of Harrison[110] on the presumptive limb rudiments in the embryos of amphibians. These disk-shaped rudiments are found laterally in the body wall. They transform into buds and differentiate into complete limbs. Needham[111] was the first to compare the steps of progressive determination of morphogenetic axes of an organ district with the progression of states of order among different mesophases. Needham wrote that "From the fundamental work of R. G. Harrison in 1921, we know that when discs of the outer wall of the body representing the buds of the future limbs are transplanted at a certain stage, it is found that the original anterior edge of the bud always produces the preaxial part of the limb. So, a limb bud of the left side, planted the right way up on the right side of the embryo, will develop into a limb with the elbow pointing forwards instead of backwards, if it is a forelimb. On the other hand, if the disc is rotated, so that the original dorsal edge of the bud is ventral, the original anterior edge will be anterior again, and a normal limb develops. Therefore, the dorso-ventral axis can be inverted without producing rearrangement, but not so the antero-posterior axis. The mediolateral axis can also be inverted at this stage with impunity, i.e., it does not matter whether the limb-bud is attached to its new situation proximally or at a distance. This means that the dorso-ventral and medio-lateral axes are still plastic, while the antero-posterior axis is determined. Later on they also become determined. Here then we have successive stages of dimensional determination. Just as the liquid crystal passes through stages of rigidity in one, two, or three dimensions, so the limb-bud passes through stages of determination in one, two, and three dimensions. The analogy here may of course be superficial, but it is sufficiently striking to warrant attention."

Harrison and Needham assumed that these transformations involve changes in the orientation of ultramicroscopic elements. Experiments have been attempted to show such changes by direct observations of X-ray diffraction patterns given by limb buds.[112] The presence of water and yolk prevents this, however. There is much current research on self-assembly, dispersion, and reassembly of microtubules into their subunits and back again. However, the relation between microtubule alignments and embryonic determination has been for the most part neglected. It

[110] R. G. Harrison, *J. Exp. Zool.* **32**, 1 (1921).

[111] J. Needham, "Biochemistry and Morphogenesis," 1st ed., Vol. 1. Cambridge Univ. Press, London and New York (2nd ed., 1950).

[112] R. G. Harrison, W. T. Astbury, and K. M. Rudall, *J. Exp. Zool.* **85**, 339 (1940).

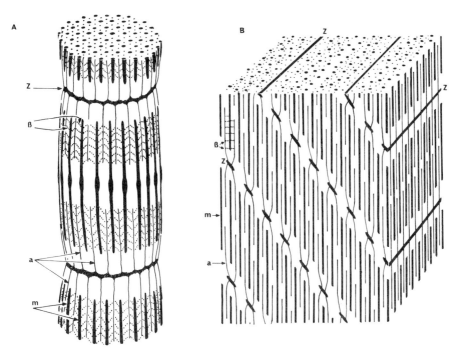

FIG. 16. A: Schematic representation of a contraction unit (sarcomere) in a cross-striated muscle. Thick myosin filaments m interdigitate with thin filaments a. Small bridges β β interconnect the two sets of filaments. Thin filaments are attached to Z elements. This pattern is repeated all along the muscle fiber. B: Similar diagram for an oblique striated muscle. The symmetries of these two systems are these of a smectic A in its B form (hexagonal packing) and of a smectic C. Contraction is the result of the relative sliding of the two sets of filaments involving ATP splitting at the level of the cross-bridges.

would be interesting to evaluate Needham's conceptions by a careful examination of microtubule distribution in various differentiating systems, particularly in the case of the vertebrate limb. It seems to us that recent ultrastructural studies, especially those concerning muscle differentiation, offer more appropriate illustrations of Needham's idea, sketching a similarity in the progression of organogenesis and the successive degrees of order which appear in a series of mesomorphic phases.

The liquid-crystal aspect of muscle fibers has been studied by Elliot and Rome.[113] These authors suggest that the system of interpenetrating sets of filaments made out of the two main proteins of myofibrils, actin and myosin (Fig. 16), is a liquid-crystal superphase in the sense defined

[113] G. F. Elliot and E. M. Rome, *Mol. Cryst. Liq. Cryst.* **8**, 383 (1969).

1 μm

FIG. 17. Longitudinal section of an oblique-striated muscle in the contracted state (*Haplosyllis depressa*; Bouligand[116]).

by Zocher and Török.[114] The interfilament distance is controlled by the balance of several possible forces: van der Waals, electrical double-layer repulsion, hydration. We shall now briefly summarize certain of our own considerations. Between smooth and striated muscles there is a structural difference resembling that found between nematics and smectics. One distinguishes cross-striated muscles (arthropods, vertebrates, other groups) and oblique-striated muscles (annelids, molluscs, etc.) as in

[114] H. Zocher and C. Török, *Acta. Crystallogr.* **22**, 151 (1967).

smectics, one has the two forms A and C. An example of oblique-striated muscle is shown in Fig. 17. In cross-striated and oblique-striated muscles the filaments may show a random distribution or may form a more or less developed hexagonal packing as in smectic B (see Huxley,[115] Bouligand[116]). Myogenesis, that is to say the differentiation of muscle fibrils, has been studied in the main structural types cited above. In the case of insects, at first, the field shows microtubules which have a general alignment.[117] The symmetry of such a system is similar to that of a nematic. It is in the next stage that the filaments which make up the myofibrils are assembled and introduce a periodicity analogous to that seen in smectics.

It is likely that progress in the theory of the liquid-crystal state will be a source of inspiration in morphogenetic research. It appears probable that a principle similar to that of least encumbrance can account in many cases for the alignment of microtubules, as happens in the transition "isotropic–liquid-nematic phase" according to the Flory and Onsager conceptions (see review in de Gennes[118]). It is also reasonable to look for analogies between the mechanisms which align the centers of gravity of smectic molecules along parallel and equidistant surfaces and those responsible for the formation of a striation in a muscle cell. The structural elements at play here vary by several orders of magnitudes; one is considering small molecules as well as the supramolecular constructions that are the actin and myosin filaments. Given the close similarity between the geometry of the various types of muscles and that of the nematic and smectic structures, one is led to look for analogies between the generating mechanisms.

VI. Conclusions

Cells and tissues are made of membranes, secretions, and fibrous networks which often are liquid crystals or close analogs. Polymerization and secondary chemical reactions of hardening lead to pseudomorphoses showing the geometrical and optical properties of the homologous mesophases. There are classical examples of self-assembly in biological materials: dispersion and reassembly of protein subunits (capsomeres) in the tobacco mosaic virus, and dissolution and rearrangement of globular proteins forming the bacterial flagella or the actin

[115] H. E. Huxley, *Sci. Am.* **199**, (5)67 (1958).
[116] Y. Bouligand, *J. Microsc. (Paris)* **5**, 305 (1966).
[117] J. Auber, *J. Microsc. (Paris)* **8**, 197 (1969).
[118] P. G. de Gennes, "The Physics of Liquid Crystals," Vol. 1. Cambridge Univ. Press, London and New York, 1974.

294 Y. BOULIGAND

filaments in muscles.[119-121] It appears that another mechanism of self-assembly exists in living systems and it is the process allowing molecules to build a liquid crystal. The parallel alignment and the relative position of molecules (depending on the smectic, nematic, or cholesteric arrangement) may appear in true mesophases and in more or less dense gels. Self-assembly of unit membranes has been observed,[122] and this process—is related to the mesomorphic nature of the cell membranes. More generally, throughout tissues, cells, and organelles the molecular organization is often very close to that of liquid crystals and it seems likely that the study of this state of matter can offer interesting insights into morphogenetic processes in living systems.

ACKNOWLEDGMENTS

The author has had fruitful discussions with Dr. Charvolin, and he appreciates considerable help from G. D. Mazur.

[119] S. Asakura, G. Eguchi, and T. Iino, *J. Mol. Biol.* **10,** 42 (1964).
[120] H. Fraenkel-Conrat, *Sci. Am.* **194**(6), 42 (1956).
[121] K. Maruyama, *Biochim. Biophys. Acta* **102,** 542 (1965).
[122] W. Stoeckenius, *Symp. Int. Soc. Cell Biol.* **1,** 349 (1962).

Author Index

Numbers in parentheses are reference numbers and indicate that an author's work is referred to although his name is not cited in the text. Numbers in italics show the page on which the complete reference is listed.

A

Abdolall, K., 257
Adams, J., 77
Agarwal, V. K., 143
Aggawal, S., 2, 9(7)
Agren, G., 252
Alder, C. J., 89
Alexander, S., 6, 8
Allison, S., 228
Allport, 2, 9(7)
Armstrong, R. S., 124
Arora, S. L., 47
Asakura, S., 294
Ashcroft, N., 12, 13(13)
Astbury, W. T., 290
Auber, J., 293
Axmann, A., 142
Azerad, R., 233, 239(49)

B

Bader, P., 232, 247(48)
Baessler, H., 140
Barnik, M. I., 203, 204, 205
Barsukov, L. I., 236
Bartolino, R., 203
Bata, L., 144
Baud, C. A., 272
Baur, G., 90, 107, 144
Beament, J. W. L., 274
Beard, R. B., 140
Becher, 15
Becker, E. D., 250
Beevers, M. S., 141
Belle, J., 246

Bennett, S., 266
Berendsen, M. J. C., 233
Berge, P., 154, 203(14)
Bergelson, L. D., 236
Bernal, J. D., 273, 274
Berreman, D. W., 78, 90, 107
Bertolotti, M., 203
Bessis, M., 272, 273
Billard, J., 41, 42(6), 43(6), 45(6)
Birbeck, M. S. C., 282
Birdsall, N. J. M., 228
Blinč, R., 236, 243
Blinov, L. M., 203, 204, 205
Bloom, M., 234, 254, 256, 257
Bloom, W., 272
Bobylev, J. P., 203
Böttcher, C. J. F., 112, 113(8), 116, 117, 119, 130(8), 131(8)
Boggs, J. M., 239
Boix, M., 78, 163
Bonnevile, M. A., 269
Bordewijk, P., 114, 116, 117(10), 118, 130, 131, 132, 134(31), 145
Bothorel, P., 246
Boulanger, Y., 257
Bouligand, Y., 1, 260, 264, 274(32,33,34), 279, 280, 282, 284, 287, 292, 293
Bourges, M. C., 270
Boven, J., 27, 28(3), 29(3), 31(3), 49(3)
Brachet, J., 267
Breternitz, V., 136
Brochard, F., 18, 81, 96, 97(57), 105, 107, 152, 153, 158, 179(13), 202(13), 256, 272
Budler, K. W., 257
Bücher, H. K., 133, 137(36), 143(36), 144, 200, 201(48a)

295

Subject Index

A

AB block copolymers, 2–3
  bulk and surface energies in, 8–9
  dense side (A) in, 6
  dilute micelles of, 9–11
  semidilute side (B) in, 6–8
  sheet structures of, 5–9
AB micelles, 9–11
  model for, 4
p-n-Alkylacetophenone, preparation of, 27, 56
p-n-Alkylanilines, preparation of, 26–32
p-Alkylbenzoylchlorides, preparation of, 56–59
p-n-Alkylbenzoic acid, preparation of, 56–58
p-n-Alkyloxyacetanilides, preparation of, 37–38
p-n-Alkyloxyanilines, preparation of, 36–39
p-p'-Alkyloxyazoxybenzene, preparation of, 52–54
p-n-Alkyloxybenzaldehydes, preparation of, 24–26
p-n-Alkyloxybenzoylchlorides, preparation of, 59–60
p-Alkyloxybenzylidene-p'-alkylanilines, in liquid crystal synthesis, 24–36
p-n-Alkyloxybenzylidene-p'-n-alkylanilines, preparation of, 32
p-n-Alkyloxyphenols, preparation of, 64–65
p-n-Alkylphenols, preparation of, 61–62
p-Aminocinnamates, preparation of, 41–43
Analog simulation, of linear instabilities in nematic liquid crystals, 205–208
Azobenzenes, permittivities of, 125
p-p'-disubstituted Azobenzenes, 124–126
p-Azoxyanisole, 116
  dielectric relaxation data for, 142
  permittivities of, 122–124
p-p'-Azoxyanisole, 48

Azoxybenzenes
  asymmetrically substituted, 128
  dipole moment of, 123–124
p-p'(di-n-alkyl)-Azoxybenzenes, preparation of, 49
p-p'-disubstituted Azoxybenzenes, preparation of, 48–54
p-Azoxyphenetole, 116

B

Backflow effect, in thermal convection, 158
Bénard–Rayleigh instability, 154
Bénard–Rayleigh thermal convection, in isotropic fluids, 153–156
p-substituted Benzylidene-p'-n-alkyloxyaniline, preparation of, 36–40
p-substituted Benzylidene-p'-aminocinnamates, preparation of, 41, 44
Biological bilayers, 268–270
Biological materials, electron microscope examination of, 267–268
Biological membranes, liquid crystals and, 265–267
Block copolymers
  excluded volume effects in, 14–15
  flexible, 2–3
Block copolymer/solvent systems, scaling laws for, 15–16

C

CBOOA, preparation of, 40
Cell membranes, liquid character and morphology of, 265–267, 271–272
Cells and tissues, ultrastructural organization of, 267–274
Chain deformations, in lipid motions, 244–249